Research Progress in
FISHERIES SCIENCE

Research Progress in FISHERIES SCIENCE

William Hunter III
Researcher, National Science Foundation, U.S.A.

Apple Academic Press

Research Progress in Fisheries Science

First Published in the Canada, 2011
Apple Academic Press Inc.
3333 Mistwell Crescent
Oakville, ON L6L 0A2
Tel. : (888) 241-2035
Fax: (866) 222-9549
E-mail: info@appleacademicpress.com
www.appleacademicpress.com

The full-color tables, figures, diagrams, and images in this book may be viewed at www.appleacademicpress.com

ISBN 978-1-926692-65-4

William Hunter III

Printed at Thomson Press (I) Ltd.
9 8 7 6 5 4 3 2 1

Cover Design: Psqua

Library and Archives Canada Cataloguing in Publication Data
CIP Data on file with the Library and Archives Canada

CONTENTS

INTRODUCTION

The control of fisheries and fish production has been going on around the world for hundreds of years. For example, the early Maori of New Zealand had strict rules in their traditional fishing activities about not taking more than could be eaten and about throwing back the first fish caught. In another example, a few hundred years later, in the early 1800s in the North Norwegian fishery off the Lofoten Islands, a law was passed to avoid taking too many fish from one specific area.

Fisheries are harvested for their value (whether commercial, recreational, or subsistence). Close to 90 percent of the world's wild fishery catches come from oceans and seas (as opposed to inland waters), and most marine fisheries are based near coasts. This is not only because harvesting in relatively shallow waters is easier than in the open ocean but also because fish are much more abundant near the coastal shelf, due to coastal upwelling and the available nutrient abundance. However, there are also productive open ocean fisheries. Most fisheries are wild fisheries, but increasingly fisheries are farmed.

Today fisheries science has become an academic discipline with practical applications. It is a multidisciplinary science that draws on oceanography, marine biology, marine conservation, ecology, population dynamics, economics, and management to provide an integrated picture of fisheries. In some cases, new disciplines have emerged from this field, as in the case of bioeconomics.

In the twenty-first century, fisheries science, with all its branches, has become ever more valuable to human society. We must find ways to protect our planet's fish, both as a food source and as a vital part of the natural world. Careers in this field have growing importance in a world where oil spills and ongoing pollution are all too real threats to the world's fisheries.

— William Hunter III

Impact of Climate Change on the Relict Tropical Fish Fauna of Central Sahara: Threat for the Survival of Adrar Mountains Fishes, Mauritania

Sébastien Trape

ABSTRACT

Background

Four central Sahara mountainous massifs provide habitats for relict populations of fish. In the Adrar of Mauritania all available data on the presence and distribution of fish come from pre-1960 surveys where five fish species were reported: Barbus pobeguini, Barbus macrops, Barbus mirei, Sarotherodon galilaeus, and Clarias anguillaris. Since 1970, drought has had a severe impact in the Adrar where rainfall decreased by 35%. To investigate whether the relict populations of fish have survived the continuing drought, a study was carried out from 2004 to 2008.

Methodology/Principal Findings

An inventory of perennial bodies of water was drawn up using a literature review and analysis of topographical and hydrological maps. Field surveys were carried out in order to locate the bodies of water described in the literature, identify the presence of fish, determine which species were present and estimate their abundance. The thirteen sites where the presence of fish was observed in the 1950s -Ksar Torchane, Ilij, Molomhar, Agueni, Tachot, Hamdoun, Terjit, Toungad, El Berbera, Timagazine, Dâyet el Mbârek, Dâyet et-Tefla, Nkedeï- were located and surveyed. The Ksar Torchane spring -type locality and the only known locality of B. mirei- has dried up at the height of the drought in 1984, and any fish populations have since become extinct there. The Timagazine, Dâyet el Mbârek and Dâyet et-Tefla pools have become ephemeral. The Hamdoun guelta appears to be highly endangered. The fish populations at the other sites remain unchanged. Four perennial pools which are home to populations of B. pobeguini are newly recorded.

Conclusion/Significance

The tropical relict fish populations of the Adrar mountains of Mauritania appear to be highly endangered. Of thirteen previously recorded populations, four have become extinct since the beginning of the drought period. New fish population extinctions may occur should low levels of annual rainfall be repeated.

Introduction

The presence of fish in the Sahara, evidence of the world's hottest desert humid past during the Holocene period, was recognised in the first Saharan exploration expeditions in the early 20th century [1]–[4]. The work undertaken in the 1950s expanded the inventory of perennial bodies of water–springs, gueltas or pool–which are home to this relict fauna [5]–[8]. More recently, systematic revisions, especially for the Barbus genus, and re-examination of the collected materials, have clarified the species identification and biogeographical affinities of the fish species present in the Sahara [9], [10].

Four of the six main central Saharan mountainous massifs are home to fish: the Ahaggar and Tassili-n-Ajjer in southern Algeria, the Adrar in Mauritania and the Tibesti in Chad [3]–[10]. There are no fish in the Aïr massif of Niger and the Iforas massif of Mali. On the southern fringes of the Sahara, relict fish fauna also survive in the Tagant of Mauritania and in the Ennedi of Chad. In total, sixteen species and sub-species of fish have managed to survive in the central Sahara, most commonly in a very low number of watering places, sometimes in a single spring or guelta (Table 1). These species are the following: Barbus macrops Boulenger, 1911, Barbus pobeguini Pellegrin, 1911, Barbus deserti Pellegrin, 1909, Barbus bynni occidentalis Boulenger, 1911, Barbus callensis biscarensis Boulenger, 1911, Labeo parvus Boulenger, 1902, Labeo niloticus (Forsskål, 1775), Raiamas senegalensis (Steindachner, 1870), Clarias anguillaris (Linnaeus, 1758), Clarias gariepinus (Burchell, 1822), Epiplatys spilargyreius (Duméril, 1861), Hemichromis bimaculatus Gill, 1862, Sarotherodon galilaeus galilaeus (Linnaeus, 1758), Sarotherodon galilaeus borkuanus (Pellegrin, 1919), Astatotilapia desfontainesi (Lacépède, 1803) and Tilapia zillii (Gervais, 1948) [3]–[5], [8], [10].

Table 1. Relict fish species of Central Sahara: distribution in Sahara and adjacent Sahelian basins.

Family and species	Distribution	
	Regions of Sahara	Adjacent basins
Cyprinidae		
Barbus c. biscarensis	Mouydir/Ahaggar/Tassili	Atlas wadis (Maghreb)
Barbus b. occidentalis	Tibesti	Senegal/Niger/Chad
Barbus deserti	Tassili	Endemic
Barbus macrops	Adrar/Ahaggar/Tibesti	Senegal/Niger/Chad
Barbus pobeguini	Adrar	Senegal/Niger
Labeo niloticus	Tibesti	Nile
Labeo parvus	Tibesti	Senegal/Niger/Chad
Raiamas senegalensis	Tibesti	Senegal/Niger/Chad
Clariidae		
Clarias anguillaris	Adrar	Senegal/Niger/Chad/Nile
Clarias gariepinus	Tassili/Tibesti	Senegal/Niger/Chad/Nile
Cyprinodontidae		
Epiplatys spilargyreius	Borkou	Senegal/Niger/Chad/Nile
Cichlidae		
Hemichromis bimaculatus	Tassili/Borkou	Senegal/Niger/Chad/Nile
Sarotherodon g. galilaeus	Adrar	Senegal/Niger/Chad/Nile
Sarotherodon g. borkuanus	Tibesti/Borkou	Endemic
Astatotilapia desfontainesi	Mouydir	Atlas wadis (Algeria/Tunisia)
Tilapia zillii	Mouydir/Ahaggar/Tassili/Tibesti	Senegal/Niger/Chad/Nile

The species richness and affinities of the relict populations of Saharan fish differ significantly depending on the massif. In the Ahaggar, three species are present, including one Paleoarctic species, B. c. biscarensis, and two Afrotropical species, B. macrops and T. zillii. In the Tassili-n-Ajjer and its Mouydir extension, the relict fauna comprises seven species: B. deserti, B. c. biscarensis and A. desfontainesi, which are Paleoarctic; and C. gariepinus, C. anguillaris, H. bimaculatus and T. zillii, which are Afrotropical. In the Adrar of Mauritania, four Afrotropical species are present: B. pobeguini, B. macrops, S. g. galilaeus and C. anguillaris. In the

Tibesti and its Borkou extension, there are nine Afrotropical species: B. macrops, B. bynni occidentalis, L. parvus, L. niloticus, R. senegalensis, C. gariepinus, E. spilargyreius, S. galilaeus borkuanus and T. zillii.

The presence of fish in the Adrar mountains of Mauritania has been known since 1913, when Pellegrin first mentioned the specimens of C. senegalensis (= C. anguillaris) and T. galilaea (= S. galilaeus) collected by Chudeau near Atar [1]. In 1937, Pellegrin added B. pobeguini and B. deserti (= B. macrops) collected by Monod in various Adrar gueltas [3]. Between 1951 and 1955, Monod published three detailed papers on the freshwater fish species of Mauritania [5]–[7]. These papers are still the only sources of original data on the inventory and population of water bodies which provide habitats for fish in the Adrar region (Figure 1). They indicated the presence of B. deserti (= B. macrops) in Ilîj, Agueni and Tachot (Seguellîl wadi basin) and in the Toungad guelta (El Abiod wadi basin), the presence of C. senegalensis (= C. anguillaris) in the same gueltas, with the exception of those in Ilîj and Tachot, the presence of T. galilaea (= S. g. galilaeus) in Molomhar and of B. pobeguini in all the above mentioned gueltas and in the Ksar Torchane spring (Tengharâda wadi, a tributary of the Seguellîl wadi), in the Terjit springs (Terjit wadi, a tributary of the El Abiod wadi) and in several gueltas or pools in the three southern Adrar basins which are completely separate from the above basins: in El Berbera and Timagazine (Timagazine-Timinit wadi basin), Dâyet el Mbârek and Dâyet et-Tefla (Nbéïké wadi) and Nkedeï (Nkedeï wadi). In 1952, Estève [11] described a new subgenus and species of Barbus in the Ksar Torchane spring: B. (Hemigrammocapoeta) mirei, characterised by an incomplete lateral line. However, their validity was doubted by Monod [6] and not recognised by Lévêque who considered B. mirei as synonym of B. pobeguini [10]. Additional data on the aquatic fauna of the Adrar region have been provided by Villiers [12] and Dékeyser & Villiers [13]. Reviews of Saharan aquatic fauna, including the Adrar region, have been carried out by Dumont [14], [15], Lévêque [10] and Le Berre [16].

Since 1970, a drought of historically unprecedented proportions persists in West Africa, with a reduction in annual recorded rainfall of 20 to 30% in the Sudanese savannah and up to 40% in the Sahel region on the southern fringes of the Sahara [17]–[20]. The rainfall deficit was also very marked in the Adrar region, with the average recorded rainfall in Atar decreasing by 35%, from 106.8 mm in 1921–1969 to only 69.8 mm in 1970–2007 [21]. Furthermore, some years were almost completely dry, with recorded rainfall in Atar of less than 20 mm in 1971, 1977, 1982 and 1996 and only 7 mm in 1983, which could be insufficient to sustain the few perennial bodies of water in the Adrar region. As all available data on the presence and distribution of fish comes from pre-1960 surveys, it became important to investigate to what extent these fish populations were able to resist such a long, intense period of drought. In order to assess the current state of ichthyologic fauna, a series of surveys was carried out from 2004 to 2008. The results of thesc surveys show that the relict fish fauna of Adrar mountains is highly endangered and that several extinctions of fish populations has already occurred.

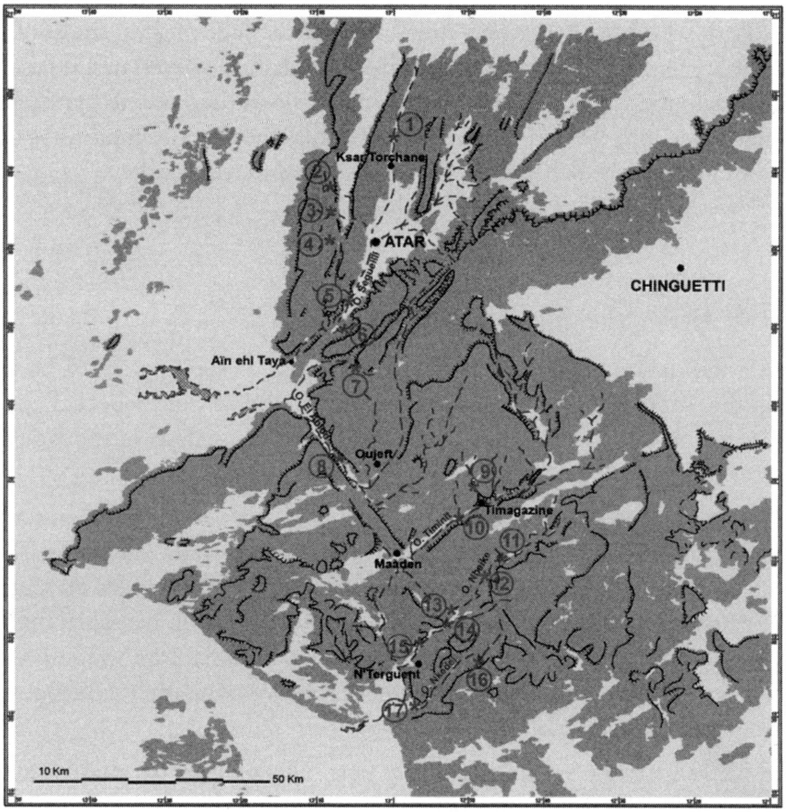

Figure 1. Map of Adrar mountains, Mauritania. Rocky areas are shown in grey, sandy areas in yellow. Numbers indicate gueltas, ponds and springs where the presence of fish was reported in the literature and/or observed during the present study. 1 : spring of Ted at Ksar Torchane ; 2 : Ilîj guelta ; 3 : Molomhar gueltas ; 4 : Agueni ponds ; 5 : Tachot guelta ; 6 : Hamdoun guelta ; 7 : Terjit springs ; 8 : Toungad guelta ; 9 : El Berbera guelta ; 10 : Timagazine wadi pond ; 11 : Dâyet el Mbârek pond ; 12 : Agmeïmine guelta ; 13 : Dâyet et-Tefla pond ; 14 : Douâyât Lemgasse pond ; 15 : Dâyet Legleïg pond ; 16 : Nkedeï guelta and pond ; 17 : Glât el Bil pond.

Materials and Methods

Study Area

The Adrar is a low central massif located between 19°–21° N and 11°–14° W, with as its highest point the 850 m peak of Teniagouri. It is mainly composed of a series of sandstone plateaux, edged by steep cliffs, which are located between 200 m and 600 m in altitude. To the west, the Adrar dominates the sandy or rocky plains of the Inchiri and Amsaga, which lie at around 100 metres above sea level. To the east and north, it is bordered by the Majâbat, Ouarane and Oumm Aghouäba dunes. To the south, it is separated from the Tagant plateaux by the Khat depression. The hydrographic network of the Adrar comprises a high number of fossil valleys, where the water only flows on a few days of the year, most often in August or September, when there is sufficient rainfall. However, despite the scarcity of rainfall, the sandstone nature of the relief allows the water to permeate and perennially seep into the various gueltas and pools, creating the conditions for the survival of a relatively rich fauna of relict fish despite the gradual drying of the Sahara since the Holocene period [22], [23].

Methods

During a first phase an inventory of the perennial or semi-perennial bodies of water in the Adrar region of Mauritania–both with or without reported presence of fish–was drawn up from a literature review and an analysis of topographical, geological and hydrological maps of the region.

In a second phase, a series of field surveys were carried out between October 2004 and June 2008 in order to locate the bodies of water mentioned in the literature and to look for the presence of fish, identify the species present and, where possible, estimate their abundance. A search for fish in perennial bodies of water where their presence was not mentioned in the literature was also carried out. Whenever local populations were present near watering places, they were also interviewed about the presence of fish.

The fish were captured using landing nets, nets or lines. In most cases they were returned to the water after being identified, but vouchers of each species and locality were kept and their identification verified using the keys drawn up by Lévêque [24], Teugels [25] and Teugels & Thys Van den Audenaerde [26]. Subaquatic visual explorations were also carried out at Ksar Torchane, Ilîj, Molomhar, Agueni, Tachot, Hamdoun, Terjit and El Berbera.

The study was approved by the review board of the IRD special programme "Action Thématique Interdépartementale Evolution Climatique et Santé."

Results

All bodies of water where the presence of fish had been indicated were identified, located in the field and surveyed. Several pools and gueltas not previously mentioned in the literature were also surveyed. The report on these surveys follows:

Ksar Torchane (Spring)

The Ksar Torchane spring (20°43′654 N/13°00′593 W) is located in the hamlet of Ted, on the edge of the Tengharâda wadi, a tributary of the Seguellîl wadi, at the northern limit of the Ksar Torchane palm grove (Figure 2A). It is the most northerly of all the sites where freshwater fish have been reported in Mauritania. The fifteen specimens of the type series of B. (Hemigrammocapoeta) mirei Estève, 1952, came from this spring. The presence of B. pobeguini has also been reported [5].

Figure 2. Gueltas and springs from the Seguellîl wadi basin. A: The former spring of Ted at Ksar Torchane; B: Ilîj guelta.; C: The first guelta of Molomhar; D: Pond at Agueni; E: Tachot guelta; F: Hamdoun guelta; G: Terjit springs; H: Toungad guelta.

The Ted spring was continually dry during the surveys carried out in October 2004, April 2005, October 2006, April 2007 and May 2007. The well in the bed of the wadi which was previously fed by the spring was flowing at the time of each survey but no fish were observed, neither from the surface nor during a sub-aquatic exploration in April 2007. The height of the water was then 2.5 m, with a water volume of 12 m3. According to the primary school teacher from Ted, who is originally from the hamlet, the fish disappeared in early 1984 when the spring and the well both ran dry for the first time and a vast cement basin adjoining the spring was also found dry. The fort overlooking the spring was at the time the headquarters of a small military garrison which enlarged the well using explosives in 1984 before having to definitively leave the site that same year. The well then continued to dry up, and new enlargement works using explosives were carried out in 1990.

Ilîj (Guelta)

The Ilîj guelta (20°38′046 N, 13°08′490 W) is located at the foot of a cliff in the canyon of the Ilîj wadi, a tributary of the Seguellîl wadi (Figure 2B). The presence of B. pobeguini and B. macrops was reported [5].

The guelta was flowing at the time of the surveys in April and May 2007 and water was seeping from the cliff at a number of locations. Several thousand B. macrops and B. pobeguini were present. The freshwater jellyfish Limnocnida tanganyicae was observed.

Molohmar (Gueltas)

The three perennial Molohmar gueltas are several dozen metres apart in a narrow canyon in the Oumm Lemhar wadi, a tributary of the Ilîj wadi, which they block completely (20°35′229 N, 13°08′794 W) (Figure 2C). They have the greatest species richness, with four species reported: B. pobeguini, B. macrops, C. anguillaris and S. g. galilaeus [5], [12], [13].

In the first, and deepest, guelta, the water was six metres deep in May 2007. Several thousand B. pobeguini, B. macrops and S. g. galilaeus were present in each. C. anguillaris and L. tanganyicae were also observed.

Agueni (Palm Grove Pools)

The Agueni palm grove is located in the bed of the Taghadem wadi, a tributary of the Seguellîl wadi. It comprises several small perennial pools fed by infiltration through the sandy soil (Figure 2D). These pools are known to be home to three species of fish: B. pobeguini, B. macrops and C. anguillaris [5].

Sixteen small pools were flowing in June 2008 at the foot of the palm groves, over a distance of around 300 m (upstream pool: 20°31′380 N, 13°08′714 W; downstream pool: 20°31′326 N, 13°08′556 W). The largest pool was 10 m long and 3 m wide; the deepest reached a depth of one metre. B. pobeguini and B. macrops were observed in most of the pools surveyed. C. anguillaris was not observed; however, according to the residents of the palm grove, it is still abundant. An ephemeral downstream guelta (20°31′204 N, 13°08′207 W) was around 100 m long in September 2007, but only 3 m long in April 2007. It was also home to B. pobeguini and B. macrops.

Tachot (Guelta)

The Tachot guelta (20°24′410 N, 13°06′465 W) is located at the foot of a cliff in the bed of the Seguellîl wadi. Monod had observed barbs there without being able to capture them and had presumed that they were B. pobeguini [5].

The water surface of this vast guelta was around 300 m2 and reached a maximum depth of 50 cm in April 2007 (Figure 2E). Around 3,000 Barbs were present, around a third of which were B. macrops and two-thirds were B. pobeguini.

Hamdoun (Guelta)

The Hamdoun guelta (20°19′380 N, 13°08′550 W) is located in the Fârech wadi, at the foot of a small cliff, several metres below the main Atar-Nouakchott road (Figure 2F). The presence of B. pobeguini, B. macrops and C. anguillaris was reported [5], [12], [13]. Other gueltas are located downstream, spread over a distance of around one kilometre, down to the confluence with the Seguellîl wadi, but all are ephemeral. Upslope of the road, there are also several gueltas where the presence of fish has been reported despite the gueltas being ephemeral [5], [12].

Eleven surveys were carried out between October 2004 and June 2008. B. pobeguini, B. macrops and C. anguillaris were regularly observed in the perennial guelta and in several of the ephemeral downstream gueltas when they were flowing. No fish were ever observed in the gueltas above the road which overlooks the perennial guelta. In October 2004, October 2006, September 2007 and November 2007, the Fârech wadi flowed over the road and into the perennial guelta after falling for four metres. It was in July 2007 that the lowest water level was measured in the perennial guelta: its maximum depth was 35 cm and the volume of water was only 1.7 m3. The water was too cloudy to measure the fish population accurately, but it was certainly lower than 200 specimens (240 barbs and seven catfish were counted two months previously when the volume of water was 4 m3).

Terjit (Springs)

The Terjit springs (20°15′045 N, 13°05′200 W), one hot (32°C) and the other cold, are several metres apart in a gorge and form a small stream which flows around a hundred metres before filtering into the sand (Figure 2G). Only the presence of B. pobeguini was reported [5], [12], [13].

B. pogeguini was present on each occasion in small numbers (probably under 300 specimens) at the four surveys that were carried out between October 2006 and May 2007.

Toungad (Guelta)

The Toungad guelta (20°03′771 N, 13°07′263) is located at the foot of a cliff in the El Abiod wadi basin (Figure 2H). The presence of B. pobeguini, B. macrops and C. anguillaris was reported [7].

Only B. pobeguini was observed in a rapid survey in April 2005 when the water was cloudy.

El Berbera (Guelta)

The El Berbera guelta (19°59′181 N, 12°49′374 W) is located in a canyon which opens into the Timinit wadi 4 km further downstream (Figure 3A). Monod observed barbs without being able to capture them and presumed that they were B. pobeguini [5].

There were several thousand specimens of B. pobeguini in May 2005.

Timagazine Wadi (Pool)

In October 1952, Monod observed the presence of B. pobeguini in a pool in the Timagazine wadi (upper Timinit wadi) located a little downstream of El Berbera Khenig [6]. The precise location of this pool is uncertain.

No perennial pools were found in the Timagazine wadi downstream of El Berbera Khenig. In September 2007, young B. pobeguini were present in a small ephemeral pool 7 km downstream of El Berbera Khenig (19°55′864 N, 12°52′716 W) (Figure 3B). The fish in this pool came undoubtedly from the El Berbera guelta.

Figure 3. Gueltas and ponds from the Timinit wadi basin. A: El Berbera guelta; B: Ephemeral pond in Timagazine wadi; C: The former perennial pond of Dâyet el Mbârek; D: Agmeïmine guelta; E: The former perennial pond of Dâyet et-Tefla; F: Douâyât Lemgasse pond; G: Dâyet Legleïg pond; H: Nkedeï guelta.

Dâyet el Mbârek (Pool)

The Dâyet el Mbârek pool (= Dhwiyat Lembarrek) is located in the upper Nbéïké wadi at the intersection of two valleys (19°48′720 N, 12°48′459 W). The presence of B. pobeguini was reported [5], [6].

This pool, located in a palm grove, was dry in June 2008 (Figure 3C). The residents of the palm grove claimed that it had been perennial until 1990 when it ran dry for the first time, leading to the extinction of the fishes. Since that date, the pool has been largely filled in by sand and only flows for around one month after each flood.

Agmeïmine (Guelta)

The Agmeïmine guelta (19°45′353 N, 12°50′062 W) is located in the bed of the Nbéïké wadi, close to the Doueïr palm grove. The presence of fish in this perennial guelta was not mentioned in the literature.

In December 2007, B. pobeguini were present in this guelta and in the temporary Dâyet el Mercrouda pool (19°45′711 N, 12°49′551 W) located 1.3 km upslope in the Doueïr palm grove. In June 2008, the visible surface of the Agmeïmine guelta was only 0.8 m2 and most of its volume of water was concealed by large rocks (Figure 3D).

Dâyet et-Tefla (Pool)

The Dâyet et-Tefla pool (19°43′121 N, 12°52′012 W) is located at the foot of a cliff in a meander of the Nbéïké wadi. The presence of B. pobeguini was reported [6].

This ancient pool, filled in by alluvial deposits and overgrown with vegetation, was dry in December 2007 and June 2008 (Figure 3E). The L'ehreïjatt guelta (19°43′279 N, 12°51′968 W), located around 300 m upstream, was flowing and home to B. pobeguini in December 2007, but was dry in June 2008.

Douâyât Lemgasse (Pool)

The Douâyât Lemgasse pool (19°42′736 N, 12°52′490 W) is located in the bed of the Nbéïké wadi at the foot of a cliff. The presence of fish in this perennial guelta was not mentioned in the literature.

B. pobeguini were present in this pool in December 2007 and June 2008, but the water was too cloudy to estimate their abundance. In June 2008 the pool was 1.5 m deep, 60 m in length and 3 m wide (Figure 3F). Around 600 m downstream, the Douâyât Begherbane pool (19°42′445 N, 12°52′534 W) contained B. pobeguini in December 2007 but was dry in June 2008.

Dâyet Legleïb (Pool)

The Dâyet Legleïb pool (19°38′654 N, 12°57′873 W) is located in the bed of the Nbéïké wadi, at the intersection of two valleys. The presence of fish in this perennial pool was not mentioned in the literature.

B. pobeguini were present in this pool in June 2008 but the water was too cloudy to estimate their abundance. The pool was 1.1 m deep, 62 m in length and 2.5 m wide (Figure 3G).

Nkedeï (Guelta and Pool)

The Nkedeï guelta (19°37′912 N, 12°48′996 W) is at the foot of a cirque cliff (Figure 3H). The Nkedeï pool (19°38′017 N, 12°49′150 W) is located in the bed of the wadi, around 250 m downstream (Figure 4A). The presence of B. pogeguini was reported both in the guelta and in the pool [5].

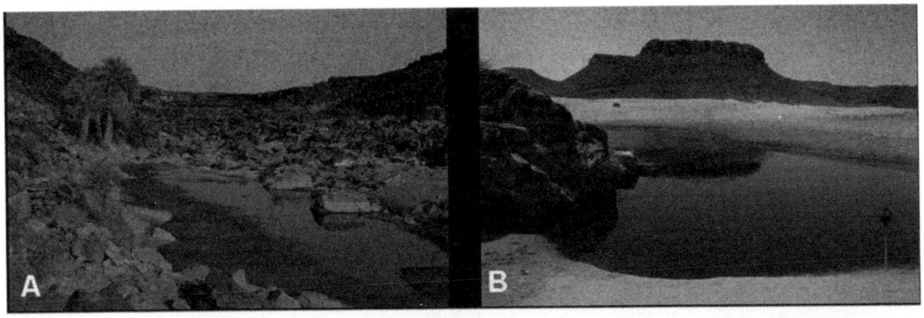

Figure 4. Gueltas and ponds from the Timinit wadi basin (continued). A: Nkedeï pond; B: Glât el Bil pond.

In December 2007, the guelta and the pool provided a habitat for several thousand B. pobeguini. In June 2008, the pool was flowing and home to B. pobeguini but the guelta was completely dry. Monod [5] claimed that the guelta seemed perennial in the 1950s but it was sometimes reduced to several puddles fed by seepage at the foot of the cliff where the fish congregated.

Glât el Bil (Pool)

Glât el Bil is a large pool located in the bed of the Nkedeï wadi (19°31′498 N, 12°57′636 W), just before it reaches a large sandy plain where in periods of flood the Timinit, Nbéïké and Nkedeï wadis join (Figure 4B). It divides into five parts when the water level falls. The presence of fish in this perennial pool is not mentioned in the literature.

In December 2007 and June 2008 the pool provided a habitat for several thousand B. pobeguini. At its largest points in June 2008, it was 110 m in length, 20 m wide and 3.2 m deep.

Discussion

Four species of fish reported in the Adrar mountains of Mauritania in the 1950s were found in the present study: B. macrops, B. pobeguini, S. g. galilaeus and C. anguillaris (Figure 5). However, the population of B. (Hemigrammocapoeta) mirei Estève, 1952, which had been described from the Ksar Torchane spring is now extinct. This was the only locality where this species and subgenus were known. Although Monod cast doubt on their validity [6], he nevertheless recognised the specific morphological characteristics (incomplete lateral line) of this population of barbs and recommended further studies in order to understand their origin. With the drying-up of the spring and the extinction of this population, this has now become impossible. The information gathered on the ground clearly indicates that the further deterioration in the rainfall deficit in 1982 and 1983 (19 mm and 7 mm of rainfall in Atar respectively), occurring after twelve years of drought, caused the spring to run dry definitively in early 1984.

Figure 5. Adrar mountains fishes. A : Barbus macrops, Hamdoun guelta ; B : Barbus pobeguini, Hamdoun guelta ; C : Sarotherodon g. galilaeus, Molomhar guelta ; D : Juvenile Clarias anguillaris, Hamdoun guelta.

Three of the pools and gueltas in the Timinit and Nbéïké wadis where Monod [5], [6] had reported the presence of B. pobeguini became ephemeral from 1984 and 1990 onwards, having been perennial before. This was probably due to the reduced underground flow of these wadis which in turn was caused by the ongoing drought. The four pools in the Nbéïké and Nkedeï wadis where the presence of B. pobeguini is mentioned for the first time in this study are thought by the populations of the neighbouring oases to have always been perennial. It is likely that the population of B. pobeguini for which they provide a habitat is a long-standing one but that it escaped previous investigations.

Of the perennial bodies of water which provide a habitat for fish and which have proved resistant to the ongoing drought, the Hamdoun guelta appears to be the most endangered. In July 2007 its volume was just 1.7 m3 with a maximum depth of 35 cm. Two months previously the guelta was still fed by a trickle of water seeping from the length of the cliff wall and providing approximately 0.7 litres of water per minute, but this seep had almost completely dried up in July and was in danger of not compensating for the very high rate of evaporation caused by the heat at this time of the year. With three species of fish present, the Hamdoun guelta presents the second greatest species richness after the Molomhar guelta. The equilibrium maintained between C. anguillaris on the one hand, and B. pobeguini and B. macrops on the other, is noteworthy. The seven adult catfish measuring 22 to 35 cm in length found in the guelta in 2006–07 could easily have killed off the two small populations of barbs - certainly fewer than two hundred individuals - which survived in July 2007. A few pieces of bread thrown into the guelta were immediately fought over by the three species. The rises in water level caused by the rains of August and September 2007, which gave rise to abundant batches of eggs, allowed the barb populations of the guelta to at least double in size within one month (310 specimens counted in September 2007) and to colonise a large number of ephemeral downstream gueltas. The absence of any fish population upstream of the road overlooking the perennial guelta, contrary to observations in the 1950s [5], [12], confirms the increased vulnerability of the Hamdoun guelta. It should also be noted that a single road accident involving one of the many lorries which use this dangerous mountain road could cause a pollution incident which would definitively wipe out the fish population in the guelta.

The Molomhar gueltas are still the most noteworthy aquatic system in the Adrar region of Mauritania and West Central Sahara. It is the only system which is home to four species of fish, and the only system in the whole central Sahara where Sarotherodon g. galilaeus has been able to survive since the Holocene period. Along with the neighbouring Ilîj guelta and several Tibesti gueltas in Chad, it is also one of the few Sahara sites where a freshwater jellyfish, Limnocnida tanganyicae, has been observed (Figure 6) [27], [28]. The reduced recorded rainfall does not seem to have affected the dry period water level, given that a similar depth of 6 m in the first guelta had been measured in February 1951 [12]. Two kilometres downstream of the gueltas, at the entrance to the Oumm Lemhar wadi canyon, an underground water harvesting system has been installed in order to provide water for the town of Atar but it does not seem to have had a major impact on the perennial gueltas. Several thousand specimens of S. g. galilaeus, B. pobeguini and B. macrops were present in the gueltas. The size of the population of C. anguillaris could not be estimated, but it is certainly much less numerous.

Figure 6. The freshwater jellyfish Limnocnida tanganyicae, Molomhar guelta.

Two large hydrographic networks cover the Adrar region of Mauritania: to the north and west, the Seguellîl wadi network, which occasionally comes into contact with the El Abiod wadi; to the south and east, the Timinit wadi network, into which flow the Nbéïké and Nkedeï wadis. These two large networks were probably separated between 3,500 and 6,000 years before present, and they were also probably in contact with the Senegal River basin in the Holocene and/or Pleistocene periods [14], [29], [30]. During the Holocene, there were large tropical lakes in several regions of Sahara, their whole catchment was probably active, and a diversified fossil fish fauna was collected all over the area [29], [30]. Only B. pobeguini have been able to survive in the Timinit wadi basin, perhaps because of high sand-bank and the reduced volume of most of the perennial bodies of water in this basin in dry periods. B. pobeguini is a small species (with a maximum total length of 53 mm among 500 adult specimens measured in the Adrar), capable of surviving in a few liters of water.

In 1951, Monod [5] was already underlining the fact that the residual bodies of water in the Sahara constituted an unrivalled natural laboratory, some of them completely isolated representing all that remains of immense but now fossil hydrographical basins. An in-depth genetic comparison of these fish populations with those in the Senegal and Niger basins has yet to be completed. It is currently in progress and should improve our knowledge of when the different basins separated and state precisely whether the levels of divergence reached are compatible with the recognition of distinct specific or subspecific entities.

Despite a drought which has lasted for nearly forty years with an intensity which is without known historical precedent, the relict species of tropical fish in the mountains of the Adrar region of Mauritania continue to survive. However, they appear to be highly endangered. Of the thirteen sites mentioned in the literature, four are no longer home to fish, at least on a permanent basis, and one appears to be highly endangered. In the basin of the Timinit, Nbéïké and Nkedeï wadis, three of the five perennial pools or gueltas previously known to provide a habitat for B. pobeguini have become ephemeral. In the Seguellîl wadi basin, the Ksar Torchane spring dried up definitively in 1984, bringing about the extinction of one of the most noteworthy Saharan populations of barb. New fish population extinctions may occur should very low levels of annual recorded rainfall be repeated.

Acknowledgements

I am thankful to Jean-Jacques Albaret, IRD Montpellier, who participated in the December 2007 survey and examined part of the specimens collected. Valuable assistance during field surveys was provided by Jean-François Trape, IRD Dakar, and the guides Kassem Sabar and Yacoub Boukheir from Azougui. Robert Vernet provided helpful comment on an earlier draft of the manuscript.

Author Contributions

Conceived and designed the experiments: ST. Performed the experiments: ST. Analyzed the data: ST. Contributed reagents/materials/analysis tools: ST. Wrote the paper: ST.

References

1. Pellegrin J (1913) Les vertébrés des eaux douces du Sahara. C R Ass Fr Avanc Sci 42: 346–352.

2. Pellegrin J (1921) Les poissons d'eaux douces de l'Afrique du Nord française. Maroc, Algérie, Tunisie, Sahara. Mém Soc Sc Nat Maroc, I, n°2, 216 p.
3. Pellegrin J (1937) Les poissons du Sahara occidental. Ass Fr Av Sc Marseille 337–338.
4. Lhote H (1942) Découverte d'un barbeau au Hoggar. Bull Soc Centr Agric Pêche 49: 19–22.
5. Monod T (1951) Contribution à l'étude du peuplement de la Mauritanie. Poissons d'eau douce. Bull IFAN 13: 802–812.
6. Monod T (1954) Contribution à l'étude du peuplement de la Mauritanie. Poissons d'eau douce (2e note). Bull IFAN 16: 295–299.
7. Monod T (1955) Contribution à l'étude du peuplement de la Mauritanie. Poissons d'eau douce (3e note). Bull IFAN sér A 17: 590–591.
8. Daget J (1959) Note sur les poissons du Borkou-Ennedi-Tibesti. Travaux de l'Institut de Recherches Sahariennes, Alger 173–181.
9. Lévêque C (1989) Remarques taxinomiques sur quelques petits Barbus (Pisces, Cyprinidae) d'Afrique de l'Ouest (première partie). Cybium 13: 165–180.
10. Lévêque C (1990) Relict tropical fish fauna in Central Sahara. Ichthyol Explor Freshwaters 1: 39–48.
11. Estève R (1952) Poissons de Mauritanie et du Sahara oriental. Bull Mus Hist Nat Paris 24: 176–179.
12. Villiers A (1953) Contribution à l'étude du peuplement de la Mauritanie. Note sur la faune aquatique et ripicole de l'Adrar mauritanien. Bull IFAN 15: 631–646.
13. Dekeyser PL, Villiers A (1956) Notations écologiques et biogéographiques sur la faune de l'Adrar. Dakar : Mémoire IFAN, n° 54, 222 p, 25 pl.
14. Dumont HJ (1982) Relict distribution patterns of aquatic animals : another tool in evaluating late pleistocene climate changes in the Sahara and Sahel. Palaeœcology of Africa 14: 1–24.
15. Dumont HJ (1987) Sahara. In: Burgis MJ, Symoens JJ, editors. African wetlands and shallow water bodies. Paris: ORSTOM. pp. 79–154.
16. Le Berre C (1989) Faune du Sahara. 1. Poissons-Amphibiens-Reptiles. Paris: Raymond Chabaud - Lechevallier.
17. Druyan L (1989) Advances in the study of Subsaharan drought. J Climato 9: 77–90.
18. Hulme M (1992) Rainfall changes in Africa: 1931–60 to 1961–90. Int J Climatol 12: 685–699.

19. Hulme M, Doherty R, Ngara T, Ncws M, Lister D (2001) African climate change: 1900–2100. Clim Res 17: 145–168.

20. L'Hôte M, Mahé G, Somé B, Triboulet JP (2002) Analysis of a Sahelian annual rainfall index updated from 1896 to 2000; the drought still goes on. Hydrol Sc J 47: 563–572.

21. Station météorologique d'Atar. Relevés pluviométriques années 1921 à 2007.

22. Vernet R (1995) Climats anciens du Nord de l'Afrique. Paris: L'Harmattan.

23. Gasse F (2006) Climate and hydrological changes in tropical Africa during the past million years. C R Palevol 5: 35–43.

24. Lévêque C (2003) Cyprinidae. In: Paugy C, Lévèque C, Teugels GG, editors. Poissons d'eaux douces et saumâtres de l'Afrique de l'Ouest/The fresh and brackish water fishes of West Africa, vol 1. Paris: IRD. pp. 322–436. In :.

25. Teugels GG (2003) Clariidae. In: Paugy C, Lévèque C, Teugels GG, editors. Poissons d'eaux douces et saumâtres de l'Afrique de l'Ouest/The fresh and brackish water fishes of West Africa, vol 2. Paris: IRD. pp. 144–173.

26. Teugels GG, Thys Van den Audenaerde D (2003) Cichlidae. In: Paugy C, Lévèque C, Teugels GG, editors. Poissons d'eaux douces et saumâtres de l'Afrique de l'Ouest/The fresh and brackish water fishes of West Africa, vol 2. Paris: IRD. pp. 520–600.

27. Dekeyser PL (1955) A propos d'une méduse mauritanienne. Notes Africaines 65: 28–31.

28. Dumont HJ (1994) The distribution and ecology of the fresh- and brackish-water medusae of the world. Hydrobiologia 272: 1–12.

29. Talbot MR (1980) Environmental responses to climatic change in the West African Sahel over the past 20 000 years. In: Faure W, Faure H, editors. The Sahara and the Nile. Rotterdam: Balkema. pp. 37–62.

30. Riser J, Petit-Maire N (1986) Paleohydrographie du bassin d'Araouane à l'Holocène. Rev Geol Dynam Geogr Phys 27: 205–212.

Selection of Reference Genes for Expression Studies with Fish Myogenic Cell Cultures

Neil I. Bower and Ian A. Johnston

ABSTRACT

Background

Relatively few studies have used cell culture systems to investigate gene expression and the regulation of myogenesis in fish. To produce robust data from quantitative real-time PCR mRNA levels need to be normalised using internal reference genes which have stable expression across all experimental samples. We have investigated the expression of eight candidate genes to identify suitable reference genes for use in primary myogenic cell cultures from Atlantic salmon (Salmo salar L.). The software analysis packages geNorm, Normfinder and Best keeper were used to rank genes according to their stability across 42 samples during the course of myogenic differentiation.

Results

Initial results showed several of the candidate genes exhibited stable expression throughout myogenic culture while Sdha was identified as the least stable gene. Further analysis with geNorm, Normfinder and Bestkeeper identified Ef1α, Hprt1, Ppia and RNApolII as stably expressed. Comparison of data normalised with the geometric average obtained from combinations of any three of these genes showed no significant differences, indicating that any combination of these genes is valid.

Conclusion

The geometric average of any three of Hprt1, Ef1α, Ppia and RNApolII is suitable for normalisation of gene expression data in primary myogenic cultures from Atlantic salmon.

Background

Skeletal muscle myogenesis involves numerous steps including the proliferation, migration and fusion of myoblasts to form myotubes; the onset of myofibrillargenesis, and the maturation and hypertrophy of muscle fibres [1,2]. Myogenesis in teleost fish has several unique features compared to mammals, including the production of myotubes throughout much of adult life [3]. The in vitro culture of fish myogenic cells is an attractive system for studying the formation and differentiation of myotubes and examining the effects of various regulatory molecules on gene expression under precisely controlled conditions [4,5]. Furthermore, since traditional gene "knockouts" are unavailable in fish, cell culture provides a viable alternative for functional assays.

A pre-requisite for the quantitative measurement of gene expression is the identification of suitable reference genes to normalise the data [6,7]. Reference genes are required to normalise for differences in RNA input and mRNA/rRNA ratios between samples [8]. Also, differences in reverse transcription efficiencies between samples can occur due to the presence of inhibitors carried over from the RNA purification [8], and the presence of PCR inhibitors can affect the number of cycles required to reach the quantification cycle value [9]. As gene expression patterns change in response to many stimuli, stable expression of reference genes needs to be confirmed for each experimental system. For example, genes identified as being stable in whole muscle samples, may not be suitable as reference genes in myogenic cell culture due to the vast changes in cell metabolism and structure that occur during the transition from myoblast to myotube. Previous myogenic cell culture experiments using the C2C12 cell line have relied on Actb

[10] and Gapd [11] as internal reference genes, however, the validity of these genes is questionable as Gapd [12,13] and Actb [14] expression has been shown to vary considerably.

A number of computer based analysis packages have been developed which analyse gene expression patterns and allow for the identification of stable reference genes. Vandesomple et al [15] designed the widely used geNorm package which uses a pairwise analysis of gene expression to identify stable reference genes. Likewise, Bestkeeper [16] performs a pairwise comparison, whereas Normfinder [17] uses a mathematical model to estimate overall expression variation of candidate reference genes, but also the variation between sample groups. Vandesomple et al [15] demonstrated that use of a single reference gene can lead to aberrant gene expression values, and now it is widely accepted that using several reference genes for normalisation is preferable.

Currently there is no information available on reference gene stability in fish myogenic cell cultures. In this paper we examine the stability of eight potential reference genes during the transition from single nucleated myoblasts to multinucleated myotubes in myogenic cell cultures derived from Atlantic salmon, one of the most commercially important aquaculture species.

Results

Cell cultures were visualised using confocal microscopy and the phenotype of cells determined at 2 d, 5 d, 8 d, 11 d and 14 d (Figure 1). The myogenic nature of the cell culture was confirmed by the presence of the myogenic marker desmin (Figure 1 a, e, i, m, q) and the presence of multi-nucleated myotubes visualised by Alexa Fluor 568-phalloidin stained actin filaments (Figure 1 b, f, j, n, r) and nuclei stained with sytox green (Figure 1 c, g, k, o, s). At 2 d, all cells were mononucleic (Figure 1a–d), which then fused to form small myotubes at 5 d (Figure 1e–h) and 8 d (Figure 1i–l) and then as the culture progressed, large myotubes (Figure 1m–p) and sheets of large multi-nucleated myotubes at 14 d (Figure 1q–t).

Each of the candidate reference genes tested gave amplification from cDNA derived at each time point of the cell culture, while the no template control (NTC) and minus reverse transcription controls (-RT) gave no signal. The specificity of each primer was by confirmed by the presence of a single band on agarose gel electrophoresis and the presence of a single peak in the dissociation curve analysis which exactly matched the dissociation curve of a plasmid standard of known sequence. Amplification of the correct product was confirmed in each case through the sequence analysis of cloned PCR products.

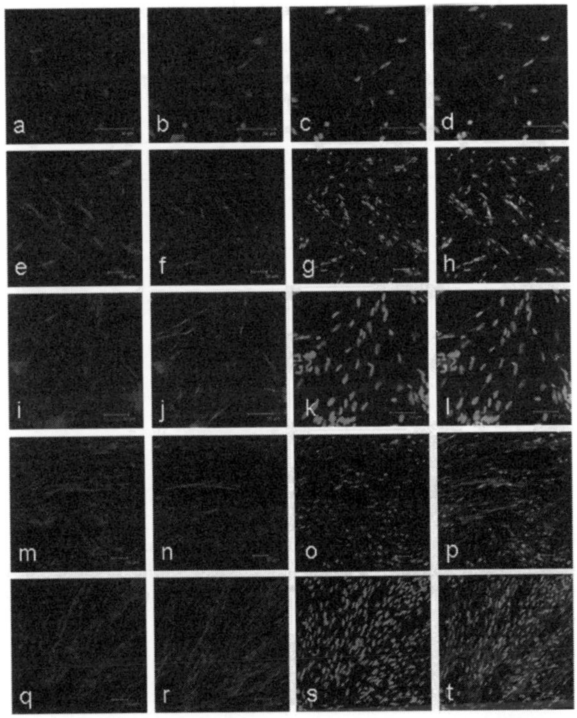

Figure 1. Growth and differentiation of myogenic cells extracted from Salmo salar fast myotomal muscle. Growth is shown at 2 d (a-d), 5 d (e-h), 8 d (i-l), 11 d (m-p) and 14 d (q-t) after cell extraction. Myogenic cells were identified by positive desmin staining (a,e,i,m,q). Actin counterstained with phalloidin (b,f,j,n,r) and nuclei stained with sytox green (c,g,k,o,s) also confirmed the presence of multinucleated myotubes shown in the overlay (d,h,l,p,t). Scale bars represent 50 μm.

Reference Gene Stability

The cell culture undergoes many structural and metabolic changes during the transition from mononucleic cells to multi-nucleated myotubes. We therefore chose to analyse the expression of genes from early time points as well as time points after the culture has produced multi-nucleated myotubes (17 d and 20 d). The analysis of reference gene stability can thus be performed in three phases, the first in developing myotubes (2 d – 11 d), the second in established myotubes (11 d – 20 d) and the third covering all time points. Based on the raw expression data (Figure 2), Sdha and Pgk show higher levels of variance than the remaining genes and appear the least stable. Figure 3 shows the raw expression values obtained for each gene at each of the time points sampled. For several of the genes, there is higher intergroup variation indicating that these genes are differentially regulated during the progression of the cell culture. For example, 18SrRNA, Pgk, and

Actb all have higher Cq values at 2 d and 5 d when the culture is predominantly mononucleic cells than they do at later time points when myotubes have formed. Sdha has high inter-assay and intra-assay variation and is clearly unsuitable as a reference gene in Atlantic salmon myogenic culture.

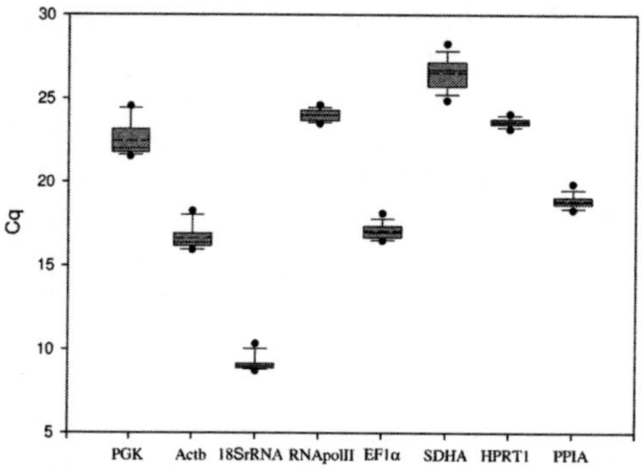

Figure 2. Expression values for all genes at all time points from Atlantic salmon myogenic cell culture. The raw quantification cycle (Cq) values (n = 42) are represented by box and whisker diagram (box represents quartiles). The mean value is indicated by the dashed line and the 5th and 95th percentiles are indicated by the dots above and below each plot.

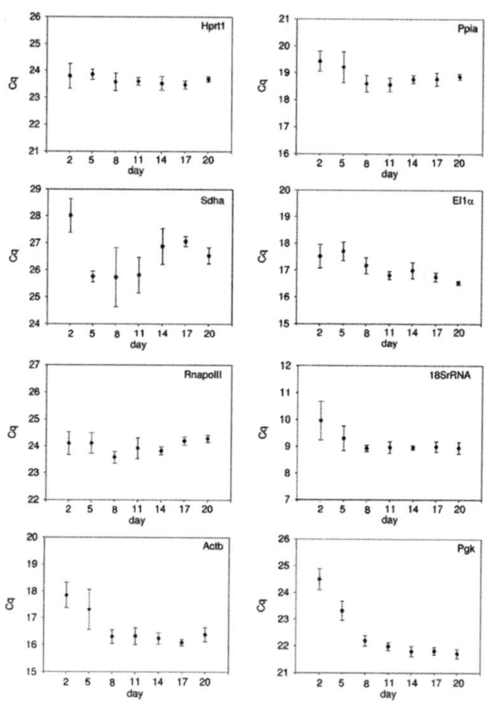

Figure 3. Individual expression profiles for each candidate reference gene at each day of culture. Values shown are raw Cq values represented as mean ± SD (n = 6).

As inspection of raw Cq values alone is insufficient for determining gene expression stability, the data obtained were further analysed using three software packages Bestkeeper, geNorm and Normfinder. Each package uses a different algorithm to determine the most stable reference gene, and as no single method has been accepted as the most appropriate for identifying stable gene expression, all three packages were used for analysis.

geNorm Analysis

Data analysis using geNorm was performed two ways. The first method used the absolute values derived from a plasmid standard curve as input, the second used the delta Cq method, with the PCR efficiencies based on a dilution series of pooled cDNA samples. The results from the three geNorm analyses covering all time points (absolute value method), developing myotubes and established myotubes are shown in figure 4. When all samples were analysed, the genes were ranked in an identical order using both analysis methods from most to least stable:

Hprt1>RNApolII>Ppia>Ef1α>18SrRNA>Actb>Pgk>Sdha.

Analysis of days 2–11 using both analysis methods revealed the same order of stability as when all days were analysed except for Ef1 and Ppia swapping order. When days 11–20 are analysed using the absolute method, the order changes from most to least stable:

Pgk>Actb>Hprt1>Ppia>Ef1>18SrRNA>RNApolII>Sdha.

When the delta Cq method was used, the order changed to:

Pgk>Hprt1>Actb>Ppia>Ef1>18SrRNA>RNApolII>Sdha.

The change in order from days 2–11 to days 11–20 likely reflects the changes in metabolism and structure that occur during differentiation and growth process as myotubes form. In all three analyses, the M values obtained for Hprt1, Ppia and Ef1α were quite similar, ranging from 0.23–0.36. Based on the similar M values, it would appear that any combination of Hprt1, Ppia and Ef1α would be suitable for normalisation.

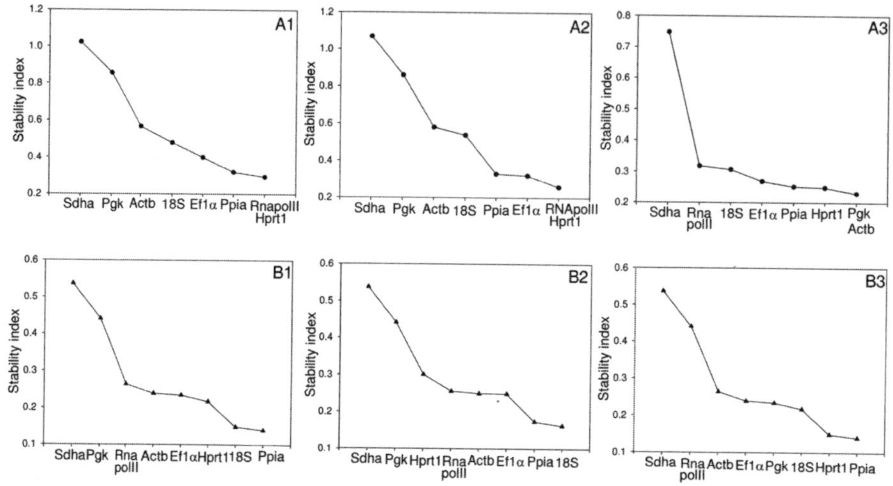

Figure 4. Stability indices calculated with geNorm (A) and Normfinder (B). Stability indices are shown for all time points (1), developing myotubes at days 2–11 (2) and in established myotubes at days 11–20 (3). Stability of gene expression is inversely proportional to the stability index, so least stable genes are to the left and the most stable to the right for each graph.

Normfinder Analysis

The stability of candidate reference genes was also analysed using Normfinder (Figure 4). The overall rank of genes from most to least stable for all time points was:

Ppia>18SrRNA>Hprt1>Ef1α>Actb>RNApolII>Pgk>Sdha.

In developing myotubes the genes were ranked:

18SrRNA>Ppia>Ef1α>Actb>RNApolII>Hprt1>Pgk>Sdha,

and in established myotubes:

Ppia>Hprt1>18SrRNA>Pgk>Ef1α>Actb>RNApolII>Sdha.

It is noteworthy that with the exception of Sdha and Pgk, all genes had stable expression in all three analyses, with stability indices between 0.3 and 0.08.

Bestkeeper Analysis

Using the initial statistics produced by Bestkeeper (Figure 2), the genes were ranked in the following order from most to least stable:

Hprt1>RNApolII>Ppia>Ef1α>Sdha>Actb>18SrRNA>Pgk

when all time points were examined. For 2d-11d, the genes were ranked:

Hprt1>RNApolII>Ef1α>Ppia>Actb>18SrRNA>Sdha>Pgk and for 11d-20d:

Pgk>Hprt1>Ppia>RNApolII> Ef1α>Actb>18SrRNA>Sdha.

All candidate reference genes examined were positively correlated with each other (Table 1), with the highest correlations found between Actb/Pgk (r = 0.914) and Ppia/Actb (r = 0.874). Correlations between the remaining genes ranged from 0.198 for RNApolII/Ef1α to 0.814 for 18SrRNA/Actb (Table 1). The low level of correlation between many of the genes is due to the small inter and intra-group variation observed for the majority of the genes. From the initial statistics, the four least stable genes were removed from further analysis. The algorithm used in Bestkeeper then calculates the correlation of each gene with the Bestkeeper Index which is calculated as the geometric mean of the candidate reference genes. The candidate reference genes were ranked in order based on their correlation with the Bestkeeper index based on the four most stable genes. For all time points the genes were ranked in order most to least stable:

Ppia>Ef1α>Hprt1>RNApolII

whereas in developing myotubes the order was:

Ppia>Ef1>RNApolII>Hprt1

and in established myotubes the ranking was Ef1α>Ppia>Pgk>Hprt1.

Table 1. Correlations between candidate reference genes expression patterns

	Pgk	*Actb*	*18SrRNA*	RNApol II	*EF1 α*	*Sdha*	*Hprt1*	*Ppia*
Actb	0.914	-	-	-	-	-	-	-
18SrRNA	0.755	0.814	-	-	-	-	-	-
RNApolII	0.229*	0.449	0.382	-	-	-	-	-
EF1 α	0.779	0.788	0.603	0.198*	-	-	-	-
Sdha	0.393	0.343	0.517	0.325	0.076*	-	-	-
Hprt1	0.514	0.612	0.488	0.57	0.656	0.281*	-	-
Ppia	0.732	0.874	0.672	0.664	0.705	0.372	0.64	-

Pearson correlation coefficients (r) based on the quantification cycle (Cq) value between reference genes. Probability greater than 0.05 is indicated by an asterisk.

Normalisation

Based on the results from the three analysis methods, four genes, EF1α, Ppia, Hprt1 and RNApolII are consistently stable. In order to assess the stability of the normalisation factors obtained, we first compared the normalised expression of

Des to various combinations of the geometric average of two genes (Figure 5). Six normalisation factors were derived by calculating the geometric averages of the following gene combinations: A: HPRT1, PPIA; B: RNApolII, HPRT1; C: EF1α, HPRT1; D: EF1α, Ppia; E: RNApolII, EF1α; F: RNApolII, Ppia. We found that there were significant differences (ANOVA $P < 0.05$) in Des gene expression at day 11 (B v D), at day 17 (A v B, D, and F; B v C, D and E; C v E and F; D v E and F) and at day 20 (B v C, D and E; D v F). We therefore examined normalisation using geometric average of three genes (Figure 6). Four normalisation factors were derived by calculating the geometric averages of the following gene combinations: A: EF1α, RNApolII, Hprt1; B: EF1α, Ppia, Hprt1; C: EF1α, RNApolII, Ppia; D: Ppia, Hprt1, RNApolII. No significant differences were observed (ANOVA $p = 0.05$) when the normalised data at each time point were compared between the different normalisation factors, indicating that using the geometric average of any three of these genes is suitable for normalisation.

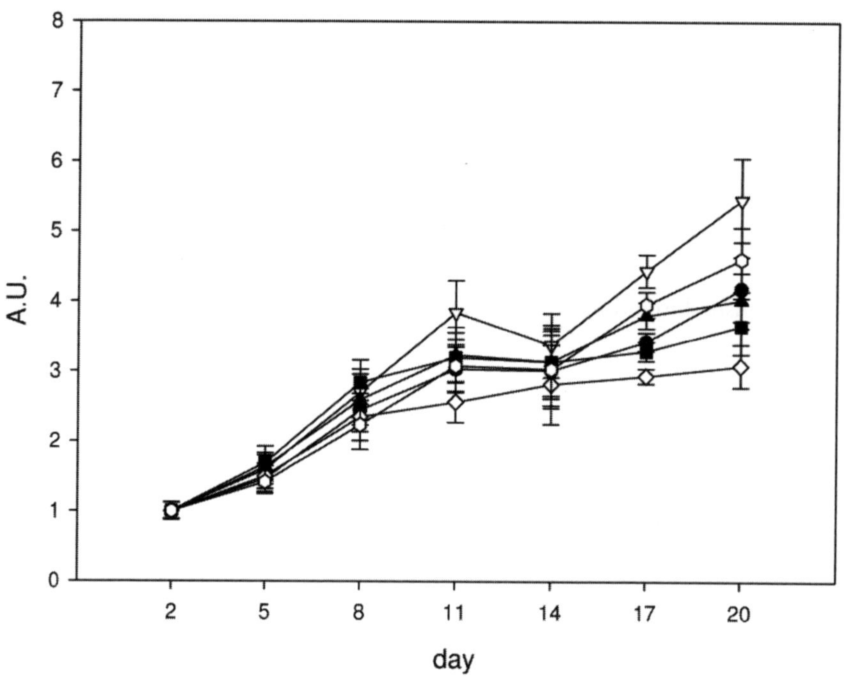

Figure 5. Normalisation of desmin mRNA expression to various combinations of the geometric average for two genes. Data shown are all calculated relative to day 2, so that day 2 values are equal to 1 arbitrary unit (A.U.). Six normalisation factors were derived by calculating the geometric averages of the following gene combinations: A: Hprt1, Ppia (closed circle); B: RNApolII, Hprt1 (open triangle); C: EF1α, Hprt1 (closed square); D: EF1α, Ppia (open diamond); E: RNApolII, EF1α(closed triangle); F: RNApolII, Ppia (open circle). Values shown are the mean normalised value ± S.E. (n = 6).

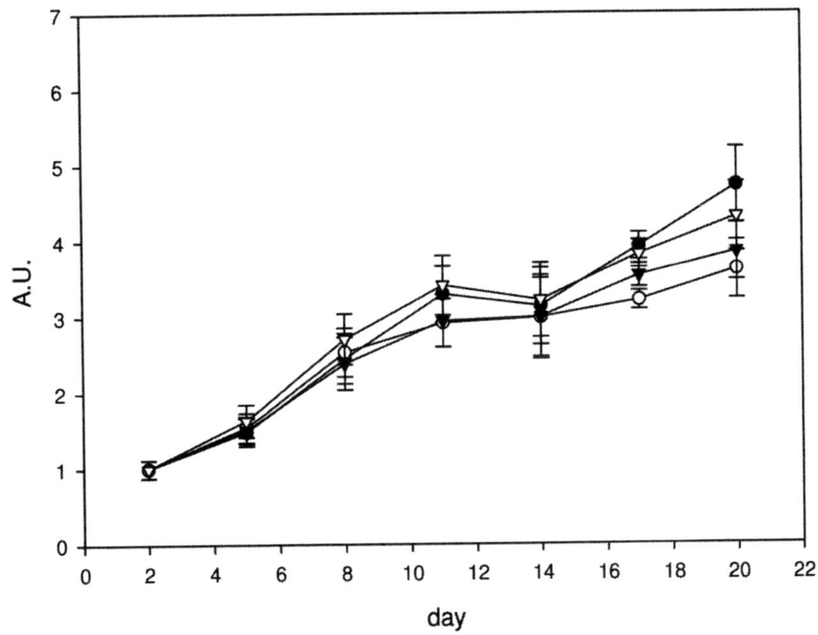

Figure 6. Normalisation of desmin mRNA expression to various combinations of the geometric average of three genes. Data shown are all calculated relative to day 2, so that day 2 values are equal to 1 arbitrary unit (A.U.). Four normalisation factors were derived by calculating the geometric averages of the following gene combinations: A: EF1α, RNApolII, Hprt1 (closed circle); B: EF1α, Ppia, Hprt1 (open circle); C: EF1α, RNApolII, Ppia (closed triangle); D: Ppia, Hprt1, RNApolII (open triangle). Values shown are the mean normalised value ± S.E. (n = 6).

Discussion

In this study, we examined the expression of eight candidate reference genes for normalisation of quantitative real-time PCR data from a primary culture of Atlantic salmon myogenic cells. The identification of genes with stable expression in all samples of an experiment is crucial as it is necessary to normalise for variability between samples introduced during the production of the cDNA [6,7]. As a universal reference gene with stable expression in all experimental systems is not available, suitable reference genes for each experiment need to be determined.

Myogenic cell culture is characterised by distinct phases where cells first proliferate, and then fuse to form multinucleated myotubes [2]. We therefore identified genes that were stable early time points where the majority of the cells are mononucleic and forming small myotubes (culture days, 2–11) those stable in established myotubes (days 11–20) and those most stable for the entire culture period. When the raw Cq values obtained at each time point are compared

(Figure 3), it is clear that for the majority of the genes examined, stable expression is observed once myotubes have become established in the culture, whereas higher intra and intergroup variation is observed when myotubes are developing. For example, Pgk, a glycolytic enzyme, appears to be upregulated as myotubes start to form, and then has stable expression in established myotubes as indicated by the low inter and intra-group variation (Figure 3).

To identify stably expressed genes, analysis packages such as geNorm and Bestkeeper perform a pairwise comparison of gene expression across the various samples in an experiment. Therefore it is crucial that genes used are not co-regulated or present on the same pathway, as co-regulated genes will likely have similar expression patterns and would therefore appear to be stably expressed in any biological experiment. For this reason we chose genes involved in a number of different biological processes (Table 2), such as nucleotide recycling (Hprt1), peptide isomerisation (Ppia), glycolysis (Pgk), citric acid cycle (Sdha), ribosome assembly (18SrRNA), transcription (RNApolII), translation (Ef1α) and cytoskeleton structure (Actb).

Table 2. GenBank accession numbers and function of selected reference gene candidates

Gene symbol	Accession number	Gene Name	Function
EF1 α	BG933853	Eukaryotic elongation factor 1a	Translation
RNApolII	BG936649	RNA polymerase 2	Transcription
18SrRNA	AJ427629	18S ribosomal RNA	Component of ribosome
Ppia	DY727143	Prolylpeptidyl isomerase A	Peptide isomerisation
Pgk	DW536646	Phosphoglycerate kinase	Glycolysis
Actb	G933897	Beta actin	Cytoskeleton
Hprt1	EG866745	Hypoxanthine phosphoribosyl transferase 1	Purine salvaging
Sdha	GE769149	Succinate dehydrogenase complex subunit A flavoprotein	Oxidation of succinate

Based on the M values obtained in geNorm (Figure 4), the stability index from Normfinder (Figure 4) and the descriptive statistics produced by Bestkeeper (Figure 2), it would appear that several of the genes used in this study are suitable for normalisation of gene expression data from salmon myogenic cell cultures. For example, Pfaffl et al [16] recommends using genes that have a standard deviation for the Cq values less than one for calculating a Bestkeeper index. In our study, all genes examined had standard deviation less than one. This is also reflected in the slight changes in the order of gene stability obtained from each of the three software packages. The least stable gene identified by all analysis methods was Sdha. Sdha has been used as a reference gene in a number of studies using different tissues [18,19], however its high inter and intra-group variation make it unsuitable for normalisation in salmon myogenic cell cultures.

Results obtained from geNorm identify Hprt1 and RNApolII as the most stable genes when all time points were examined, however, the M values obtained for Ppia and Ef1α are quite similar and thus these genes are also likely to be suitable for normalisation. The same set of genes was found to be stable in developing myotubes, but differed in established myotubes where Pgk/Actb were found to be the most stable, although the Hprt1 and Ppia also had low M values and can be considered stably expressed.

The most stable genes identified for all time points by Normfinder ranked in descending order were Ppia>18SrRNA>Hprt1>Ef1α. Both Ppia and Hprt1 have been reported to give stable expression in mouse C2C12 myotubes [20,21] and Ef1α has been reported to have stable expression in some Salmon tissues [22]. As the 18S and 28S ribosomal RNAs are highly abundant and account for the vast majority of RNA, it is unsurprising that 18SrRNA is found to be stable across the samples as equal amounts of RNA were reverse transcribed. However, Vandesomple et al [15] criticise the use of 18SrRNA as a housekeeping gene due to its high abundance making baseline subtraction difficult. Also, transcription of rRNA and mRNA occur via RNA polymerase I and II respectively which may lead to imbalances in the two mRNA fractions as reported by Solanas et al [23]. The similar stability indices obtained for Ppia, Hprt1 and Ef1α, identify all of these genes as suitable for normalisation.

Similar to the results of geNorm and Normfinder, Bestkeeper analysis revealed the most stable genes to be Ppia, Ef1α, Hprt1 and RNApolII. Interestingly, Actb, which has been used as a reference gene in numerous studies [10,24], was found to be the third least stable gene in this analysis, having high intra-group variation in developing myotubes as well as high inter-group variation when comparing developing and established myotubes. These differences in Cq values between developing and established myotubes indicate that Actb is differentially regulated during differentiation of Atlantic salmon myogenic cells, as reported in chicken and mouse myoblast culture [25,20] and is therefore unsuitable as reference gene for myogenic culture. Interestingly, RNApolII and Hprt1, which were identified as the most stable genes in geNorm (Figure 4) had a correlation coefficient of only 0.57, which was lower then for many of the other genes (Table 1). The selection as the most stable genes in geNorm is likely a reflection of the low intra and inter-group variation observed for both of these genes (Figure 3).

Vandesomple et al [15] recommend using the geometric average of three reference genes for accurate normalisation. To assess the suitability of the reference gene candidates, we first normalised the expression of Des to combinations of the geometric average of two reference genes from Ppia, Hprt1, Ef1α and RNApolII (Figure 5). We found significant differences in Des expression at days 11, 17 and 20 when comparing results from different combinations of reference genes. However, when three genes were used, there were no significant differences between any of the combinations of reference genes (Figure 6) indicating that all four genes are suitable for normalisation when the geometric average of three genes is used.

Conclusion

To the best of our knowledge, this is the first study examining gene expression stability in myogenic culture of a teleost species and thus provides a useful platform for gene expression studies using this system. The data provided in this paper may also be useful in guiding researchers performing myogenic cell culture in other teleost species. We recommend using a three gene normalisation factor using the geometric average of any combination of EF1α, Ppia, RNApolII and Hprt1.

Methods

Isolation of Myogenic Satellite Cells

Myosatellite cells were isolated using a method similar to that described by Koumans et al [26]. Juvenile Atlantic salmon (Salmo salar L) 30 ± 6 g (mean ± s.d., N = 10) were used for each culture. As the experimental animals had not undergone gonadal development, the gender of the fish was not determined. Fast myotomal muscle was dissected under sterile conditions and placed in extraction media consisting of Dulbecco's modified eagle's media (DMEM) 9 mM NaHCO3, 20 mM HEPES (pH 7.4) with 15% (v/v) horse serum and 1 × antibiotics (100 units/ml penicillin G, 100 μg/ml streptomycin sulfate, 0.25 μg/ml amphotericin B) (Sigma, Gillingham, Dorset, UK) at a ratio of 1 gram of muscle per 5 ml extraction media. The tissue was then minced with a sterile scalpel before centrifugation at 300 g for 5 min, and two washes with DMEM without horse serum. The muscle pieces were digested with collagenase (0.2% m/v in DMEM, Type 1a, Sigma, Gillingham, Dorset, UK) for 70 minutes at room temperature in the dark, before centrifugation at 300 g for 5 minutes. The resulting pellet was washed twice with DMEM before being passed through a pipette repeatedly to separate cells.

Samples were further digested with trypsin (0.1% in DMEM) for 20 minutes at room temperature. The resulting cell suspension was centrifuged (300 g, 1 min). The supernatant was poured into 20 × vol of extraction media containing serum to inhibit trypsin activity. The pellet was further digested by a second treatment with trypsin for 20 min at room temperature, before centrifugation at 300 g 1 min. The supernatant was poured into 20 × volume of extraction media. The extraction media containing the cell suspension was centrifuged 300 g, 20 min. Cell pellets were re-suspended in 30 ml of basal medium before mechanical trituration through 10 ml and 5 ml pipettes until cells are separated. The cell suspension was then passed through 100 μm and 40 μm nylon cell strainers (BD Biosciences San Jose, CA, USA) and centrifuged 20 min 300 g. The cells were resuspended in basal media, cell number determined using hymaecytometer, and then diluted to give approximately 1.5 × 106 cells/ml.

Cell Culture

All cell culture methods were performed using Aseptic technique in a Microflow 2 Advanced biosafety cabinet (Bioquell Ltd, Andover, UK). 6 well cell culture plates (Greiner Bio-One Ltd, Stroudwater, UK) were treated with a 100 ug/ml poly-lysine solution (Sigma, Gillingham, Dorset, UK) at 4 μg/cm^2 for 5 minutes at room temperature, then aspirated before 2 washes with sterile water and allowed to air dry. 1 ml of laminin (Sigma, Gillingham, Dorset, UK) in DMEM at 20 μg/ml was applied to each well and incubated at 18°C overnight prior to plated cells. Cell culture was performed using complete medium (DMEM, 9 mM NaHCO3, 20 mM HEPES (pH 7.4), supplemented with 10% foetal calf serum (Sigma, Gillingham, Dorset, UK) and 1 × antibiotics (Sigma, Gillingham, Dorset, UK) which was changed daily.

Immunofluorescence of Culture Cells

Cells were grown on glass coverslips treated with poly-L-lysine and laminin as described above. Samples were washed 2 × in PBS, fixed in 4% (m/v) paraformaldehyde for 20 min at room temperature, washed 2 × 5 mins in PBS, permeabilised with 0.2% triton X-100 PBS for 5 minutes, washed 2 × in PBS and then blocked in 5% NGS, 1.5% BSA, 0.1% triton X-100 PBS for 1 hour at room temperature. All antibody steps were performed in PBST (1% BSA, 0.1% triton X-100 in PBS). Desmin antibody (Sigma, Gillingham, Dorset, UK) was diluted 1:20 in PBST and incubated overnight at 4°C, washed 3 × in PBS. A 1: 400 dilution of anti-rabbit Alexa Fluor 405 antibody (Invitrogen, Carlsbad, CA, USA) in PBST was incubated for 1 hour at room temperature, and washed 3 × in PBS. Cells were then counterstained for actin with Alexa Fluor Phalloidin 568 (Invitrogen, Carlsbad, CA, USA) and nuclei with Sytox green (Invitrogen, Carlsbad, CA, USA) as per manufacturer's recommendations. Cells were imaged using a Leica TCS SP2 confocal microscope.

Quantitative Real Time PCR Experiments

The following procedures were performed as to comply with the MIQE guidelines [27].

RNA Extraction and cDNA Synthesis

RNA was immediately extracted from duplicate wells of 3 separate cell cultures. RNA extraction and genomic DNA removal was performed using a RNeasy plus kit (Qiagen Inc., Chatsworth, CA, USA) as per manufacturer's recommendations. RNA was concentrated by ethanol precipitation and quantified using a NanaoDrop 1000 spectrophotometer (Thermo Fisher Scientific, Waltham, MA, USA). Only RNA with an A260/280 ratio between 1.8 and 2.1 and an A260/230 above 1.9 was used for cDNA synthesis. For samples where enough RNA was obtained (excludes day 2), the integrity of the RNA was confirmed by gel electrophoresis. Residual genomic DNA was removed using the genomic DNA wipeout buffer included in the Quantitect reverse transcription kit (Qiagen Inc., Chatsworth, CA, USA). 800 ng of RNA was reverse transcribed into cDNA for 30 min at 42°C using a Quantitect reverse transcription kit (Qiagen Inc., Chatsworth, CA, USA) as per manufacturer's recommendations.

Quantitative PCR

qPCR was performed using a Stratagene MX3005P QPCR system (Stratagene, La Jolla, CA, USA) with Brilliant II SYBR (Stratagene, La Jolla, CA, USA). cDNA used in qPCR was first diluted 80-fold with nuclease free H_2O. Each qPCR reaction mixture contained 7.5 μl 2 × Brilliant II SYBR green master mix (Surestart Taq DNA polymerase, 2.5 mM $MgCl_2$), 6 μl cDNA (80-fold dilution), 500 nM each primer and RNase free water to a final volume of 15 μl. Amplification was performed in duplicate in 96 well plates (Stratagene, La Jolla, CA, USA) with the following thermal cycling conditions: initial activation 95°C for 10 minutes, followed by 40 cycles of 15 s at 95°C, 30 s at 60°C, and 30 s at 72°C. Control reactions included a no template control (NTC) and no reverse transcription control (-RT). Dissociation analysis of the PCR products was performed by running a gradient from 60 to 95°C to confirm the presence of a single PCR product. Products were also sequenced to confirm identity. A 4-fold dilution series made from known concentrations of plasmid containing the PCR inserts was used to calculate absolute copy numbers for each of the genes examined. PCR efficiencies for input into Bestkeeper were calculated from a dilution series (1/20, 1/40, 1/80, 1/160, 1/320, 1/640) of cDNA

Standards for calculating absolute copy number for each gene were prepared by cloning the PCR product from each primer pair into a T/A pCR4-TOPO vector (Invitrogen, Carlsbad, CA, USA) and transformation of chemically competent TOP10 Escherichia coli cells (Invitrogen, Carlsbad, CA, USA). Individual colonies were grown and plasmids purified using Fastprep plasmid purification method (Eppendorf, Hamburg, Germany). The concentration of each plasmid was calculated based on absorbance at 260 nm, and a dilution series produced for calculation of copy number via qPCR.

Primer Design

Primers were designed using NetPrimer (Premier BioSoft, Palo Alto, CA, USA) to have Tm of 60°C, and where possible, were designed to cross an exon-exon junction to avoid amplification of contaminating genomic DNA. To determine exon-intron junction sites, genomic sequences for orthologous genes from Danio rario, Gasterosteus aculeatus, Oryzias latipes, Takifugu rubripes and Tetraodon nigroviridis were retrieved from Ensembl http://www.ensembl.org/index.html, and compared to the Salmo salar cDNA sequences using the Spidey software tool http://www.ncbi.nlm.nih.gov/spidey/. Primers were designed across conserved exon-intron junctions and used at a final concentration of 500 nM. 18SrRNA, at 500 nM gave poor amplification efficiencies, however this was improved using a final concentration of 1.5 μM. The primers used for qPCR are listed in table 3 and have been submitted to rtprimerdb http://www.rtprimerdb.org/ [28].

Table 3. qPCR primer sequences, and amplification parameters.

Gene	Primer sequence (5'-3')	Amplicon size (bp)	Tm (°C)	E (%) plasmid	R^2 plasmid	E (%) cDNA	R^2 cDNA
EF1α	f: GAATCGGCTATGCCTGGTGAC r: GGATGATGACCTGAGCGGTG	141	86.0	96.0	0.998	99.5	0.999
RNApolII	f: CCAATACATGACCAAATATGAAAGG r: ATGATGATGGGGATCTTCCTGC	157	84.8	95.3	0.996	98.5	0.999
18SrRNA	f: TCGGCGTCCAACTTCTTA r: GCAATCCCCAATCCCTATC	189	86.5	95.6	0.995	94.5	0.999
Ppia	f: CATCCCAGGTTTCATGTGC r: CCGTTCAGCCAGTCAGTGTT	203	85.9	96.4	0.998	96.5	0.999
Pgk1	f: CTCGGTGATGGGGCTTAGG r: TCATTGGTGGAGGCGACA	160	87.0	98.1	0.999	99.5	0.999
Actb	f: TGACCCAGATCATGTTTGAGACC r: CTCGTAGATGGGTACTGTGTGGG	146	83.8	93.2	0.997	100	0.999
Hprt1	f: CCGCCTCAAGAGCTACTGTAAT r: GTCTGGAACCTCAAACCCTATG	255	81.8	92.1	0.996	90.0	0.997
Sdha	f: CATGTTACCAAGGGCTGCAT r: GTGTCAGATGATATCTCAACCCAG	207	85.8	99.0	0.997	95.5	0.998
Des	f: GTCCATCTGGATCTGCACCT r: GGCTGCTTTCAGAGCTGATG	169	82.8	99.5	0.998	99.0	0.996

PCR efficiencies, amplicon melting temperature (Tm) and correlation coefficients of standard curves (from plasmid and cDNA) for candidate reference genes and desmin.

Data Analysis

The stability of candidate reference genes was determined using geNorm [15], Normfinder [17] and Bestkeeper [16]. Input data for geNorm and Normfinder were absolute values derived from a plasmid standard curve with the data for geNorm transformed as per author's guidelines. Input for Bestkeeper was the Cq values, and the PCR efficiencies calculated from a dilution series (1/20, 1/40, 1/80, 1/160, 1/320, 1/640) of cDNA. Normfinder Analysis of inter and intra group variation was performed on all data, days 2–11 and days 11–20. Statistical analysis was performed with Minitab (Minitab Inc).

Abbreviations

DMEM: Dulbecco's modified eagle's media; PBS: phosphate buffered saline; Actb: actin beta; Ef1α: eukaryote elongation factor 1 alpha; RnapolII: Rna polymerase 2; Gapd: glyceraldehyde-3-phosphate dehydrogenase; Hprt1: hypoxanthine phosphoribosyl transferase 1; Ppia: prolylpeptidyl isomerase A; Pgk: phosphoglycerate kinase; Sdha: succinate dehydrogense; 18SrRNA: 18S ribosomal RNA; Des: desmin.

Author Contributions

NB performed the experimental work and wrote the first draft of the manuscript. IJ contributed to study design and writing of the manuscript. Both authors read and approved the final manuscript.

Acknowledgements

This work was supported by a grant from the Biotechnology and Biological Research Council grant (BB/D015391/1). The authors thank Daniel Garcia de la serrana Castillo for guidance in cell culture methods and Dr Jorge Fernandes for EF1α, RNApolII and Actb primers.

References

1. Horsley V, Pavlath GK: Forming a multinucleated cell: molecules that regulate myoblast fusion. Cells Tissues Organs 2004, 176:67–78.
2. Richardson BE, Nowak SJ, Baylies MK: Myoblast fusion in fly and vertebrates: New genes, new processes and new perspectives. Traffic 2008, 9(7):1050–1059.
3. Johnston IA: Environment and plasticity of myogenesis in teleost fish. J Exp Biol 2006, 209(12):2249–2264.

4. Fauconneau B, Paboeuf G: Sensitivity of muscle satellite cells to pollutants: an in vitro and in vivo comparative approach. Aquat Toxicol 2001, 53(3–4):247–263.

5. Castillo J, Ammendrup-Johnsen I, Codina M, Navarro I, Gutierrez J: IGF-I and insulin receptor signal transduction in trout muscle cells. Am J Physiol Regul Integr Comp Physiol 2006, 290(6):R1683–R1690.

6. Bustin SA: Absolute quantification of mRNA using real-time reverse transcription polymerase chain reaction assays. J Mol Endocrinol 2000, 25(2):169–193.

7. Le Bail A, Dittami SM, de Franco PO, Rousvoal S, Cock MJ, Tonon T, Charrier B: Normalisation genes for expression analyses in the brown alga model Ectocarpus siliculosus. Bmc Molecular Biology 2008, 9(75).

8. Bustin SA: Quantification of mRNA using real-time reverse transcription PCR (RT-PCR): trends and problems. J Mol Endocrinol 2002., 29(1): UNSP 0952-5041/0902/0029-0023

9. Kontanis EJ, Reed FA: Evaluation of real-time PCR amplification efficiencies to detect PCR inhibitors. J Forensic Sci 2006, 51(4):795–804.

10. Sakiyama K, Abe S, Tamatsu Y, Ide Y: Effects of stretching stress on the muscle contraction proteins of skeletal muscle myoblasts. Biomed Res 2005, 26(2): 61–68.

11. Pearen MA, Ryall JG, Maxwell MA, Ohkura N, Lynch GS, Muscat GEO: The orphan nuclear receptor, NOR-1, is a target of beta-adrenergic signaling in skeletal muscle. Endocrinology 2006, 147(11):5217–5227.

12. Deindl E, Boengler K, van Royen N, Schaper W: Differential expression of GAPDH and beta-actin in growing collateral arteries. Mol Cell Biochem 2002, 236(1–2):139–146.

13. Fernandes JMO, Mommens M, Hagen O, Babiak I, Solberg C: Selection of suitable reference genes for real-time PCR studies of Atlantic halibut development. Comp Biochem Physiol B 2008, 150(1):23–32.

14. Selvey S, Thompson EW, Matthaei K, Lea RA, Irving MG, Griffiths LR: beta-actin – an unsuitable internal control for RT-PCR. Mol Cell Probes 2001, 15(5):307–311.

15. Vandesompele J, De Preter K, Pattyn F, Poppe B, Van Roy N, De Paepe A, Speleman F: Accurate normalization of real-time quantitative RT-PCR data by geometric averaging of multiple internal control genes. Genome Biol 2002, 3(7):RESEARCH0034.

16. Pfaffl MW, Tichopad A, Prgomet C, Neuvians TP: Determination of stable housekeeping genes, differentially regulated target genes and sample integrity:

BestKeeper – Excel-based tool using pair-wise correlations. Biotechnol Lett 2004, 26(6):509–515.

17. Andersen CL, Jensen JL, Orntoft TF: Normalization of real-time quantitative reverse transcription-PCR data: A model-based variance estimation approach to identify genes suited for normalization, applied to bladder and colon cancer data sets. Cancer Res 2004, 64(15):5245–5250.
18. Nalubamba KS, Gossner AG, Dalziel RG, Hopkins J: Differential expression of pattern recognition receptors during the development of foetal sheep. Dev Comp Immunol 2008, 32(7):869–874.
19. Silver N, Best S, Jiang J, Thein SL: Selection of housekeeping genes for gene expression studies in human reticulocytes using real-time PCR.Bmc Molecular Biology 2006, 7:33.
20. Nishimura M, Nikawa T, Kawano Y, Nakayama M, Ikeda M: Effects of dimethyl sulfoxide and dexamethasone on mRNA expression of housekeeping genes in cultures of C2C12 myotubes. Biochem Biophys Res Commun 2008, 367(3):603–608.
21. Chen IHB, Huber M, Guan TL, Bubeck A, Gerace L: Nuclear envelope transmembrane proteins (NETs) that are up-regulated during myogenesis. Bmc Cell Biology 2006, 7:38.
22. Ingerslev HC, Pettersen EF, Jakobsen RA, Petersen CB, Wergeland HI: Expression profiling and validation of reference gene candidates in immune relevant tissues and cells from Atlantic salmon (Salmo salar L.). Mol Immunol 2006, 43(8):1194–1201.
23. Solanas M, Moral R, Escrich E: Unsuitability of using ribosomal RNA as loading control for Northern blot analyses related to the imbalance between messenger and ribosomal RNA content in rat mammary tumors. Anal Biochem 2001, 288(1):99–102.
24. Alfieri RR, Bonelli MA, Cavazzoni A, Brigotti M, Fumarola C, Sestili P, Mozzoni P, De Palma G, Mutti A, Carnicelli D, et al.: Creatine as a compatible osmolyte in muscle cells exposed to hypertonic stress. J Physiol 2006, 576(2):391–401.
25. Schwartz RJ, Rothblum KN: Gene switching in myogenesis – differential expression of the chicken actin multigene family. Biochemistry 1981, 20(14):4122–4129.
26. Koumans JTM, Akster HA, Dulos GJ, Osse JWM: Myosatellite cells of Cyprinus-Carpio (teleostei) invitro – isolation, recognition and differentiation. Cell Tissue Res 1990, 261(1):173–181.

27. Bustin SA, Benes V, Garson JA, Hellemans J, Huggett J, Kubista M, Mueller R, Nolan T, Pfaffl MW, Shipley GL, et al.: The MIQE Guidelines: Minimum Information for Publication of Quantitative Real-Time PCR Experiments. Clin Chem 2009, 55(4):611–622.

28. Pattyn F, Robbrecht P, Speleman F, De Paepe A, Vandesompele J: RTPrimerDB: the real-time PCR primer and probe database, major update. Nucleic Acids Res 2006, 34:D684–D688.

Comparative Chromosome Mapping of Repetitive Sequences. Implications for Genomic Evolution in the Fish, *Hoplias malabaricus*

Marcelo B. Cioffi, Cesar Martins and Luiz A. C. Bertollo

ABSTRACT

Background

Seven karyomorphs of the fish, Hoplias malabaricus (A-G) were previously included in two major groups, Group I (A, B, C, D) and Group II (E, F, G), based on their similar karyotype structure. In this paper, karyomorphs from Group I were analyzed by means of distinct chromosomal markers, including silver-stained nucleolar organizer regions (Ag-NORs) and chromosomal location of repetitive sequences (18S and 5S rDNA, and satellite 5SHindIII-

DNA), through fluorescence in situ hybridization (FISH), in order to evaluate the evolutionary relationships among them.

Results

The results showed that several chromosomal markers had conserved location in the four karyomorphs. In addition, some other markers were only conserved in corresponding chromosomes of karyomorphs A-B and C-D. These data therefore reinforced and confirmed the proposed grouping of karyomorphs A-D in Group I and highlight a closer relationship between karyomorphs A-B and C-D. Moreover, the mapping pattern of some markers on some autosomes and on the chromosomes of the XY and X_1X_2Y systems provided new evidence concerning the possible origin of the sex chromosomes.

Conclusion

The in situ investigation of repetitive DNA sequences adds new informative characters useful in comparative genomics at chromosomal level and provides insights into the evolutionary relationships among Hoplias malabaricus karyomorphs.

Background

Although usually reported as a single taxonomic entity, Hoplias malabaricus (Characiformes, Erythrinidae) has significant karyotypic diversity and well-defined population differences concerning the diploid number, morphology of chromosomes and sex chromosome systems. Such intraspecific diversity enabled the characterization of seven main karyomorphs (A-G), in which those without differentiated sex chromosome systems (A, C, E and F) show a wider geographical distribution compared to those that have such systems (B, D and G), which highlights the derivative character of the latter [1]. Despite differences in diploid chromosome number and in the presence or absence of differentiated sex chromosome systems, the seven karyomorphs were subdivided into two major groups (I and II) based on general karyotype similarity [1]. Thus, Group I included karyomorphs A-D, while Group II included karyomorphs E-G (Table 1).

Table 1. Karyomorphs previously identified in Hoplias malabaricus according to Bertollo et al. (2000).

Karyomorphs	Chromosome number	Sex Chromosomes	Geographic occurrence
Group I: first chromosome pairs of similar sizes			
Karyomorph A	2n = 42	-	Northen to southern Brazil, Uruguay and northen Argentina
Karyomorph B	2n = 42	XX/XY	Vale do Rio Doce (Minas Gerais State) and Iguaçú River (Paraná State, Brazil)
Karyomorph C	2n = 40	-	Northen Brazil to northen Argentina
Karyomorph D	Female - 2n = 40/Male - 2n = 39	$X_1X_1X_2X_2/X_1X_2Y$	Upper Paraná hydrographic basin, Brazil
Group II: first chromosome pairs with clearly distinct sizes			
Karyomorph E	2n = 42	-	Trombetas River (Paraná State, Brazil)
Karyomorph F	2n = 40	-	Surinam to southeastern Brazil
Karyomorph G	Female - 2n = 40/Male - 2n = 41	XX/XY_1Y_2	Amazonian Rivers, Brazil

The karyotype diversity in H. malabaricus indicates the probable occurrence of distinct species, which is reinforced by the sympatry between some karyomorphs, without evidence of gene flow between them [1]. Specifically for karyomorphs A and C, and for karyomorphs A and D, the results obtained using RAPD-PCR genomic markers were also consistent with a lack of gene flow, providing additional evidence for karyomorphs as distinct evolutionary units [2].

Among Erythrinidae, a repetitive DNA class named 5SHindIII-DNA that shares similarities to 5S rDNA "true" repeats was previously isolated and characterized [3]. This sequence could not be found in the chromosomes of Erythrinus, Hoplerythrinus or Hoplias lacerdae and is therefore likely to be exclusive to H. malabaricus. Its exclusive presence in this species shows that this repetitive DNA class probably originated after the divergence of the three Erythrinidae genera and Hoplias species [4]. Ribosomal genes (18S and 5S rDNA) were also useful markers in H. malabaricus, showing significant differences among populations of this species [5,6].

Molecular organization and cytogenetic mapping of ribosomal genes and other repetitive DNA sequences have provided important contributions to the characterization of biodiversity and the evolution of ichthyofauna [7-9]. In fact, a substantial fraction of any eukaryotic genome consists of repetitive DNA sequences, including satellites, minisatellites, microsatellites and transposable elements. Despite intensive study in recent decades, the molecular forces that generate, propagate and maintain repetitive DNAs in the genome are still under discussion [10]. Possible functions of satellite DNAs have been studied in several groups of animals, evidencing that these sequences may play an important role at the chromosomal and nuclear level [8,9,11-16].

This report presents a comparative study of the H. malabaricus karyomorphs belonging to Group I (A, B, C and D) by means of distinct chromosomal markers, including silver-stained nucleolar organizer regions (Ag-NORs) and chromosomal location of repetitive sequences (18S and 5S rDNA, and satellite 5SHindIII-DNA), through fluorescence in situ hybridization (FISH), in order to evaluate the evolutionary relationships among them.

Methods

Mitotic Chromosome Preparations

Chromosome preparations were obtained from H. malabaricus specimens from different river basins belonging to karyomorphs A, B, C and D, as specified in Table 2. The animals were first injected in the abdominal region with a 0.025% aqueous solution of colchicine at a dose of 1 ml/100 g of weight. After 50–60 minutes, the specimens were sacrificed, and the chromosomal preparations were obtained from cells of the anterior kidney [17].

Table 2. Collection sites of Hoplias malabaricus, with the respective karyomorphs and sample sizes.

Locality	Karyomorph	N
Descalvado (SP) – Pântano River	A	8 – Male 6 – Female
Parque Florestal do Rio Doce (MG) – lagoons: Doce River	B	5 – Male 6 – Female
Poconé (MT) – lagoons: Bento Gomes River	C	5 – Male 8 – Female
São Carlos (SP) – UFSCar reservoir: Monjolinho Stream	D	10 – Male 7 – Female

SP = São Paulo, MT = Mato Grosso, and MG = Minas Gerais Brazilian States.

Probes

Three tandem-arrayed DNA sequences isolated from the H. malabaricus genome were used. The first probe contained a 5S rDNA repeat copy and included 120 base pairs (bp) of the 5S rRNA encoding gene and 200 bp of the non-transcribed spacer (NTS) [3]. The second probe contained a copy of the repetitive satellite 5SHindIII-DNA sequence with 360 bp composed of a 95-bp segment with similarity to the 5S rRNA gene of the first probe and a 265-bp segment similar to the NTS of the first probe [3]. The third probe corresponded to a 1,400 bp-segment of the 18S rRNA gene obtained via PCR from nuclear DNA [5].

FISH Procedure, Sequential Ag-NOR Detection and Karyotype Analysis

Fluorescence in situ hybridization (FISH) was performed on mitotic chromosome spreads [18]. The probes were labeled by nick translation with biotin-14-dATP (Bionick labeling system-Invitrogen). The metaphase chromosome slides were incubated with RNAse (40 μg/ml) for 1.5 h at 37°C. After denaturation of chromosomal DNA in 70% formamide/2× SSC at 70°C, spreads were incubated in 2× SSC for 4 min at 70°C. Hybridization mixtures containing 100 ng of denatured probe, 10 mg/ml dextran sulfate, 2× SSC, and 50% formamide in a final volume of 30 μl were dropped on the slides, and the hybridization was performed overnight at 37°C in a 2× SSC moist chamber. These hybridization conditions were previously described for the 5S rDNA and 5SHindIII-DNA probes in order to avoid possible cross-hybridization [3] Post-hybridization washes were carried out at 37°C in 2× SSC, 50% formamide for 15 min, followed by a second wash in 2× SSC for 15 min, and a final wash at room temperature in 4× SSC for 15 min. Detection of hybridized probes was carried out with 0.07% avidin-FITC conjugate (Sigma) in C buffer (0.1 M $NaHCO_3$, 0.15 M NaCl) for 1 h followed by 2 rounds of signal amplification using 2.5% anti-avidin biotin conjugate (Sigma) in blocking buffer (1.26% $NaHCO_3$, 0.018% sodium citrate, 0.0386% triton and 1% non-fat dried milk) for 30 min. Each treatment with anti-avidin biotin conjugate was followed by a treatment with avidin-FITC. The treatments with avidin-FITC and anti-avidin-biotin were conducted in a 2× SSC moist chamber at 37°C. After each amplification step, the slides were washed 3 times for 5 min each in blocking buffer at 42°C. The post-hybridization washes were performed on a shaker (150 rpm). The chromosomes were counterstained with propidium iodide (50 μg/mL) and analyzed with an Olympus BX50 epifluorescence microscope. The chromosome images were captured using CoolSNAP-Pro software (Media Cybernetic). Slides previously treated by FISH were washed with water and dehydrated with washes in 75%, 85% and 100% ethanol for 5 minutes for each concentration. After air drying, the slides were prepared for the Ag-NORs detection [19]. Approximately 30 metaphase spreads were analyzed per specimen to determine the diploid chromosome number and karyotype structure. The chromosomes were classified as metacentric (m), submetacentric (sm) or subtelocentric (st), according to arm ratios [20].

Results

Sites of 5SHindIII-DNA, 5S rDNA and 18S rDNA were clearly detected by the FISH procedures, allowing their clear identification and location in the chromosomes of H. malabaricus (Figures 1, 2, 3 and 4). These data were organized in the form of idiograms (Figure 5) to facilitate the comparative analysis between the karyomorphs.

The satellite 5SHindIII-DNA was mapped in the centromeric region of several chromosome pairs. The karyomorphs A and B presented eight chromosome pairs (nos. 1, 5, 6, 12, 13, 14, 18 and 19) carrying these sites. Additionally, 5SHindIII-DNA sites were also highlighted on chromosome pair 21 of karyomorph A and on chromosomes X and Y of karyomorph B. On the other hand, the karyomorphs C and D presented 10 chromosome pairs (nos. 1, 5, 6, 8, 9, 10, 14, 15, 17 and 19), in addition to sites located on chromosomes no. 20 of karyomorph C and X2 and Y of karyomorph D (Figures 1 and 5).

18S rDNA sites were proximal to the centromere or in the telomeric region of the chromosomes. In the latter case, bitelomeric sites, i.e., present in both telomeric regions, could also be seen. Karyomorphs A and B presented four chromosome pairs bearing such sites, three of them (nos. 5, 16 and 18) showing a conserved location in both karyomorphs. The fourth site was exclusive to karyomorph A (pair no. 21) or karyomorph B (pair no. 4). In addition, a conspicuous cistron was present on the X chromosome of karyomorph B, occupying a great extent of its long arms. In karyomorphs C and D, five chromosome pairs (nos. 5, 11, 13, 15 and 19) carried 18S rDNA sites. The chromosome 11 of karyomorph C showed correspondence with chromosome X1 of karyomorph D, both in the form, size and location of a conspicuous NOR site on the long arms, proximal to the centromere. In turn, the Y chromosome of karyomorph D also showed a proximal 18S rDNA site in the short arms (Figures 2, 3 and 5). In general, there was perfect correspondence between the number and location of 18S rDNA and Ag-NOR sites (Figure 3).

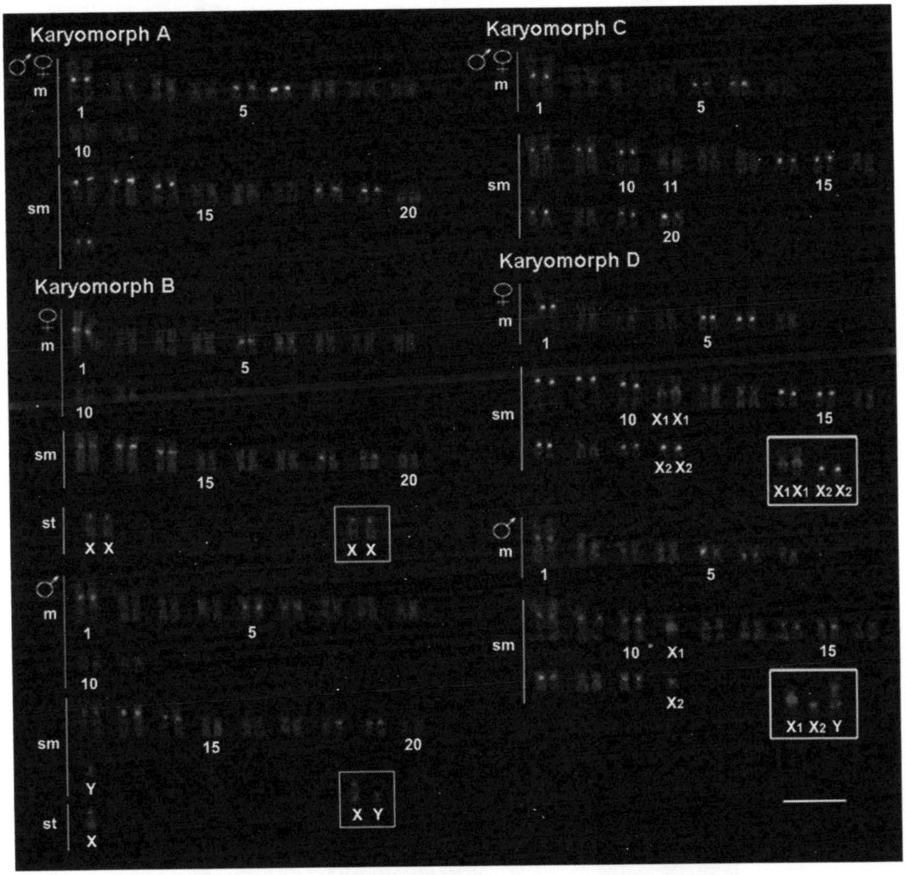

Figure 1. Karyotypes of Hoplias malabaricus (karyomorphs A-D) arranged from chromosomes probed with 5SHindIII-DNA satellite sequences (white sectionss) and counterstained with propidium iodide. The sex chromosomes of karyomorphs B and D are boxed. Bar = 5 μm.

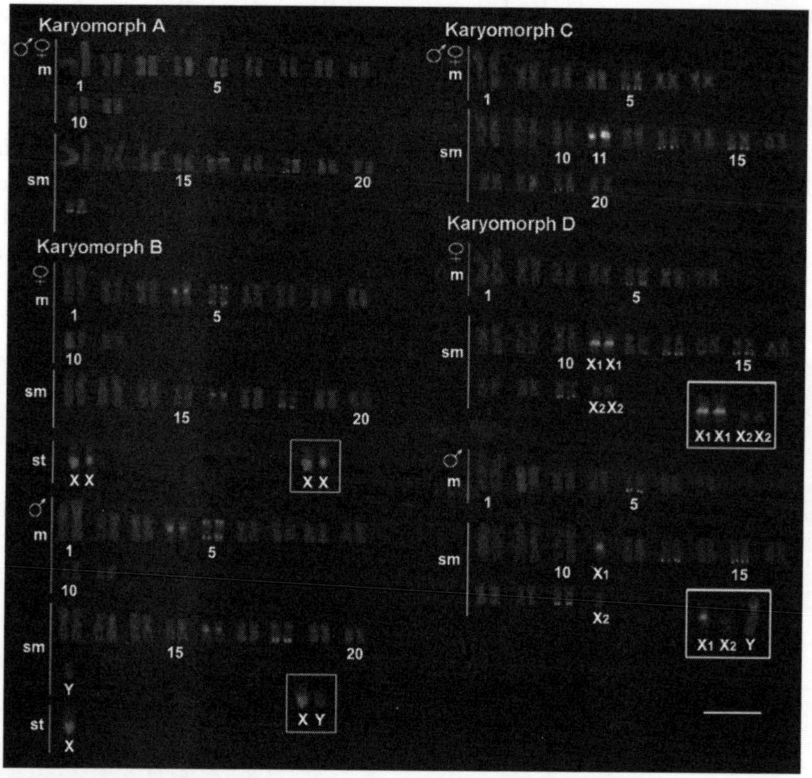

Figure 2. Karyotypes of Hoplias malabaricus (karyomorphs A-D) arranged from chromosomes probed with 18S rDNA (white sections) and counterstained with propidium iodide. The sex chromosomes of karyomorphs B and D are boxed. Bar = 5 μm.

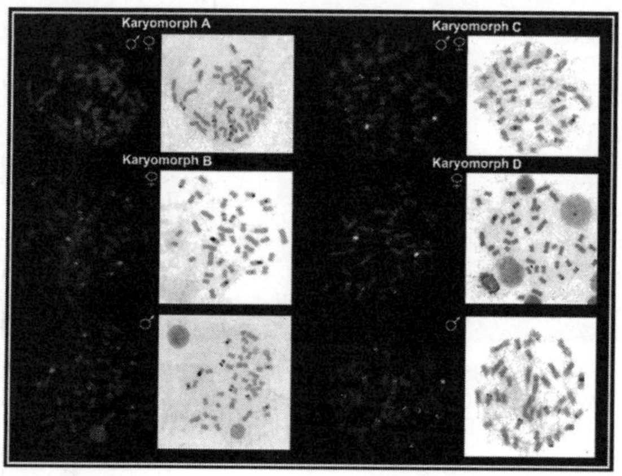

Figure 3. Hoplias malabaricus FISH metaphases (karyomorphs A-D), showing the 18S rDNA and Ag-NORs sites with sequential analysis. Note the general correspondence between the number and location of the 18S rDNA cistrons and Ag-NORs.

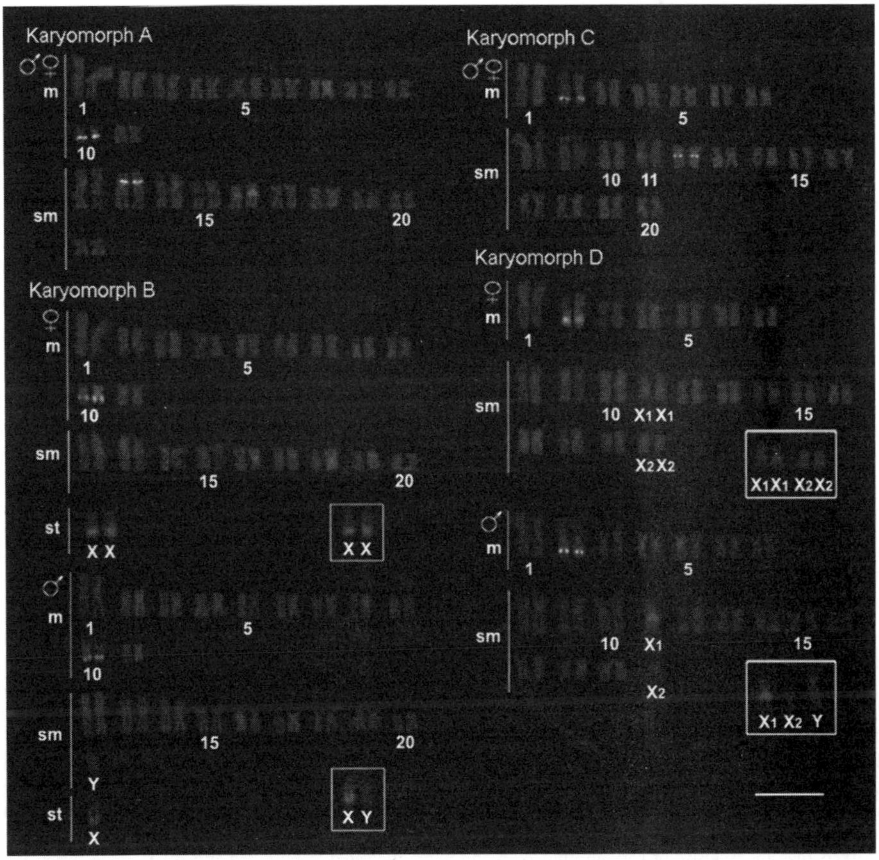

Figure 4. Karyotypes of Hoplias malabaricus (karyomorphs A-D) arranged from chromosomes probed with 5S rDNA (white sections) and counterstained with propidium iodide. The sex chromosomes of karyomorphs B and D are boxed. Bar = 5 μm.

Cytogenetic mapping of the 5S rDNA sequences showed conserved markers only in corresponding chromosomes of karyomorphs A-B (metacentric pair no. 10) or karyomorphs C and D (metacentric pair no. 2), both with interstitial sites on the long arms. However, an exclusive proximal cluster was located in the short arms of the submetacentric pair no. 13 of karyomorph A and in the long arms of the submetacentric pair no. 12 of karyomorph C (Figures 4 and 5).

Figure 6 summarizes all of the corresponding chromosomes of karyomorphs A-D, karyomorphs A-B, or karyomorphs C-D, as well as the chromosomes that showed sites exclusive to each karyomorph, considering the repetitive DNAs analyzed.

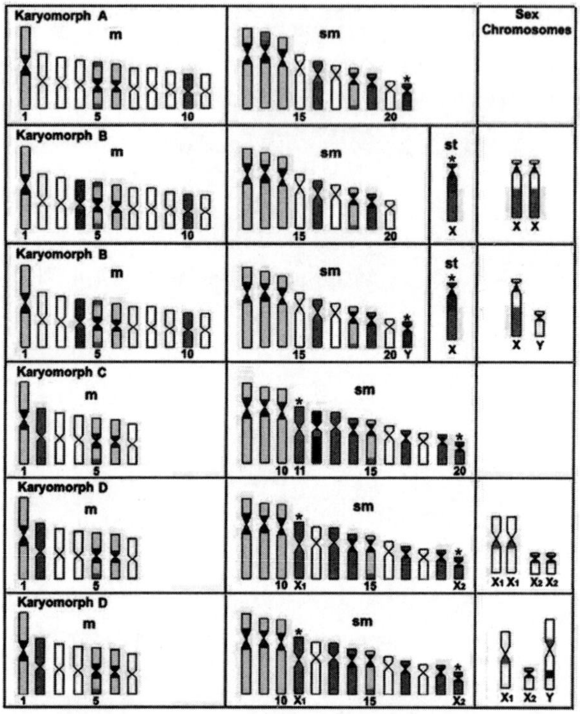

Figure 5. Representative idiogram of Hoplias malabaricus karyomorphs A-D, on the basis of repetitive DNA sequences analyzed. The sites location of the satellite 5SHindIII-DNA, 18S rDNA and 5S rDNA on the chromosomes is indicated in deep blue, green and gray, respectively. Yellow indicates the corresponding chromosomes of karyomorphs A-B-C-D; pink indicates the corresponding chromosomes of karyomorphs A-B and in orange are indicated the corresponding chromosomes of karyomorphs C-D. The chromosomes bearing markers that are exclusive of karyomorphs A, B and C are indicated in blue, red and black, respectively. The asterisks indicate probable relationships between the sex chromosomes of karyomorphs B and D with some autosome pairs of karyomorphs A and C, respectively.

Karyomorphs	Corresponding chromosomes	Specific chromosomes
A – B – C – D	1 – 5 – 6	
	{ 12 – 13 – 14 – 18 (A - B) / 8 – 9 – 10 – 15 (C - D) }	
A – B	10 – 16 – 19	
	{ 21 (A) / X - Y (B) }	
C – D	2 – 13 – 14 – 17 – 19	
	{ 11 - 20 (C) / X_1 - X_2 (D) }	
A		13(p)
B		4
C		12

Figure 6. Chromosomal pairs bearing repetitive DNA sites with correlation or specificity for Hoplias malabaricus karyomorphs. The chromosomes between curly brackets are indicated according to their position in the karyotypes.

Discussion

Despite the differences regarding the diploid number and the occurrence of differentiated sex chromosomes, the four karyomorphs possess a relatively homogeneous karyotypic structure, basically formed by meta-submetacentric chromosomes, constituting an apparently related evolutionary group – Group I – in H. malabaricus [1]. Karyomorphs A and B have 2n = 42 chromosomes, suggesting that karyomorph B was most likely derived from the emergence of a sex chromosome system XX/XY, where X corresponds to the only subtelocentric chromosome of the karyotype. Such a relationship also seems to be applicable for the karyomorphs C (2n = 40) and D (2n = 39 males/2n = 40 females), in that the latter could also have been derived by the emergence of a multiple sex chromosomes system, $X_1X_1X_2X_2/X_1X_2Y$ [1].

The cytogenetic mapping of different repetitive DNA sequences provided reliable chromosomal markers, which allowed the determination of relationships among different karyomorphs. The use of these same markers in a comparative analysis among different populations of karyomorph A also demonstrated a continuing genomic differentiation in this group, allowing the detection of recent evolutionary events, independent of great variations in karyotypes [5]. In fact, repetitive DNAs are highly dynamic throughout evolution, allowing their employment in evolutionary studies. Evaluation of all of the obtained markers shows seven corresponding chromosome pairs in the four karyomorphs, four corresponding pairs in karyomorphs A-B and seven corresponding pairs in karyomorphs C-D, along with some exclusive chromosomes (Figure 6). In addition to corroborate the inclusion of karyomorphs A-D in the same major evolutionary group (Group I), the results provide further evidence for a greater proximity between the karyomorphs A-B and karyomorphs C-D, and also identify some of their peculiarities.

The repetitive 5SHindIII-DNA sequence is a tandemly organized DNA family that shares similarity with 5S rDNA and probably originated from duplicated segments of this ribosomal DNA class [3]. This satellite DNA family was relatively common in the H. malabaricus genome, with 18 chromosomal sites in karyomorphs A-B and 22 sites in karyomorphs C-D. Although 18 sites of 5SHindIII-DNA was the common situation found in karyomorph A, a comparative analysis among different populations of this karyomorph showed 22 sites for only one of them. However, such additional sites showed no correspondence with those present in the chromosomes of karyomorphs C and D [5]. All 5SHindIII-DNA sites had an exclusive location in the centromeric region of chromosomes, consistent with previous findings for other Erythrinidae populations [4]. It is well known that centromeric regions are rich in repetitive DNAs, as seen in several organisms, including humans, mice, maize, fruit flies and yeast [21]. It is therefore likely that

5SHindIII-DNA repetitions may have some structural or functional role in H. malabaricus chromosomes as components of their centromeric DNA [3].

In higher eukaryotes, the rRNA genes are organized as two distinct multigene families, represented by 45S rDNA (18S+5.8S+28S) and 5S rDNA. Both families are composed of tandemly repeated units, with hundreds to thousands of copies. Multiple copies of the 45S rDNA correspond to the nucleolar organizing regions [22]. Although NORs are frequently telomeric in fish, they show a great variability in this group, both in position, number and size of the cistrons [7,23]. A comparative analysis between the 18S rDNA sites and the Ag-NORs showed great similarity regarding the location of the nucleolar organizing regions in H. malabaricus. Any decreases in the number of Ag-NORs, observed in some karyomorphs, can be attributed to differential gene activity among the 18S rDNA sites in the cells, since the Ag-NORs represent only those cistrons that were active in the preceding interphase [24,25]. Along with some exclusive chromosome markers for each karyomorph, corresponding chromosomes could also be observed, as is the case for metacentric no. 5 (karyomorphs A-D) and the submetacentrics nos. 18 (karyomorphs A and B) and 15 (karyomorphs C and D). The two latter chromosomes, although occupying distinct karyotype positions between karyomorphs, may correspond to the same chromosome, not only by their shape and size, but also by sharing a 5SHindIII-DNA site. In turn, in karyomorphs C and D, the metacentric no. 5 harbored a single 18S rDNA site. In karyomorphs A and B this chromosome showed bitelomeric NORs, indicating that additional events had occurred allowing the acquisition of new sites of rDNA (Figure 5). It is interesting to note that the presence of bitelomeric NORs in H. malabaricus is relatively frequent in karyomorphs A and B [6,26-28].

When compared to 18S rDNA and 5SHindIII-DNA, the 5S rDNA was a more specific marker, since corresponding chromosomes were not found among the four karyomorphs, in addition to some exclusive sites in karyomorphs A and C. In this way, it seems that karyotype differentiation in Group I did not retain any basal characteristic concerning this chromosomal marker. In this context, the 5S rDNA appears to have gone through karyotypic changes more pronounced than the 18S rDNA and 5SHindIII-DNA, since only evolutionary more related karyomorphs, i.e., A-B and C-D, show corresponding chromosomes. Coincidentally, sites of 5S rDNA were shown to be good population markers in H. malabaricus, because they showed significant differences even among populations from the same karyomorph [5].

5S rRNA genes are generally found in an interstitial position on chromosomes from the majority of fish species [29], as well as in other vertebrates [30-33], suggesting that such a pattern of distribution is not a coincidence. Furthermore, its chromosomal location is usually not syntenic with the 45S rDNA sites. Although the populations analyzed here show that the location of 18S and 5S rDNA sites were always independent, synteny was already found in a population of karyomorph A [5], highlighting again the dynamic behavior of rDNAs throughout the karyotype evolutionary process of H. malabaricus. Telomeric regions would be more conducive to genetic material transfer between chromosomes due their proximity inside the interphase nucleus [34]. This fact could be associated with the greater numerical conservation of the 5S rDNA sites in relation to 18S rDNA sites in fish, possibly due to a preferential location in interstitial and telomeric regions of the chromosomes, respectively [29].

Concerning the sex chromosome systems, the distribution of 5SHindIII-DNA and 18S rDNA also indicates a likely correlation between some chromosomes of karyomorphs A and C, with the sex chromosomes present in karyomorphs B and D, respectively. This is the case for chromosome no. 21 of karyomorph A, which shares the 18S rDNA and 5SHindIII-DNA sites with chromosomes X and Y of karyomorph B, in addition to showing a marked similarity in relation to the size and morphology of chromosome Y (Figure 5). In the same way, chromosome no. 11 of karyomorph C is similar to chromosome X1 of karyomorph D, in both the physical location of an 18S rDNA site and the correspondence between CMA3 positive regions (data not shown). It is likely that the differentiation of the XX/XY system of karyomorph B has occurred from a heterochromatinization process. This resulted in a large subtelocentric X chromosome that harbors a conspicuous heterochromatic block on the long arms, which co-locates with a NOR site and presents a polymorphic behavior [26]. On the other hand, a clear case of translocation is associated with the origin of the X1X1X2X2/X1X2Y sex chromosome system in karyomorph D, resulting in a large Y chromosome present only in males and a consequent diploid number reduction in this sex [35,36]. The data presented here suggest that the sex chromosomes of karyomorphs B and D were derived from chromosomes 21 and 11 of karyomorphs A and C, respectively. Ongoing studies with additional chromosomal markers will provide a conclusive analysis of the evolution of sex chromosome systems in H. malabaricus.

Conclusion

Repetitive DNAs were important for the genomic evolutionary process of H. malabaricus, as evidenced by the presence and distribution of these sequences on chromosomes. These findings lend further support to the idea that Group I is representative of karyomorphs that are closely related and suggest a greater evolutionary proximity between the karyomorphs A-B and karyomorphs C-D, as well as the probable chromosomal origin of the sex systems.

Author Contributions

MBC carried out the cytogenetic analyses and drafted the manuscript. CM helped in cytogenetic analysis and drafted the manuscript. LACB designed and coordinated the study, and drafted and revised the manuscript. All authors read and approved the final manuscript.

Acknowledgements

The authors would like to thank Drs. Liano Centofante and Jorge A. Dergam and MSc. Uédson Jacobina for supplying fish specimens. This work was supported by the Brazilian agencies FAPESP (Fundação de Amparo à Pesquisa do Estado de São Paulo – proc. n. 2007/05565-5) and CNPq (Conselho Nacional de Desenvolvimento Científico e Tecnológico).

References

1. Bertollo LAC, Born GG, Dergam JA, Fenocchio AS, Moreira-Filho O: A biodiversity approach in the Neotropical fish Hoplias malabaricus. Karyotypic survey, geographic distribution of cytotypes and cytotaxonomic considerations. Chromosome Res 2000, 8:603–613.
2. Dergam JA, Suzuki HI, Shibatta OA, Silva LFD, Júlio HF, Caetano LG, Black WC: Molecular biogeography of the Neotropical fish Hoplias malabaricus (Erythrinidae, Characiformes) in the Iguassu, Tibagi, and Paraná rivers. Genet Mol Biol 1998, 21:493–496.
3. Martins C, Ferreira IA, Oliveira C, Foresti F, Galetti PM Jr: A tandemly repetitive centromeric DNA sequence of the fish Hoplias malabaricus (Characiformes: Erythrinidae) is derived from 5S rDNA. Genetica 2006, 127:133–141.
4. Ferreira IA, Bertollo LAC, Martins C: Comparative chromosome mapping of 5S rDNA and 5SHindIII repetitive sequences in Erythrinidae fishes (Characiformes) with emphasis on the Hoplias malabaricus' species complex. Cytogenet Genome Res 2007, 118:78–83.
5. Cioffi MB, Martins C, Centofante L, Jacobina U, Bertollo LAC: Chromosomal variability among allopatric populations of Erythrinidae fish Hoplias malabaricus: mapping of three classes of repetitive DNAs. Cytogenet Genome Res 2009, in press.

6. Vicari MR, Artoni RF, Bertollo LAC: Comparative cytogenetics of Hoplias malabaricus (Pisces, Erythrinidae). A population analysis in adjacent hydrographic basins. Genet Mol Biol 2005, 28:103–110.
7. Almeida-Toledo LF: Cytogenetic markers in neotropical freshwater fishes. In Phylogeny and Classification of Neotropical Fishes. Edited by: Malabarba LR, Reis RE, Vari, RP, Lucena ZMS, Lucena CAS. Porto Alegre: Edipucrs; 1998:583–588.
8. Jesus CM, Galetti PM Jr, Valentini SR, Moreira-Filho O: Molecular and chromosomal location of two families of satellite DNA in Prochilodus lineatus (Pisces, Prochilodontidae), a species with B chromosomes. Genetica 2003, 118:25–32.
9. Vicari MR, Artoni RF, Moreira-Filho O, Bertollo LAC: Colocalization of repetitive DNAs and silencing of major rRNA genes. A case report of the fish Astyanax janeiroensis. Cytogenet Genome Res 2008, 122:67–72.
10. Biemont C, Vieira C: Junk DNA as an evolutionary force. Nature 2006, 443:521–524.
11. Singer MF: Highly repetitive sequences in mammalian genomes. Int Rev Cytol 1982, 76:67–112.
12. Hummel S, Meyerhhof W, Korge E, Knochel W: Characterization of highly and moderately repetitive 500 bp EcoRI fragments from Xenopus laevis DNA. Nucleic Acids Res 1984, 12:4921–4937.
13. Haaf T, Schmid M: Chromosome topology in mammalian interphase nuclei. Exp Cell Res 1991, 192:325–332.
14. Larin Z, Fricker MD, Tyler-Smith C: De novo formation of several features of a centromere following introduction of an Y alphoid YAC into mammalian cells. Hum Mol Genet 1994, 3:689–695.
15. Clabby C, Goswami U, Flavin F, Wilkins NP, Houghton JA, Powell R: Cloning, characterization and chromosomal location of a satellite DNA from the Pacific oyster, Crassostrea gigas. Gene 1996, 168:205–209.
16. Oliveira C, Wright JM: Molecular cytogenetic analysis of heterochromatin in the chromosomes of tilapia, Oreochromis niloticus (Teleostei: Cichlidae). Chromosome Res 1998, 6:205–211.
17. Bertollo LAC, Takahashi CS, Moreira-Filho O: Cytotaxonomic considerations on Hoplias lacerdae (Pisces, Erythrinidae). Brazil J Genet 1978, 1:103–120.
18. Pinkel D, Straume T, Gray J: Cytogenetic analysis using quantitative, high sensitivity, fluorescence hybridization. Proc Natl Acad Sci USA 1986, 83:2934–2938.

19. Howell WM, Black DA: Controlled silver staining of nucleolus organizer regions with a protective colloidal developer: A 1-step method. Experientia 1980, 36:1014–1015.

20. Levan A, Fredga K, Sandberg AA: Nomenclature for centromeric position on chromosomes. Hereditas 1964, 52:201–220.

21. Henikoff S, Ahmad K, Malik HS: The centromere paradox: stable inheritance with rapidly evolving DNA. Science 2001, 293:1098–1102.

22. Long EO, Dawid ID: Repeated genes in eukaryotes. Ann Rev Biochem 1980, 49:727–764.

23. Galetti PM Jr: Chromosome diversity in neotropical fishes: NOR studies. Ital J Zool 1998, 65:53–65.

24. Hsu TC, Spirito SC, Pardue ML: Distribution of 18+28S ribosomal genes in mammalian genomes. Chromosoma 1975, 53:25–33.

25. Miller DA, Dev VG, Tantravahi R, Miller OJ: Supression of human nucleolus organizer in mouse-human somatic hybrid cells. Exp Cell Res 1976, 101:235–243.

26. Bertollo LAC: The nucleolar organizer regions of Erythrinidae fish. An uncommon situation in the genus Hoplias. Cytologia 1996, 61:75–81.

27. Born GG, Bertollo LAC: An XX/XY sex chromosome system in a fish species, Hoplias malabaricus, with a polymorphic NOR-bearing X chromosome. Chromosome Res 2000, 8:111–118.

28. Born GG, Bertollo LAC: Comparative cytogenetics among allopatric populations of the fish Hoplias malabaricus. Cytotypes with 2n = 42 chromosomes. Genetica 2000, 110:1–9.

29. Martins C, Galetti PM Jr: Chromosomal localization of 5S rDNA genes in Leporinus fish (Anostomidae, Characiformes). Chromosome Res 1999, 7:363–367.

30. Vitelli L, Batistoni R, Andronico F, Nardi I, Barsacchi-Pilone G: Chromosomal localization of 18S + 28S and 5S ribosomal RNA genes in evolutionary divergent anuran amphibians. Chromosoma 1982, 84:475–491.

31. Schmid M, Vitelli L, Batistoni R: Chromosome banding in Amphibia. IV. Constitutive heterochromatin, nucleolus organizers, 18S+28S and 5S ribosomal RNA genes in Ascaphidae, Pipidae, Discoglossidae and Pelobatidae. Chromosoma 1987, 95:271–284.

32. Lucchini S, Nardi I, Barsacchi G, Batistoni R, Andronico F: Molecular cytogenetics of the ribosomal (18S + 28S and 5S) DNA loci in primitive and advanced urodele amphibians. Genome 1993, 36:762–773.

33. Mäkinem A, Zijlstra C, de Haan NA, Mellink CHM, Bosma AA: Localization of 18S plus 28S and 5S ribosomal RNA genes in the dog by fluorescence in situ hybridization. Cytogenet Cell Genet 1997, 78:231–235.

34. Schweizer D, Loidl J: A model for heterochromatin dispersion and the evolution of C band patterns. Chromosomes Tod 1987, 9:61–74.

35. Bertollo LAC, Fontes MS, Fenocchio AS, Cano J: The X1X2Y sex chromosome system in the fish Hoplias malabaricus. I. G-, C- and chromosome replication banding. Chromosome Res 1997, 5:493–499.

36. Bertollo LAC, Mestriner CA: The X1X2Y sex chromosome system in the fish Hoplias malabaricus (Pisces, Erythrinidae). II. Meiotic analyses. Chromosome Res 1998, 6:141–147.

Communities of Gastrointestinal Helminths of Fish in Historically Connected Habitats: Habitat Fragmentation Effect in a Carnivorous Catfish *Pelteobagrus fulvidraco* from Seven Lakes in Flood Plain of the Yangtze River, China

Wen X. Li, Pin Nie, Gui T. Wang and Wei J. Yao

ABSTRACT

Background

Habitat fragmentation may result in the reduction of diversity of parasite communities by affecting population size and dispersal pattern of species.

In the flood plain of the Yangtze River in China, many lakes, which were once connected with the river, have become isolated since the 1950s from the river by the construction of dams and sluices, with many larger lakes subdivided into smaller ones by road embankments. These artificial barriers have inevitably obstructed the migration of fish between the river and lakes and also among lakes. In this study, the gastrointestinal helminth communities were investigated in a carnivorous fish, the yellowhead catfish Pelteobagrus fulvidraco, from two connected and five isolated lakes in the flood plain in order to detect the effect of lake fragmentation on the parasite communities.

Results

A total of 11 species of helminths were recorded in the stomach and intestine of P. fulvidraco from seven lakes, including two lakes connected with the Yangtze River, i.e. Poyang and Dongting lakes, and five isolated lakes, i.e. Honghu, Liangzi, Tangxun, Niushan and Baoan lakes. Mean helminth individuals and diversity of helminth communities in Honghu and Dongting lakes was lower than in the other five lakes. The nematode Procamallanus fulvidraconis was the dominant species of communities in all the seven lakes. No significant difference in the Shannon-Wiener index was detected between connected lakes (0.48) and isolated lakes (0.50). The similarity of helminth communities between Niushan and Baoan lakes was the highest (0.6708), and the lowest was between Tangxun and Dongting lakes (0.1807). The similarity was low between Dongting and the other lakes, and the similarity decreased with the geographic distance among these lakes. The helminth community in one connected lake, Poyang Lake was clustered with isolated lakes, but the community in Dongting Lake was separated in the tree.

Conclusion

The similarity in the helminth communities of this fish in the flood-plain lakes may be attributed to the historical connection of these habitats and to the completion of the life-cycles of this fish as well as the helminth species within the investigated habitats. The diversity and the digenean majority in the helminth communities can be related to the diet of this fish, and to the lacustrine and macrophytic characters of the habitats. The lake isolation from the river had little detectable effect on the helminth communities of the catfish in flood-plain lakes of the Yangtze River. The low similarities in helminth communities between the Dongting Lake and others may just be a reflection of its unique water environment and anthropogenic alterations or fragmentation in this lake.

Background

Human activities greatly alter the size, shape, and spatial arrangement of natural habitats, and habitat fragmentation may influence the size of populations and dispersal pattern of individuals among populations [1], thus reducing species richness and abundance [2]. Dam construction can disrupt the connectivity of aquatic ecosystems and impede the abilities of aquatic biota to adapt to changes in environmental conditions [3], and also impact persistence of fish populations [4]. The increased habitat fragmentation and reduced local host population may threaten at least local extinction of parasites [5], especially for autogenic parasites with limited dispersal ability, which have been considered one of the main causes leading to the low number of parasite species [6,7]. However the "rescue effect" from other parasite metapopulations can prevent global extinction [8].

In the flood plain of the Yangtze (Changjiang) River, there were many lakes which were historically connected with the river. Since the 1950s, however, most of them have become isolated from the river due to the construction of dams and sluices, and many larger lakes have been subdivided into smaller ones by hydrological projects, road embankment etc [9]. The construction of these artificial barriers has inevitably obstructed the migration of fish among lakes [10]. The yellowhead catfish Pelteobagrus fulvidraco is a common fish species found in the Yangtze River [11], and is residential in being able to sexually reproduce in most of these localities where it occurs, such as in lakes, reservoirs and rivers. In previous research, Li et al. [12] investigated populations of a parasitic nematode in the intestines of P. fulvidraco in connected and isolated lakes in the flood plain of the river, but no fragmentation effect was detected at the level of genetic diversity of the nematode populations.

In order to detect if the lake fragmentation has any effect on the helminth communities of fish, the present study was designed to investigate communities of helminths in alimentary tracts of the yellowhead catfish P. fulvidraco from two connected lakes and five isolated lakes in the flood plain of the Yangtze River, China.

Methods

During February 2004, fish samples were collected from 12 sites in seven lakes, with three sites in Poyang lake, two each in Dongting, Honghu, and Liangzi lakes, and one each in Tangxun, Niushan, and Baoan lakes. Poyang and Dongting lakes represent the only two lakes which are still connected with Yangtze River. The distribution of these lakes and sampling sites were given in a previous paper by

Li et al. [12]. These lakes are shallow in depth ranging among 1.91 – 6.39 m, and vary in area among 37 – 2933 km^2 (Table 1). At least 30 yellowhead catfish were obtained from each sample site. The fork length was measured, and the stomach and intestine of each fish were examined for helminths within 24 h after sampling.

Table 1. Features of seven lakes in the flood plain of the Yangtze River, China

Lakes	Longitude	Latitude	Average depth (m)	Area (km^2)
Poyang	E115°49'~116°46'	N28°24'~29°46'	5.10	2933
Dongting	E111°53'~113°05'	N28°44'~29°35'	6.39	2432
Niushan	E114°19'~114°29'	N30°23'~30°29'	4.00	40
Honghu	E113°11'~113°28'	N29°38'~29°59'	1.91	344
Baoan	E114°39'~114°46'	N30°12'~30°18'	3.40	48
Tangxun	E114°19'~114°29'	N30°23'~30°29'	1.85	37
Liangzi	E114°21'~114°39'	N30°05'~30°18'	4.16	304

Communities of the gastrointestinal helminths were analysed at the infra- and component levels. Prevalence and abundance, as defined by Bush et al. [13], were calculated for each parasite species. Measures of component community structure are: the total number of helminth species, the Berger-Parker dominance index (d = NmaxN-1, in which Nmax represents the number of individuals in the most abundant species, and N the total number of the species in the community), and the Shannon-Wiener index (H = - ΣP1 lnP1, where P1 is the proportion of the individuals in the ith species) which describes the richness and abundance of parasites. Measures of infracommunity structures are: mean number of helminth species per fish, mean number of helminth individuals per fish, and mean Brillouin's index per fish (B = (lnN! - Σlnn1!) N-1, where n1 is the number of individuals in the ith species). Indexes are defined and calculated as in Magurran [14]. Similarities between individual fish were compared between lakes using the quantitative percentage similarity index (P = Σmin (Pxi, Pyi), where Pxi and Pyi are the proportions of parasite species i in the x and y host population, respectively), which compares similarity of two communities in number of species and parasite individuals as described by Hurlbert [15].

Correlation of lake area and diversity was analysed using correlation matrices. Analysis of variance (ANOVA) was used to examine significant difference in fish fork length among lakes. Significant difference in diversity between connected and isolated lakes was analysed statistically using a t-test. Correlation analysis of similarities and mean geographical distance between any two lakes was performed by the Mantel test. Cluster analysis of similarities of helminth communities was conducted by using an unweighted pair-group average method (UPGAM).

Results

A total of 11 species of helminths including 6 species of digeneans, 3 species of nematodes, 1 species of cestode, and 1 acanthocephalan were found in the stomach and intestine of P. fulvidraco from the seven lakes, with their infection levels listed in Table 2. The most prevalent and abundant parasite species was the nematode Procamallanus fulvidraconis, which was found in all the lakes and in 74.2% of the fish studied and comprised 69.8% of the total parasite specimens recorded. The digeneans Genarchopsis goppo and Coitocoecum plagiorchis comprised 13.4%, 6.1% of the total parasite specimens, respectively, and the remaining helminth species less than 5% each.

Table 2. Prevalence (%) and mean abundance (± SD) of helminths in Pelteobagrus fulvidraco from 7 lakes in the flood plain of the Yangtze River, China

Helminth species	**Poyang**	**Dongting**	Niushan	Honghu	Baoan	Tangxun	Liangzi
Genarchopsis goppo	63.3% 2.0 ± 2.4	3.3% 0.1 ± 0.6	63.9% 2.6 ± 4.1	6.4% 0.1 ± 0.8	70% 5.9 ± 7.4	90% 6.0 ± 4.8	28.3% 0.4 ± 0.8
Orientocreadium siluri	23.3% 0.5 ± 1.1	21.7% 0.5 ± 1.2	25% 0.8 ± 1.8	15.4% 0.4 ± 2.3	30% 0.43 ± 0.77	6.7% 0.1 ± 0.3	50% 1.7 ± 4.3
Coitocoecum plagiorchis	41.1% 2.4 ± 5.0	-	22.2% 0.4 ± 0.8	-	30% 1 ± 1.9	30% 1 ± 2.8	16.7% 0.3 ± 0.7
Echinoparyphium lingulatum	12.2% 0.5 ± 1.7	3.3% 0.1 ± 0.2	13.9% 0.3 ± 0.8	3.8% 0.1 ± 0.3	-	13.3% 0.1 ± 0.3	13.3% 0.4 ± 1.5
Dollfustrema vaneyi	-	-	8.3% 0.3 ± 1.7	-	-	-	-
Opisthorchis parasiluri	-	-	-	-	-	6.7% 0.1 ± 0.5	-
Procamallanus fulvidraconis	92.2% 10.2 ± 11.1	40% 3.4 ± 8.2	100% 12.3 ± 9.9	74.4% 6.1 ± 6.2	100% 15.8 ± 9.0	93.3% 7.6 ± 6.7	98.3% 13.0 ± 10.1
Spinitectus gigi	2.2% 0.1 ± 0.2	-	5.6% 0.1 ± 0.2	6.4% 0.1 ± 1.0	6.7% 0.1 ± 0.3	-	-
Camallanus cotti	1.1% 0.1 ± 0.1	1.7% 0.1 ± 0.1	-	-	3.3% 0.1 ± 0.4	-	-
Gangesia pseudobagri	-	3.3% 0.1 ± 0.3	-	11.5% 0.1 ± 0.5	20% 0.3 ± 0.7	40% 1.1 ± 2.2	3.3% 0.1 ± 0.8
Hebsoma violentum	-	-	5.6% 0.1 ± 0.2	-	3.3% 0.1 ± 0.2	-	15% 0.4 ± 1.3

There were no significant differences in fish fork length among the lakes (P > 0.05). The total number of helminth species ranged among 6 and 8 species in the helminth communities (Table 3). The mean number of helminth species per fish was the highest (2.80) in Tangxun lake, but lower in Dongting and Honghu lakes (0.78 and 1.18). The mean number of helminth individuals per fish was the highest (23.61) in Baoan lake, but lower in Dongting and Honghu lakes (4.20 and 6.90). The Brillouin's index was low, being 0.12 and 0.15 in Dongting and Honghu lakes, the index was greater than 0.4 in the other five lakes. The Shannon-Wiener index was apparently lower in Honghu and Dongting lakes (0.22 and 0.18, respectively) than in the other five lakes (all > 0.5). Therefore, the helminth infra-communities were poorer in the species number and individual number in Dongting and Honghu lakes, and richer in Poyang, Niushan, Baoan, Tangxun and Liangzi lakes. The value of the Berger-Parker dominance index varied from 0.47 to 0.88, with the highest observed in Dongting and Honghu Lakes (0.82 and 0.88), but the dominant species in the seven lakes was the nematode P. fulvidraconis (Table 3).

Table 3. Characteristics of helminth communities and fish length of Pelteobagrus fulvidraco from 7 lakes in the flood plain of the Yangtze River, China

Characteristics	**Poyang**	**Dongting**	Niushan	Honghu	Baoan	Tangxun	Liangzi	**Connected lakes**	Isolated lakes
Sample size	90	60	36	78	30	30	60	150	234
Mean fish length ± SD (cm)	16.12 ± 2.31	17.31 ± 1.72	16.61 ± 1.61	16.53 ± 2.20	17.20 ± 2.11	17.33 ± 2.32	17.53 ± 2.43	16.97 ± 2.37	16.76 ± 2.36
Mean no. of species ± SD	2.37 ± 0.94	0.78 ± 0.88	2.44 ± 1.11	1.18 ± 0.94	2.60 ± 1.1	2.80 ± 1.03	2.60 ± 1.20	1.73 ± 1.20	2.13 ± 1.26
Mean no. of hel ind ± SD +	15.70 ± 11.10	4.20 ± 8.90	16. 80 ± 12.2	6.90 ± 6.72	23.61 ± 12.11	16.00 ± 10.72	17.52 ± 11.74	11.09 ± 11.68	14.47 ± 11.73
Brillouin index ± SD	0.45 ± 0.30	0.12 ± 0.19	0.41 ± 0.31	0.15 ± 0.24	0.51 ± 0.25	0.65 ± 0.20	0.44 ± 0.33	0.36 ± 0.31	0.39 ± 0.32
Total no. of species	8	7	8	6	8	7	8	8	11
Shannon-Wiener index	0.58	0.18	0.53	0.22	0.62	0.83	0.57	0.48 ± 0.40	0.50 ± .041
Berger-Parker index	0.65	0.82	0.74	0.88	0.67	0.47	0.74	0.74	0.70
Dominant species	P.p. *	P.p. *	P.p. *	P.p. *	P.p. *	P.p. *	P.p. *	P.p. *	P.p. *

+ Mean number of helminth individuals
* P.p. represents *Procamallanus fulvidraconis*

There were no significant correlations between lake area and the value of Shannon-Wiener and Brillouin's indexes (P > 0.05). Significant difference in the value of Shannon-Wiener index was not found between connected lakes (0.48) and isolated lakes (0.50) (P = 0.57); but, a significant difference was detected in the mean number of species and mean number of helminth individuals between connected and isolated lakes. This may, however, reflect the low values observed in one of the two connected lakes and also the only two connected lakes still existing in the flood-plain of the river.

The similarity was the highest between Niushan and Baoan lakes (0.67), but lowest between Tangxun and Dongting lakes (0.18). The similarity between Dongting lake and others were very low, all below 0.28. Other comparisons were all above 0.40, except that between Tangxun and Honghu Lakes. From the clustering tree of similarities between helminth communities, Poyang, Niushan, Baoan and Liangzi are clustered together, and then with Tangxun and Honghu lakes; but, Dongting lake was quite separate (Figure 1). The similarity of helminth communities decreased significantly with the geographic distance between lakes (R = - 0.55, P < 0.05; Figure 2).

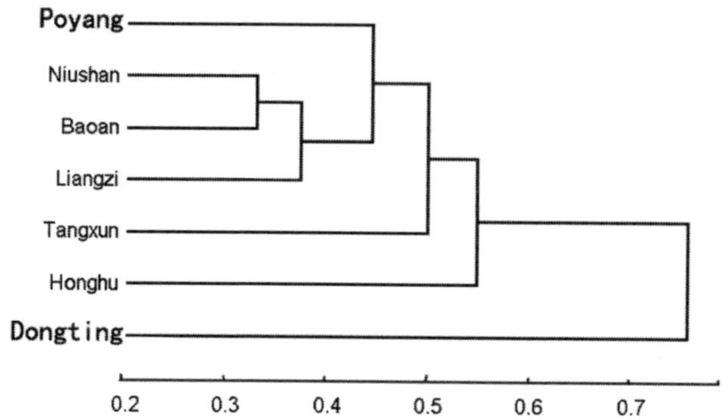

Figure 1. Clustering tree based on percentage similarity index of between helminth communities in the flood plain of the Yangtze River using unweighted pair-group average method. Number below the axis indicated the dissimilarity values.

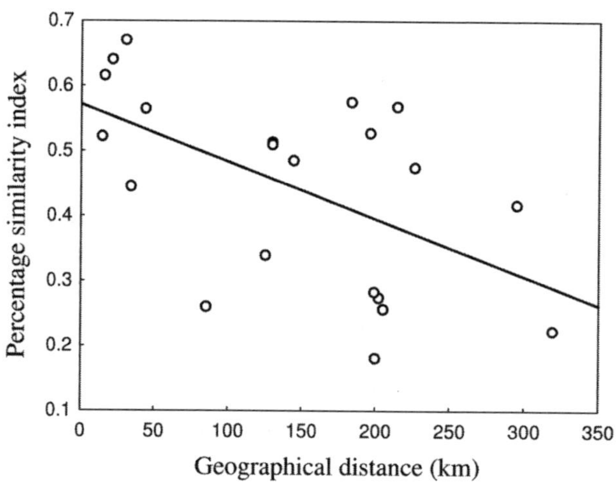

Figure 2. Relationship between geographical distance and percentage similarity index between each 2 lakes ($R = -0.55$, $P < 0.05$).

Discussion

It is interesting to note that some characteristics of the gastrointestinal helminth communities, such as the species composition, and the total number of species, were quite similar, or even the same, such as the dominant species, in the investigated connected and isolated flood-plain lakes of the Yangtze River. Furthermore, despite some differences in the mean number of helminth species per fish, the mean number of helminth individuals per fish, and diversity and similarity indexes, the helminth communities of P. fulvidraco in isolated lakes were clustered with one connected lake, i.e. Poyang Lake, with the exception of Dongting Lake. This, together with the non-significant difference in the value of Shannon-Wiener index between connected lakes and isolated lakes, may indicate that the isolation from the Yangtze River might have caused little or un-detected effects on major characteristics of the gastrointestinal helminth communities of P. fulvidraco.

The low infection level of helminths in Dongting Lake, and its low similarities with other lakes, may just be a reflection of its unique characters. Lakes in the flood-plain of the Yangtze River are in general crescent/oxbow lakes, and were once connected with the river [9]. They are eutrophic, but macrophytic if not influenced recently by urbanization or pollution [9]. Dongting Lake had been vastly reclaimed for agricultural and aquacultural purposes in the last century, with many channels constructed in the lake area in favour of transportation and flood diversion, resulting in a high degree of fragmentation in the lake area although its connection with the Yangtze River is still maintained [16,17]. Most importantly, this lake receives water from four rivers and then flows into the Yangtze River, although Poyang Lake also receives water from a river. The fragmentation, the water composition and probably the overall environment in this lake may account for the observed low infection rate of helminths, and thus the lower values of mean number of species and mean number of helminth individuals, as well as the low similarities observed between Dongting Lake and others, which may then be reflected as a separate tree in the clustering tree. It is nevertheless somewhat disappointing that only two connected lakes were included in the present study, which may also have some effect on the comparison between connected and isolated lakes.

In a study to examine the effect of lake fragmentation on parasite communities of fish, Valtonen et al. [18] also found a high degree of similarities in communities of ecto- and endo- parasites in roach (Rutilus rutilus) and perch (Perca fluviatilis) from four adjacent lakes, including two isolated lakes in central Finland. They considered that the fish and the parasite faunas may have been originally very similar and minor changes might have occurred over a period of 50 years. This may also be the case for the helminth communities of P. fulvidraco in the flood-plain lakes of the Yangtze River, although historical data on the parasites of this fish is not available. The fish can reproduce in lakes and also in rivers, although they may migrate between them, and the helminth species recorded in P. fulvidraco can all complete their life cycles in these investigated lakes [19]. The historical connection, the abundance of definitive fish hosts and the completion of life cycles of the helminth species in the flood-plain lakes of the Yangtze River may account for the observed similarity of the gastrointestinal helminth communities in P. fulvidraco in the present study. It seems possible that the observed pattern of helminth communities in P. fulvidraco in these flood-plain lakes might have been maintained historically, and reflected the historical connection of these localities.

The helminths recorded in the present study are all autogenic. Allogenic parasites have not previously been reported in this catfish [19], although body cavities and other organs of this fish were not examined in the present study. In a previous study, also carried in the flood-plain lakes, the parasites found in the common carp Cyprinus carpio were all autogenic [20]. Indeed, allogenic parasites, e.g. allogenic cestodes and/or nematodes are seldom reported in these flood-plain lakes [19], which may be a unique character for the parasite composition of fish in these waters. Autogenic species are considered to have a highly patchy distribution and to be more unpredictable in occurrence than allogenic [7,21], which may lead to the low similarities of parasite communities, as observed in the present study and reported by Esch et al. [7]. Nie et al. [20] also detected a low level of similarity between helminth communities of common carp. However, in other systems, similarities are normally higher; for example the quantitative similarity was high from 0.39 to 0.93 (mean = 0.65 ± 0.03) in whitefish in a group of lakes in Alberta, Canada, and from 0.39 to 0.94 (mean = 0.70 ± 0.04) in lakes in Finland, with the domination of specialists [22,23]. However, in the British Isles, where fish parasite communities have been well-investigated, the level of similarity between component communities is quite variable; but on average is low as determined in natural lakes and reservoirs. These communities were often dominated by generalist parasites, generally acanthocephalans, and are thus unpredictable [7,24,25]. As few data have been gathered in flood-plain lakes, it is far too early to make a predictable suggestion on the pattern of parasite communities in these lakes.

In the flood-plain lakes, several helminth parasites in P. fulvidraco may be considered as specialists. Most interestingly, the dominant nematode Procamallanus fulvidraconis is a specialist parasite found mostly in this fish, although it has been reported in other species of fish, such as Silurus asotus [19]. The higher infection rate of P. fulvidraconis observed in these lakes, with the exception of Dongting lake, may be attributed to the lacustrine environment in which zooplanktonic cyclops serving as the intermediate host for the nematode [26] may be abundant, and thus account for, at least to certain extent, the observed similarity between the flood-plain lakes. Orientocreadium siluri, Opisthorchis parasiluri, Gangesia pseudobagri, Spinitectus gigi are normally reported from P. fulvidraco and other silurid fish, and can then be considered as specialists. The three species, Genarchopsis goppo, Hebsoma violentum and Camallanus cotti, which infect a wide range of fish species, such as cyprinids and silurids may be considered as generalists. Coitocoecum plagiorchis and Dollfustrema vaneyi, which use carnivorous fish as definitive hosts, infect a wide range of fish species [19,27], may also be recognized as specialists, and Echinoparyphium lingulatum was also reported from carnivorous fish [28]. It is thus likely that component communities in P. fulvidraco are composed mainly of helminths specific to this fish or silurid fish, and if extended, to carnivorous fish.

The diversity indexes of helminth communities of the catfish were rather high in both isolated and connected lakes. Usually, the richness of helminth communities was related to the diet of host [29]. The yellowhead catfish is carnivorous and feeds mainly on invertebrates, including insect larvae, shrimps and crawfish, and also small fish [30]. Other studies also suggested that helminth communities in carnivorous fish species were generally richer than those in herbivores and omnivores [31]. The mean Brillouin's index of gastrointestinal helminth communities in the carnivorous catfish (0.39) was apparently higher than that (less than 0.16) in another omnivorous European eel Anguilla anguilla [32-34], and brown trout Salmo trutta [35,36]. Indeed, the Brillouin's index and the species richness of intestinal helminth communities were rather low (from 0 to 0.41 and 1 to 7, respectively) in the common carp Cyprinus carpio in flood-plain lakes of the river, which is omnivorous and feeds mainly on invertebrates and plant matters [20]. The majority of digeneans in the component communities can be attributed to the lacustrine and macrophytic environment which may provide habitats for mollusc intermediate hosts and thus transmission of digeneans, although detailed information on molluscan species composition and abundance is not available.

However, the similarity of helminth communities in P. fulvidraco reduced with geographical distances among lakes. The similarity of parasite communities decreasing with geographical distance is a pattern of many reported fish parasite communities [37-39]. Geographical distance may influence the probability of parasite exchanges [22,38], and there may be several explanations for such reduced similarity. The historical connection may be attributed to the observed similarity of helminth communities, but the degrees of stocking fishery in these lakes, and possible factors such flooding which may occur frequently in this flood-plain [9] may cause the variation of the helminth community similarities, or increase the similarity between adjacent lakes. Overall, the island biogeography theory may provide some explanation for the similarities of helminth communities between lakes. In particular, the lower similarity between Dongting Lake and the others might have influenced substantially the reduced similarity, and it may not be the only matter of distance, but the uniqueness of Dongting Lake as described above.

In addition, populations of these helminth species may be large enough to maintain population structure, and thus reduce the difference at both the population and community levels. Li et al. [12] reported that the nematode P. fulvidraconis, which dominated the helminth communities of P. fulvidraco in flood-plain lakes, showed no genetic difference at the level of populations also in these lakes.

It may also be possible that helminths can reinvade from other local populations of fish, with the movement of intermediate and definitive hosts. Therefore, lake isolation from the Yangtze River has hardly any effect on helminth communities of P. fulvidraco in such a short period of about half a century.

Other factors, such as trophic status have been suggested to have some effect on helminth communities of fish. Historical and recent studies have all shown that helminth component community in fish was associated with physicochemical characteristics and thus productivity of lakes [18,23,40]. Eutrophication has always been a problem in flood-plain lakes of the Yangtze River, but the increase in trophic level and artificial stocking in some lakes in the flood-plain have inevitably destroyed the submerged plantation, leading to the conversion of original, so-called macrophytic lakes to algal lakes in which the carnivorous catfish Pelteobagrus fulvidraco has almost disappeared [41]. So, the effect of eutrophication on helminth communities of fish in these lakes may be interpreted cautiously or differently as generally recognized.

Competing Interests

The authors declare that they have no competing interests.

Author Contributions

WXL conducted the field work and data analysis and drafted the manuscript. PN generated the research idea and finalized the manuscript. GTW contributed to the research plan. WJY participated in the field work. All authors read and approved the final manuscript.

Acknowledgements

This research was financially supported by the State Key Laboratory of Freshwater Ecology and Biotechnology (2006FB10, 2008FBZ06).

References

1. Fahrig L, Merriam HG: Conservation of fragmented population. Conservation Biology 1994, 8:50–59.

2. Gibb H, Hochuli DF: Habitat fragmentation in an urban environment: large and small fragments support different arthropod assemblages. Biological Conservation 2002, 106:91–100.
3. Pringle CM, Freeman MC, Freeman BJ: Regional effects of hydrologic alterations on riverine macrobiota in the New World: tropical-temperate comparisons. BioSience 2000, 50:807–823.
4. Morita K, Yamamoto S: Effects of habitat fragmentation by damming on the persistence of stream-dwelling charr populations. Conservation Biology 2002, 16:1318–1323.
5. Dobson AP, Pacala SW: The parasites of Anolis lizards of the northern Lesser Antilles. II. The structure of the parasite community. Oecologia 1992, 92:118–125.
6. Kennedy CR, Bush AO, Aho JM: Patterns in helminth communities: why are birds and fish different. Parasitology 1986, 93:205–215.
7. Esch GW, Kennedy CR, Bush AO, Aho JM: Patterns in helminth communities in freshwater fish in Great Britain: alternative strategies for colonization. Parasitology 1988, 96:519–532.
8. Bush AO, Kennedy CR: Host fragmentation and helminth parasites: hedging your bets against extinction. International Journal for Parasitology 1994, 24:1333–1343.
9. Wang S, Dou H: Chinese lakes. Beijing, China, Science Press; 1998. [In Chinese.]
10. Chang JB, Cao WX: Fishery significance of the river-communicating lakes and strategies for the management of fish resources. Resources and Environment in the Yangtze River 1999, 8:153–157. [In Chinese.]
11. Cheng Q, Zheng B: Systematic synopsis of Chinese fishes. Beijing, Science Press; 1987. [In Chinese.]
12. Li WX, Wang GT, Nie P: Genetic variation of fish parasite populations in historically connected habitats: undetected habitat fragmentation effect on populations of the nematode Procamallanus fulvidraconis in the catfish Pelteobagrus fulvidraco. J Parasitol 2008, 94:634–637.
13. Bush AO, Lafferty KD, Lotz JM, Shostak AW: Parasitology meets ecology on its own terms: Margolis et al. revisited. Journal of Parasitology 1997, 83:575–583.
14. Magurran AE: Ecological Diversity and its Measurement. Princeton University Press; 1988.

15. Hurlbert SH: The measurement of niche overlap and some relatives. Ecology 1978, 59:67–77.
16. Xiong JX: Integrity of spatial structure of wetland landscape in west Dongting Lake and its optimization. Wetland Science & Management 2008, 4:16–19. [In Chinese]
17. Xiong JX, Wu NF: Analysis of spatial structural integrity of wetland landscape in the east Dongting Lake. Environmental Science & Management 2008, 33:30–33. [In Chinese]
18. Valtonen ET, Holmes JC, Koskivaara M: Eutrophication, pollution, and fragmentation: effects on parasite communities in roach (Rutilus rutilus) and perch (Perca fluviatilis) in four lakes in central Finland. Canadian Journal of Fisheries Aquatic Sciences 1997, 54:572–585.
19. Chen CI, Eds, et al.: An illustrated guide to the fish disease and causative pathogenic fauna and flora in the Hupei Province. Beijing, Science Press; 1973. [In Chinese.]
20. Nie P, Yao WJ, Gao Q, Wang GT, Zhang YA: Diversity of intestinal helminth communities of carp from six lakes in the flood plain of the Yangtze River, China.J ournal of Fish Biology 1999, 54:171–180.
21. Esch G, Fernández J: A functional biology of parasitism: Ecological and evolutionary implications. London, Chapman & Hall; 1993.
22. Karvonen A, Valtonen ET: Helminth assemblages of whitefish (Coregonus lavaretus) in interconnected lakes: similarity as a function of species specific parasites and geographical separation. Journal of Parasitology 2004, 90:471–476.
23. Goater CP, Baldwin RE, Scrimgeour GJ: Physico-chemical determinants of helminth component community structure in whitefish (Coregonus clupeaformes) from adjacent lakes in Northern Alberta, Canada. Parasitology 2005, 131:713–722.
24. Kennedy CR: Helminth communities in freshwater fish: structural communities or stochastic assemblages? In Parasite Communities: Patterns and Processes. London, Chapman & Hall; 1990.
25. Hartvigsen R, Kennedy CR: Patterns in the composition and richness of helminth communities in brown trout, Salmo trutta, in a group of reservoirs. Journal of Fish Biology 1993, 43:603–615.
26. Li HC: The taxonomy and early development of Procamallanus fulvidraconis n. sp. Journal of Parasitology 1935, 21:102–113.

27. Wang GT: A note on the monthly changes of Dollfustrema Vaneyi in the digestive tract of the mandarin fish, Siniperca chuatsi. Acta Hydrobiological Sinica 2003, 27:108–109. [In Chinese.]

28. Wang PQ: Some digenetic treamatodes from fishes in Fujian province, China. Acta Zootaxonomica Sinica 1984, 9:122–131. [In Chinese.]

29. Dogiel VA: General Parasitology (English translation). Edinburgh, Oliver and Boyd; 1964.

30. Du JR: A study on the ingredient and emergence rate of the food and the reproduction of Pseudobagrus fulvidraco in Liangzi Lake. Chinese Journal of Zoology 1963, 2:74–77. [In Chinese.]

31. Pérez-Ponce de León G, García-Prieto L, León-Règagnon V, Choudhury A: Helminth communities of native and introduced fishes in Lake Pátzcuaro, Michoacán, México. Journal of Fish Biology 2000, 57:303–325.

32. Kennedy CR: The dynamics of intestinal helminth communities in eels Anguilla anguilla in a small stream: long-term changes in richness and structure. Parasitology 1993, 107:71–78.

33. Sures B, Knopf K, Wurtz J, Hirt J: Richness and diversity of parasite communities in European eels Anguilla anguilla of the River Rhine, Germany, with special reference to helminth parasites. Parasitology 1999, 119:323–330.

34. Kristmundsson A, Helgason S: Parasite communities of eels Anguilla anguilla in freshwater and marine habitats in Iceland in comparison with other parasite communities of eels in Europe. Folia Parasitologica 2007, 54:141–153.

35. Molloy S, Holland C, Poole R: Metazoan parasite community structure in brown trout from two lakes in western Ireland. Journal of Helminthology 1995, 69:237–242.

36. Kennedy CR, Hartvigsen RA: Richness and diversity of intestinal metazoan communities in brown trout Salmo trutta compared to those of eels Anguilla anguilla in their European heartlands. Parasitology 2000, 121:55–64.

37. Poulin R, Morand S: Geographical distances and the similarity among parasite communities of conspecific host populations. Parasitology 1999, 119:369–374.

38. Poulin R: The decay of similarity with geographical distance in parasite communities of vertebrate hosts.J ournal of Biogeography 2003, 30:1609–1615.

39. Oliva1 ME, González MT: The decay of similarity over geographical distance in parasite communities of marine fishes. Journal of Biogeography 2005, 32:1327–1332.

40. Wisniewski WL: Characterization of the parasite fauna of an eutrophic lake. Acta Parasitologica Polonica 1958, 6:1–64.

41. Yan G, Ma J, Qiu D, Wu Z: Succession and species replacement of aquatic plant community in East lake. Acta Phytoecologica Sinica 1997, 21:319–327. [In Chinese.]

Theoretical Analysis of Pre-Receptor Image Conditioning in Weakly Electric Fish

Adriana Migliaro, Angel A. Caputi and Ruben Budelli

ABSTRACT

Electroreceptive fish detect nearby objects by processing the information contained in the pattern of electric currents through the skin. The distribution of local transepidermal voltage or current density on the sensory surface of the fish's skin is the electric image of the surrounding environment. This article reports a model study of the quantitative effect of the conductance of the internal tissues and the skin on electric image generation in Gnathonemus petersii (Günther 1862). Using realistic modelling, we calculated the electric image of a metal object on a simulated fish having different combinations of internal tissues and skin conductances. An object perturbs an electric field as if it were a distribution of electric sources. The equivalent distribution of electric sources is referred to as an object's imprimence. The high conductivity of the fish body lowers the load resistance of a given object's imprimence, increasing the electric image. It also funnels the current generated by the electric organ in such a way that the field and the imprimence of objects in the vicinity of the rostral electric fovea are enhanced. Regarding skin conductance, our results show that the actual value is in the optimal range for transcutaneous voltage modulation by nearby objects. This result suggests that "voltage" is the answer to the long-standing question as to whether current or voltage is the effective stimulus for electroreceptors. Our analysis shows that the fish body should be conceived as an object that interacts with nearby objects, conditioning the electric image. The concept of imprimence can be extended to other sensory systems, facilitating the identification of features common to different perceptual systems.

Introduction

Electroreceptive fish detect nearby objects by processing the information contained in the pattern of electric currents through the skin. In weakly electric fish, these currents result from a self-generated field, produced by the electric organ discharge (EOD). Local transepidermal voltage or current density is the effective stimulus for electroreceptors. The distribution of voltage or current on the sensory surface of the fish's skin is the electric image of the surrounding environment [1–3]. From this image, the brain constructs a representation of the external world. Therefore, to understand electrolocation it is necessary to know the image-generation strategy used by electrolocating animals.

Theoretical analysis of image generation has yielded realistic models that predict with acceptable accuracy the electrosensory stimulus [4–12]. One general conclusion of previous reports is that the skin conductance and the conductivity difference between the internal tissues of the fish and the water are the main factors shaping the electric image: the seminal paper by Lissmann and Machin [13] started a long-lasting controversy about the roles of these factors. Lissmann and Machin argued that if "… the fish has approximately the same conductivity as the water and that it does not appreciably distort the perturbing field (i.e., does not produce an image of the image), the potential distribution around the fish due to the perturbing field can be calculated." However, several reports [3,7,14] have indicated that the internal conductivity of freshwater fish is high with respect to the surrounding water, and that the high conductance of internal tissues is critical for enhancing the local EOD field as well as for generating the centre-surround opposition pattern that characterizes electric images and that is coded by primary afferents [15].

Experimental studies in pulse gymnotids have confirmed theoretical predictions, showing that the high conductivity of the fish body funnels the self-generated current to the perioral region, where an electrosensory fovea has been described on the basis of electroreceptor density, variety, and central representation [16]. This funnelling effect enhances the stimulus at the foveal region. In addition, two different types of skin have been described in some electric fish of the family Mormyridae: the low-conductance mormyromast epithelium where electroreceptors are present, and the high-conductance non-mormyromast epithelium where electroreceptors are absent [7,17]. The mormyromast epithelium is found on the head in front of the gills, as well as along the dorsum of the back and along the ventral surface of the trunk. The non-mormyromast epithelium is found along the sides of the trunk. This heterogeneity of skin conductance introduces another factor shaping physical electric images.

This article describes a realistic modelling study of the effect of the internal and skin conductance on electric image generation in G. petersii. We have calculated the electric image of a metal object on a simulated fish having different magnitudes of conductances for internal tissues and skin. While the high conductivity of the fish body enhances the electric image by a combination of mechanisms, the skin conductance appears to optimize the transcutaneous voltage modulation by nearby objects. In contrast, transcutaneous current increases monotonically with skin conductivity. These results suggest that transcutaneous voltage is the critical proximal stimulus for electroreceptors.

We generalize two concepts: "object perturbing field" and "imprimence," introduced early in electroreception research [13], to other sensory systems. An object perturbs an electric field as if it were adding a new field to the basal one. This perturbing field can be considered as equivalent to a certain distribution of electric sources. This distribution is referred to as an object's imprimence.

Results

Electric fields and images generated by metal objects were described in previous reports (reviewed by [3]). In Figure 1, we present results obtained with a realistic fish model and a metal cube, as a reference for the following simulations. Figure 1A shows the basal field, i.e., the field generated by the EOD in the absence of objects. Since all the components of the scene are purely resistive elements, the electric field generated by the EOD has a constant spatial distribution and thus can be described with a static analysis. Therefore, the EOD has been represented by a DC current flowing caudal to rostral along the electric organ (EO). The isopotential lines run closely parallel to the skin, and the distance between them diminishes close to the tip of the "barbillon," a finger-like extension of the lower lip found in some mormyrid fish. This indicates that field strength and, consequently, current magnitudes are larger at the tip of the barbillon, due to edge effects. The barbillon may be thought of as an "electrosensory fovea" [16,17]. Figure 1B shows the distortion of the basal field (i.e., the object perturbing field) produced by the presence of the cube. This distortion depends on the characteristics of the object and the basal electric field in its neighbourhood. The object perturbing field shown in Figure 1 could also be produced by a set of dipoles oriented almost perpendicular to the fish skin at the point closest to the cube (object imprimence) [13]. The electric image is the difference between the current densities through the skin in the presence and the absence of the object (Figure 1C). Note that the currents increase (positive values) in the skin facing the cube and decrease in a larger surrounding region, producing a "Mexican hat"-type image that can be seen in the graph of Figure 1D, which shows a profile of the image along a line on the sagittal plane. Thus the object image is distributed over a large part of the sensory surface and is not restricted to just the area of skin facing it.

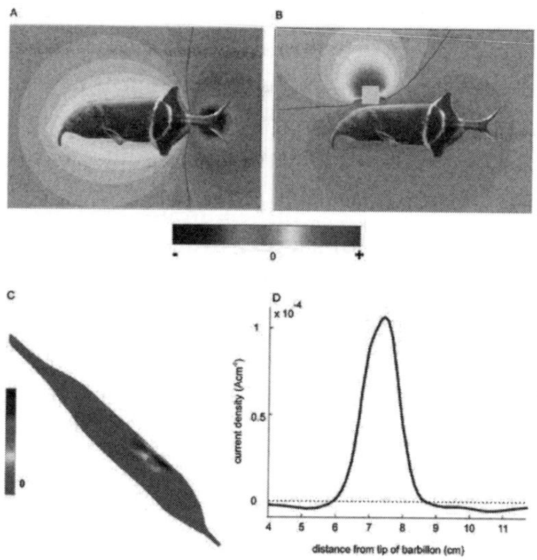

Figure 1. Image Generation in a Fish with Realistic Internal Conductivity and Homogeneous Highly Conductive Skin

(A) The colored background represents the difference in voltage between each point surrounding the fish and an infinitely distant point, using a non-linear arctangent color scale (used to highlight values close to zero) shown in the color bar below for the basal field (in the absence of objects). The black line shows the zero equipotential surface, which is perpendicular to the axis of the EO equivalent dipole distribution.

(B) A similar colored representation shows the perturbing field (i.e., the field in the presence of the object minus the basal field) produced by a metal cube (1 cm3) close to the skin (0.5 mm). The black line shows the zero equipotential surface, which is perpendicular to the axis of the object equivalent dipole distribution.

(C) Electric image of the metal object depicted in a color map on the modelled realistic fish from a scorci view.

(D) Electric image along the intersection of the skin with the sagittal plane, illustrating its "Mexican hat" profile.

To study the effect of the skin and internal conductances on the generation of the electric image, we departed from the situation proposed by Lissmann and Machin (1958), in which all fish tissues have the same conductivity as the water. Secondly, we studied the effect of changing internal conductivity, while maintaining a skin conductance that was very high and therefore of negligible effect. Thirdly, for an internal conductivity similar to that experimentally determined, we studied the effect of changing skin conductance as if it were uniform along the fish surface. Finally, we compared results obtained with homogeneous skin conductances and those obtained with the heterogeneous distribution of the skin conductances that is present in G. petersii.

Images as a Function of Fish Internal Conductivity

We have proposed that the low resistivity of the fish body is a very important factor for the shaping of the electric image. To assess its contribution, we simulated electric images for fish having a high skin conductance but with different internal conductivities.

We first modelled a fish with an internal conductivity equal to that of the water, as assumed by Lissmann and Machin [13]. This is described as a "transparent fish." In this case, the conductivity of the surrounding medium is homogeneous except for the object. Thus, the images calculated as the distribution of the current density across a virtual sensory epithelium are the perturbing fields at the surface of the skin. The basal field generated by the EO is similar to the field of a dipole in a homogeneous medium (Figure 2A). Consequently, and in contrast to the real situation, the isopotentials lines do not run closely parallel to the skin, and the field at the tip of the barbillon does not show an edge effect. Figure 2B shows the perturbing field (the field in the presence of the object minus the basal field) produced by a metal cube close to the skin. The imprimence of the object is equivalent to a certain distribution of dipoles located at the object site, no longer oriented perpendicular to the skin, in contrast to the naturally realistic case (see Figure 1B).

Comparison of Figures 1 and 2 shows that the direction of the field is nearly parallel to the transparent fish body, and nearly perpendicular to the real fish body. This indicates that the conductivity of the fish body distorts the field produced by the EO. It is worth noting that the internal conductivity of the fish not only funnels the current rostrally but also exerts an effect on the field direction generated at the object location: as a consequence, the electric image is more symmetric. Comparison of image profiles along a sagittal plane (Figures 1D and 2D) shows an enhancement of the image amplitude produced by the presence of the fish body. The body exerts this effect in two ways: a) by increasing the local field in the vicinity of the object, therefore increasing the perturbing field and its imprimence, and b) by introducing an impedance gradient at the site of the sensory surface. Previous research has shown that the amplitude of the image generated by a dipole increases up to two times when the fish/water conductivity ratio increases [18]. To test this mechanism, we calculated the image of a dipole perpendicular to the skin of a transparent fish (Figure 3A, with the negative pole facing the skin), comparing it with the image of the same dipole on a fish with normal internal impedance (Figure 3B). While the waveform remains similar as shown in the current profiles along the skin intersecting with the sagittal and coronal planes, the amplitude of the profile for the realistic fish is twice that for the transparent fish (Figure 3C).

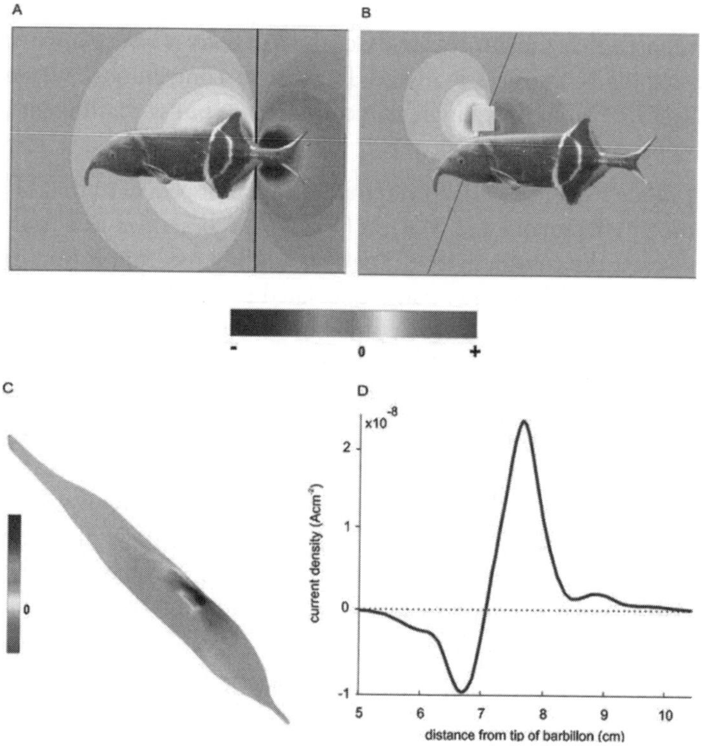

Figure 2. Image Generation in a Fish with Internal Conductivity like That of Water and with a Homogeneous Highly Conductive Skin

The black bars show the zero equipotential surfaces as in Figure 1.

(A) Basal field (in the absence of objects).

(B) Perturbing field produced by the same scene as in Figure 1B.

(C) Electric image of the metal object depicted in a color map on the modelled transparent fish from a scorci view.

(D) Electric image along the intersection of the skin with the sagittal plane.

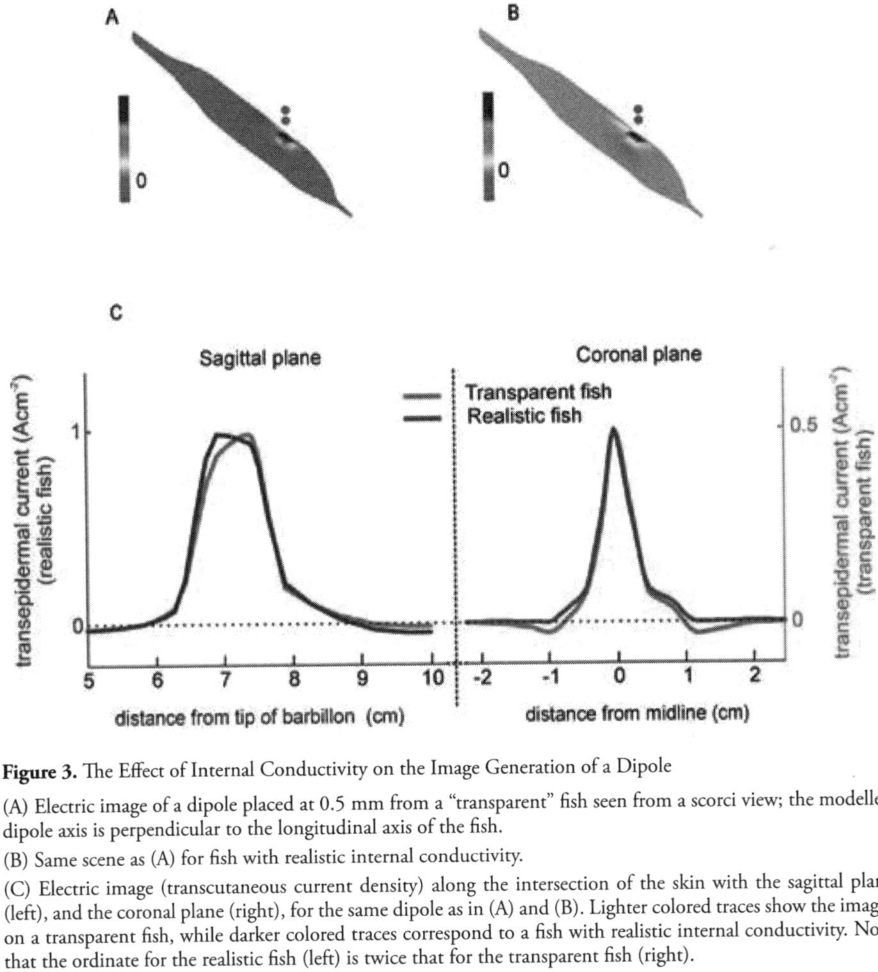

Figure 3. The Effect of Internal Conductivity on the Image Generation of a Dipole

(A) Electric image of a dipole placed at 0.5 mm from a "transparent" fish seen from a scorci view; the modelled dipole axis is perpendicular to the longitudinal axis of the fish.

(B) Same scene as (A) for fish with realistic internal conductivity.

(C) Electric image (transcutaneous current density) along the intersection of the skin with the sagittal plane (left), and the coronal plane (right), for the same dipole as in (A) and (B). Lighter colored traces show the images on a transparent fish, while darker colored traces correspond to a fish with realistic internal conductivity. Note that the ordinate for the realistic fish (left) is twice that for the transparent fish (right).

Figure 4A shows the normalized electric image of a metal cube calculated for fish with different body conductivities. In order to maintain a constant electric source, the tail region was modelled as an independent compartment with realistic internal conductivity. As shown in the normalized images (Figure 4A), both edges shift rostrally with a predominant shift of the rostral border, so that the image becomes wider as body conductivity increases. In addition, the shape of the profile, which initially consists of two main deflections (one caudal positive and one rostral negative), becomes more symmetric, resembling a Mexican hat. The amplitude of the image is an increasing function of body conductance (Figure 4B). These changes are correlated with an increase in the magnitude and a change in the direction of the basal field around the object.

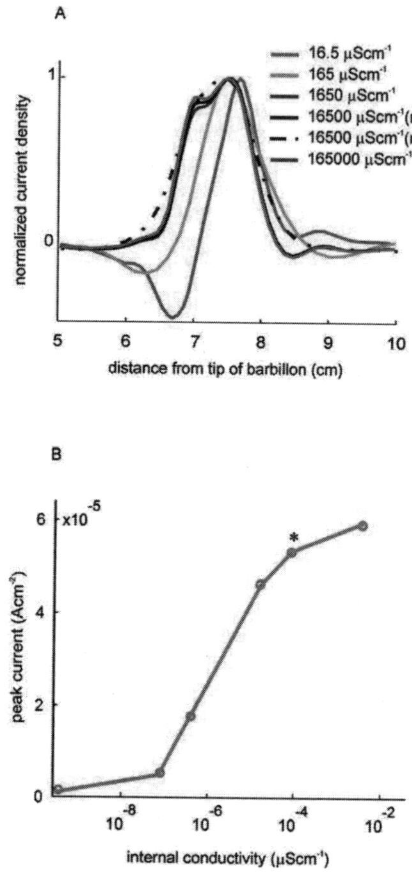

Figure 4. The Effect of Internal Conductivity on Electric Image Generation

(A) Normalized electric images of the same metal cube (identical position) on fish with different internal conductivities. Red: 16.5 μScm−1 (the same as water conductivity), cyan: 165 μScm−1, blue: 1,650 μScm−1, black: 16,500 μScm−1 (normal conductivity), magenta: 165,000 μScm−1. The skin is modelled for all cases, with a homogeneous conductivity of 500,000 μScm−1. The dashed line shows the case of a fish with realistic internal conductivity and skin conductivity distribution. rl, realistic internal conductivity; rlh, realistic internal conductivity, heterogeneous skin distribution.

(B) Peak amplitude of the electric image of a metal cube (1 cm3) placed at 0.5 mm from the fish, as a function of body internal conductivity. The difference in the peak amplitude of the electric image corresponding to the realistic internal conductivity fish shown in this figure and that shown in Figure 1 is due to the use of two compartment bodies (see Materials and Methods).

The Effect of Skin Conductance

To assess the contribution of the skin to image formation, we studied the effect of different uniform skin conductances for a fish with normal internal conductivity. For very low skin conductivity, the transepithelial currents produced by the EO are negligible (Figure 5): the current short-circuits inside the fish because it cannot flow through the skin. The transepithelial current increases with the skin conductivity, approaching an asymptotic value (red trace in Figure 5A and 5C). Since transcutaneous voltage is the quotient of the current density divided by skin conductance, voltage increases differently with skin conductance, rising to a maximum and then decreasing (blue trace in Figure 5B and 5C). The value of skin conductance at which voltage reaches a maximum, 100 μScm^{-2}, is close to the actual measured value for the mormyromast epithelium. This suggests that electroreceptors operate in a voltage detection mode rather than in a current detection mode. The normalized curves in Figure 5D show that the image is smoother and wider as the skin conductance decreases. Continuous traces correspond to uniform skin conductances, where the cyan one is the closest to the mormyromast epithelium value. Realistic electric images were calculated as a reference, using the distribution of conductivities determined experimentally (dotted traces) [7].

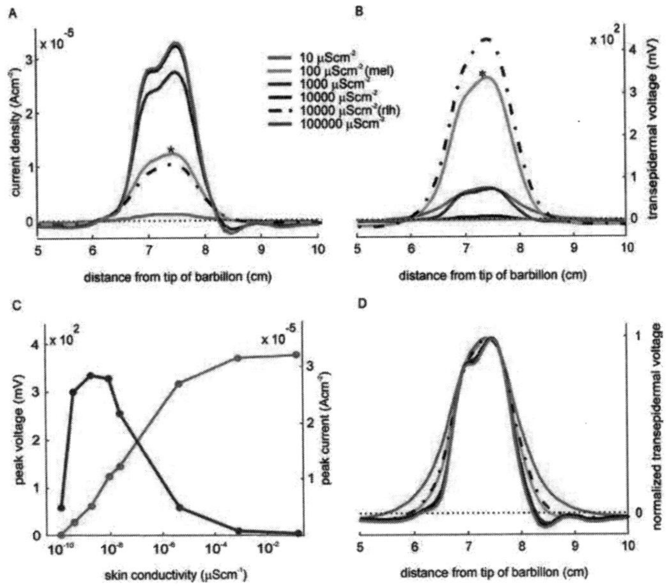

Figure 5. The Effect of Skin Conductivity on Electric Image Generation

(A) Transcutaneous current density (electric image) of a metal cube (1cm3) placed at 0.5 mm from the skin, modelled on skin with different conductivities. Red: 10 μScm−2, cyan: 100 μScm−2 (similar to mormyromast epithelium), blue: 1,000 μScm−2, black: 10,000 μScm−2, magenta: 100,000 μScm−2. All these fish have an internal conductivity of 3,300 μScm−1. Dashed line shows the case of a fish with realistic internal conductivity and skin conductivity distribution.

(B) Transcutaneous voltage calculated from the transcutaneous current densities shown in (A), using the same color code.

(C) Current peak (right axis, red trace) and voltage peak (left axis, blue trace) as a function of skin conductivity for fish with homogeneous skin.

(D) Normalized plot for (B), using same color code. mel, mormyromast epithelium-like conductivity; rlh, realistic internal conductivity, heterogeneous skin distribution.

Discussion

Animals extract information from the environment and from their own bodies by analyzing changes in the patterns of energy impinging on their sensory surfaces. In that sense, it can be affirmed that to see is to reconstruct visual scenes from a light pattern on the retina or to hear is to extract auditory scenes from sound patterns at the cochlea [19]. Similarly, electric sensing is to reconstruct electric scenes from the pattern of electric currents through the skin.

In electrosensory perception, each object generates a signal that results from the deformation that its presence causes in an electric field. This deformation is a virtual field, called "object perturbing field" by Lissmann and Machin [13]. The object perturbing field is not directly measurable, but computable as the electric field in the presence of the object minus the electric field in its absence, also called "basal field." As any electric field, the object perturbing field can be considered as caused by an electric source, which is equivalent to the presence of the object.

The "imprimence" of an object, an expression also coined by Lissmann and Machin [13] referring to the electric sources equivalent to the object, not only generates an image but also a change in the field that interacts with other objects. Thus, the effect of a given object not only generates its own image but also modifies the images of other objects [10]. There are theoretical and experimental reasons indicating that the fish body is also an object, and that this is of particular importance since it is an object that is always present as a major determinant of sensory imaging [5,7,18,20]. This leads to the proposition that the fish body, by its presence and movements, constitutes a critical pre-receptor mechanism that conditions sensory signals [16,21]. We, therefore, discuss here the effect of relevant components of the fish's body on image generation.

The Effect of the Fish's Internal Conductivity

The imaging process consists of two steps: imprimence generation (yellow boxes in Figure 6) and image generation (purple boxes in Figure 6). The simplest example occurs in a "transparent" fish, isoconductive with water. The electric image (green arrow in Figure 6A) is the difference between the electrosensory stimulus generated in the presence of an object (light-blue arrow in Figure 6A) and the electrosensory stimulus generated in the absence of that object (dark-blue arrow in Figure 6A). The latter is referred to as the basal stimulus because it is caused by the basal field. Since, in the case of the transparent fish, the basal field is not distorted by the fish body, the electric image results from the projection on the skin of a field perturbation induced only as a consequence of interaction of the object with the basal field (green arrows in Figure 6A). Then, in a transparent fish, image formation can be described as a simple process consisting of two steps: a) the generation of a field by the EO and b) the deformation of this field by the presence of the object.

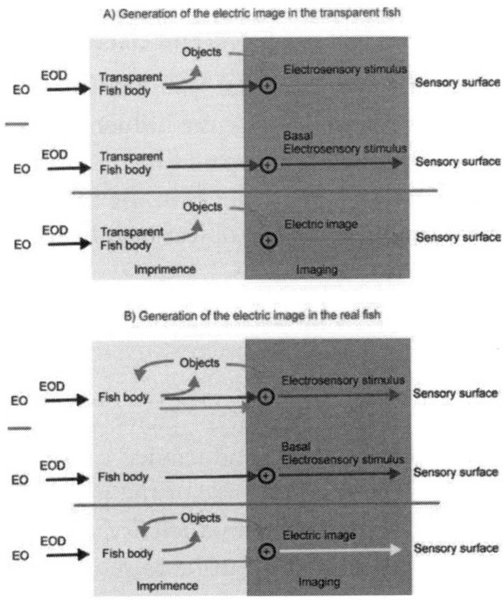

Figure 6. Schematic Representation of Electric Image Generation First row, generation of stimulation in the presence of the object; second row, basal stimulation in the absence of objects; third row, sensory image.

(A) Fish with water-like internal conductivity. Imprimence generation (yellow boxes) precedes image generation (purple boxes). A field perturbation (green arrows) is induced as a consequence of the object interaction with the basal field (dark-blue arrows). The electric image is the difference between the perturbing (light-blue arrow) and the basal fields at the skin.

(B) Fish with realistic internal conductivity. The interaction of the body with the field perturbed by the object (red arrows) introduces another component (orange arrow) to the electrosensory stimulus (magenta arrow). The electric image (yellow arrow) is the electrosensory stimulus minus the basal field (blue arrow, representing the sum of the effects of the fish body and the object in the presence of each other). (See Discussion for explanation.)

However, in nature, the basal field is different from that produced by the EO in a homogeneous medium, because it is affected by the inextricable presence of the fish body. Similarly, the object perturbing field is also affected by the fish's body. This interaction (red arrows) produces two extra components that add to the basal electrosensory stimulus (dark-blue arrow): the perturbing field of the object (green arrow) and the perturbing field of the fish body (orange arrow). This resulting field acting on the skin is the electrosensory stimulus (magenta arrow). To calculate the electric image, we subtracted the effect of the basal stimulus (fish body alone, blue arrow). Thus, the electric image (yellow arrow) results from the addition of the perturbing field of the fish's body in the presence of the object (orange arrow) plus the perturbing field of the object in the presence of the fish's body (green arrow). When the object is large enough and surrounds the fish, its effect becomes very important, having a strong influence on the overall pattern of current flow. This is the case when the fish chooses to stay in confined spaces that are frequently its preference in the natural habitat, or in the tube-shaped shelters commonly used in captivity. The fish's positioning of its body in this manner strongly affects the electric images of objects and electrosensory responses [22].

When the object is relatively small or far from the fish body, the loop between the object and the fish body opens, because the influence of the field of the object on the fish body becomes negligible compared to the basal field. Consequently, the scheme of image generation is the same as in the case of the transparent fish. However, the basal field illuminating the object is different than that in the case of a transparent fish and so is the image.

The Effect of Skin Conductivity

The skin conductance is the other important factor shaping the electric image. A homogeneous decrease of the skin conductance causes: a) a decrease of the transepithelial current density, b) an increase of the transepithelial voltage up to a maximum at the range of natural skin conductivity, c) a decrease of the relative slope of the flanks of the image, and d) an increase of the centre region of the "Mexican hat" profile.

For measuring electrosensory stimulus, either local field (equivalent to current flow) or transcutaneous voltage has often been used indiscriminately [1,20,23]. However, our results indicate that current density and voltage are not equivalent stimuli. The transepithelial change in voltage caused by an object is the maximum within the range of skin conductances that are actually measured in the mormyromast epithelium of G. petersii (70–500 µScm–2 [7]), suggesting that the low conductance observed in the mormyromast epithelium might be an adaptation for optimizing voltage sensing. This low conductance of the mormyromast epithelium is caused by a thin layer of tightly packed epithelial cells [24], which makes the mormyromast epithelium up to ten times more resistant than the non-mormyromast epithelium [7]. If receptors electrically shunt their low conductance non-sensitive surroundings, transcutaneous voltage could be considered to be the meaningful parameter of the stimulus. Experimental measurements testing this hypothesis should be done.

Our study also shows a consistent decrease in the relative slope of the flanks of the image and an increase in the centre region of the "Mexican hat" profile with increasing skin resistance (see Figure 5D). In this study, restricted to single conductive objects close to the skin, these changes are rather small, indicating that the main factor for determining object image shape is the internal conductance of the fish body.

The Generality of the Concepts of Object Perturbing Field and Imprimence

Object recognition is an important issue in all sensory systems (including electrical perception), but it is not well understood. The comparative study of different sensory systems leads to general concepts and a language that could potentially be shared by researchers in different systems. In this paper, we focus on peripheral imaging mechanisms, a subject common to sensory systems. We focus in particular on the way in which pre-receptor mechanisms and interactions between different objects in a given scene shape the image.

We emphasize two concepts that were introduced early on in electroreception research [13]: object perturbing field and imprimence. An object perturbs an electric field as if it were a distribution of electric sources. The equivalent distribution of electric sources is referred to as an object's imprimence. Object perturbing field is a concept that relates to reflections and refractances in vision, echoes in audition, etc. In the same way that objects of different impedance than the water modify an electric field, objects in a visual scene modify the illumination. Similarly, echoes and resonances produced by objects modify the distribution of sound in an auditory scene. For example, the sound of a pulsed string on a guitar is greatly modified by the resonance of the box, giving the sound a characteristic timbre. The concept of imprimence can be extended in the same way. Objects producing reflexions, refractances, echoes, and resonances can be considered new sources of energy.

The imprimence produced by the animal's own body acts as a pre-receptor mechanism. The fish body can be considered as an object that interacts with other objects in the scene, generating an imprimence that through the perturbing field modifies the basal field of the scene and, consequently, the imprimence of the other objects. Many species of fish (including Mormyrids) hear underwater due to the imprimence produced by air-filled sacs such as the swimbladder [25]. In a less fundamental way, our body changes the visual image by interfering with and reflecting light, modifying the images of nearby objects. In addition, the interactions between objects and the perturbations of the fields by the imprimences of other objects are used to extract information from a scene. For example, the imprimence of the external ear modifies the incoming sound, allowing for the computation of the altitude of a source [26].

Animal senses explore nature using a limited number of types of energy and receptors with limited dynamic ranges. This constrains and conditions the representation of external reality according to the capabilities of each animal. Humans circumvent these limitations by creating artificial systems, such as radar or sonar, which expand the repertoire of representable qualities of objects. The concepts of imprimence and perturbing field may be applied to the design of artificial sensory systems. It is a common practice to deal with interactions between objects and the perturbations of the fields by the imprimences of other objects as undesirable interference. Nevertheless, evolution has developed neural operations that use images resulting from object interferences as a source of information, in some cases using this to infer object attributes. In these cases, interference between objects may increase the amount of available information contained in the image. Development of the theory of peripheral imaging is a necessary step for the design of computational procedures, allowing the extraction of a larger amount of information from the same signals.

Conclusions

The electric image of an object results from the projection on the skin of a virtual field caused by the presence of an object, in a given electrosensory scene.

The fish's large internal conductance (compared with water) causes a rostral funnelling of the current. This leads to an increase in the imprimence of objects close to the rostral regions of the fish and, consequently, to an increase in the amplitude of their images.

The large difference in conductivity between the inside and outside of the fish forces the field to be almost perpendicular to the sensory surface and, consequently, makes the shape of the image more symmetrical.

An object modifies the field of other objects immersed in the same global field. The fish body itself is a major object, inherent to the process of image generation. Thus, a global field results from the reciprocal interaction between the fish body and nearby objects.

The conductance of the skin changes the shape of the image only slightly, but drives the amplitude (considered as the distribution of transepithelial voltages) close to its maximum, for a given set of other electrical parameters. This result suggests that the high resistance of the mormyromast surface, a property conferred by a thin layer of tightly packed epithelial cells, may serve to optimize object images.

The use of a realistic computational model has allowed us to settle the controversy about the relative importance of the internal and skin conductivities in the determination of the magnitude and shape of the electric image, an issue that has been debated since the seminal paper by Lissmann and Machin [13].

We propose that the concepts of perturbing field and imprimence [13] may be usefully applied to the analysis of other sensory systems and the design of artificial ones.

Materials and Methods

The Model

Simulations were run using a program written to simulate the electric image in weakly electric fish (i.e., the currents through the fish skin), which uses the Boundary Element Method (BEM [27], as proposed by Assad [4], and has been described previously [10,28]). This program allows the determination of the electric field and the electrosensory image in a given environment (scene), calculating the currents through the skin. A scene may include objects (other than the fish) of different conductivity, shape, and size, and is defined by setting the geometry and location of one or more electric fish and objects. Water, internal, and skin conductivities can be specified as required. When the skin conductivity is not homogeneous, different regions can be defined using a graphic interface. Complex shapes, including the fish body, are approximated by a surface composed by triangles. Although the fish shape is kept constant throughout this article, the model allows its modification if required. Once the scene is determined, the potentials and current densities through the skin of the fish and through the objects are calculated. The graphic presentations were made by Matlab standard subroutines.

Changes in Internal Conductivity and Skin Conductance

We studied the effect of the skin and internal conductivity on the electric image in the presence of a metallic (high conductance) cube placed symmetrically to the sagittal plane and facing the dorsal skin 0.5 mm away. Water conductivity was kept at 16.5 μScm^{-1}.

To assess the influence of the internal conductivity of the fish body, different values ranging from that equal to surrounding water conductivity (in which case the fish may be considered transparent) to 16,500 μScm−1 were examined, including the value experimentally determined (3,300 μScm−1). In order to maintain a constant electric source, tail and body regions were modelled as independent compartments, maintaining the tail with a realistic internal conductivity while applying different values for the body. In these cases, the conductance across the model skin was set low enough to be considered irrelevant.

To study the influence of the skin conductance, we explored the effect of different skins with homogeneously distributed conductances ranging from 10–100,000 μScm−2 and a natural-like skin with heterogeneous conductance distribution. The internal conductivity in this case was close to that experimentally determined (3,300 μScm−1).

Two singular conditions were used for comparison purposes: a) when the fish model has experimentally determined conductances (where the fish body exerts its normal effect on the electric image); and b) when it has water-like conductances (i.e., where the fish body exerts no effect on the electric image).

Acknowledgements

The authors would like to thank Dr. Kirsty Grant and the anonymous reviewers for their helpful comments and suggestions of improvement. This work was partially financed by the Comision Sectorial de Investigación Científica (CSIC), Universidad de la República, Montevideo, Uruguay (fellowship for AM and equipment), and a grant for international cooperation from the French Ministère des Affaires Etrangères, (ECOS-Sud U03B01).

Author Contributions

AM is a postgraduate student (PEDECIBA, Uruguay) whose thesis is being advised by AAC and RB. AAC and RB conceived and designed the experiments, and contributed reagents/materials/analysis tools. AM performed the experiments. AM, AAC, and RB analyzed the data and wrote the paper.

References

1. Bastian J, editor (1986) Electrolocation. In: Bullock TH, Heiligenberg W, editors. Electroreception. New York: Wiley & Sons. 722 p.
2. Bell CC (1989) Sensory coding and corollary discharge effects in mormyrid electric fish. J Exp Biol 146: 229–253.
3. Budelli R, Caputi A, Gomez L, Rother D, Grant K (2002) The electric image in Gnathonemus petersii. J Physiol Paris 96: 421–429.
4. Assad C (1997) Electric field maps and boundary element simulations of electrolocation in weakly electric fish. Pasadena (California): California Institute of Technology.

5. Budelli R, Caputi AA (2000) The electric image in weakly electric fish: Perception of objects of complex impedance. J Exp Biol 203: 481–492.

6. Caputi A, Budelli R (1995) The electric image in weakly electric fish: I. A data-based model of waveform generation in Gymnotus carapo. J Comput Neurosci 2: 131–147.

7. Caputi A, Budelli R, Grant K, Bell C (1998) The electric image in weakly electric fish: II. Physical images of resistive objects in Gnathonemus petersii. J Exp Biol 201: 2115–2128.

8. Heiligenberg W (1973) Electrolocation of objects in the electric fish Eigenmannia rhamphichthyidae Gymnotoidei. J Comp Physiol 87: 137–164.

9. Lissmann HW (1958) On the function and evolution of electric organs in fish. J Exp Biol 35: 156–191.

10. Rother D, Migliaro A, Canetti R, Gomez L, Caputi A, et al. (2003) Electric images of two low resistance objects in weakly electric fish. Biosystems 71: 169–177.

11. Rasnow B (1996) The effects of simple objects on the electric field of Apteronotus leptorhynchus. J Comp Physiol A 178: 397–411.

12. Assad C, Rasnow B, Stoddard PK (1999) Electric organ discharges and electric images during electrolocation. J Exp Biol 202: 1185–1193.

13. Lissmann HW, Machin KE (1958) The mechanisms of object location in Gymnarchus niloticus and similar fish. J Exp Biol 35: 457–486.

14. Caputi AA, Aguilera PA, Castello ME (2003) Probability and amplitude of novelty responses as a function of the change in contrast of the reafferent image in G. carapo. J Exp Biol 206: 999–1010.

15. Gomez L, Budelli R, Grant K, Caputi AA (2004) Pre-receptor profile of sensory images and primary afferent neuronal representation in the mormyrid electrosensory system. J Exp Biol 207: 2443–2453.

16. Castello ME, Aguilera PA, Trujillo-Cenoz O, Caputi AA (2000) Electroreception in Gymnotus carapo: Pre-receptor processing and the distribution of electroreceptor types. J Exp Biol 203: 3279–3287.

17. von der Emde G, Schwarz S (2002) Imaging of objects through active electrolocation in Gnathonemus petersii. J Physiol Paris 96: 431–444.

18. Sicardi EA, Caputi A, Budelli R (2000) Physical basis of distance discrimination in weakly electric fish. Physica A 283: 86–93.

19. Bregman AS (2001) Auditory scene analysis: The perceptual organization of sound. Cambridge (Massachusetts): MIT Press. 773 p.

20. Aguilera PA, Castello ME, Caputi AA (2001) Electroreception in Gymnotus carapo: Differences between self-generated and conspecific-generated signal carriers. J Exp Biol 204: 185–198.

21. Caputi AA (2004) Contributions of electric fish to the understanding sensory processing by reafferent systems. J Physiol Paris 98: 81–97.

22. Pereira AC, Centurion V, Caputi AA (2005) Contextual effects of small environments on the electric images of objects and their brain evoked responses in weakly electric fish. J Exp Biol 208: 961–972.

23. von der Emde G (1990) Discrimination of objects through electrolocation in the weakly electric fish, Gnathonemus petersii. J Comp Physiol A 167: 413–421.

24. Quinet P (1971) Etude systématique des organes sensoriels de la peau des Mormyriformes (Pisces, Mormyriformes). Annls Musée Royal Afrique Centrale, Tervuren (Belgium) 190: 1–97.

25. Yan HY, Curtsinger WS (2000) The otic gasbladder as an ancillary auditory structure in a mormyrid fish. J Comp Physiol [A] 186: 595–602.

26. Hudspeth AJ, Konishi M (2000) Introduction—Auditory neuroscience: Development, transduction, and integration. Proc Natl Acad Sci U S A 97: 11690–11691.

27. Hunter P, Pullan A (2002) FEM/BEM notes. Available: http://lola.unimo.it/~fonda/DISPENSA_Tb_html/Programma/Prog_TB_4/Dipolo/fembemnotes.pdf . Accessed 15 June 2005.

28. Rother D (2003) Simulación de imágenes eléctricas en peces eléctricos de descarga débil. Montevideo: PEDECIBA-Universidad de la República. 45 p.

Defining Global Neuroendocrine Gene Expression Patterns Associated with Reproductive Seasonality in Fish

Dapeng Zhang, Huiling Xiong, Jan A. Mennigen, Jason T. Popesku, Vicki L. Marlatt, Christopher J. Martyniuk, Kate Crump, Andrew R. Cossins, Xuhua Xia and Vance L. Trudeau

ABSTRACT

Background

Many vertebrates, including the goldfish, exhibit seasonal reproductive rhythms, which are a result of interactions between external environmental stimuli and internal endocrine systems in the hypothalamo-pituitary-gonadal axis. While it is long believed that differential expression of neuroendocrine genes contributes to establishing seasonal reproductive rhythms, no systems-level investigation has yet been conducted.

Methodology/Principal Findings

In the present study, by analyzing multiple female goldfish brain microarray datasets, we have characterized global gene expression patterns for a seasonal cycle. A core set of genes (873 genes) in the hypothalamus were identified to be differentially expressed between May, August and December, which correspond to physiologically distinct stages that are sexually mature (prespawning), sexual regression, and early gonadal redevelopment, respectively. Expression changes of these genes are also shared by another brain region, the telencephalon, as revealed by multivariate analysis. More importantly, by examining one dataset obtained from fish in October who were kept under long-daylength photoperiod (16 h) typical of the springtime breeding season (May), we observed that the expression of identified genes appears regulated by photoperiod, a major factor controlling vertebrate reproductive cyclicity. Gene ontology analysis revealed that hormone genes and genes functionally involved in G-protein coupled receptor signaling pathway and transmission of nerve impulses are significantly enriched in an expression pattern, whose transition is located between prespawning and sexually regressed stages. The existence of seasonal expression patterns was verified for several genes including isotocin, ependymin II, GABAA gamma2 receptor, calmodulin, and aromatase b by independent samplings of goldfish brains from six seasonal time points and real-time PCR assays.

Conclusions/Significance

Using both theoretical and experimental strategies, we report for the first time global gene expression patterns throughout a breeding season which may account for dynamic neuroendocrine regulation of seasonal reproductive development.

Introduction

Fundamental to the survival of most organisms, be they yeast, plants, fishes, or mammals, are biological rhythms with periodic (daily, monthly or annual) changes in behavior and physiology [1]. The daily circadian rhythm is exemplified by opening/closing of flowers or the daily sleep cycle in humans. The menstrual cycle of women is a typical monthly rhythm whereas circannual rhythms include bird migrations, hibernation in frogs and mammals. Marked reproductive seasonality in numerous vertebrate classes, including fish, ensures that reproduction and subsequent development of offspring is coordinated with optimal environmental and nutritional conditions. It has long been accepted that external environmental influences such as photoperiod [2], [3], [4] and temperature [4], [5], [6] exert dominant roles in biological rhythms, and internal neuroendocrine systems such as the pineal gland, hypothalamus and pituitary coordinate these signals [7], [8], [9], [10]. In this study, we use theoretical and experimental approaches to better understand global genomic regulation of the neuroendocrine system during seasonal reproduction.

The bony fishes or teleosts represent more than half of all vertebrates. Numerous characteristics of goldfish (Carassius auratus), a member of the family Cyprinidae, one of the largest vertebrate families, make this species an excellent model for neuroendocrine signaling and the control of seasonal reproduction [11], [12]. Similar to many teleosts, goldfish employ a seasonal reproductive strategy [5] characterized by distinct stages. In temperate regions of the Northern hemisphere the exact timing of these stages in goldfish depends on geographical location but can be generally classified as follows: gonadal regression (June-October), recrudescence or redevelopment (October-March), prespawning (April-May) and spawning or breeding season (May or early June).

Seasonal reproductive cyclicity including annual regeneration of gonadal tissues is mediated by a hormone regulatory pathway primarily involving gonadotropin-releasing hormone (GnRH), luteinizing hormone (LH), FSH, growth hormone (GH), melatonin, and sex steroid hormones [8], [9], [10]. The main hypophysiotrophic systems controlling LH and GH are located in the hypothalamus and the preoptic region of the telencephalon, which have been well-described and characterized in goldfish [8], [13]. Transduction of photoperiodic signals via retino-hypothalamo-pineal pathways results in melatonin secretion patterns typically reflecting the length of the dark period [14], [15], [16]. Melatonin influences the neuroendocrine system via its receptors which are widely distributed in the teleost brain [17], [18], [19]. Multiple GnRH forms [20] activate anterior pituitary receptors to stimulate the expression and release of LH, follicle-stimulating hormone (FSH) and GH, which are all involved in controlling seasonal gonadal development and sex steroid synthesis. Moreover, synchronization of fish reproductive behaviors in the breeding season involves a well-defined sex pheromone response system, culminating in surge release of LH that induces ovulation in females and sperm release in males [21]. Testosterone (T) and estrogen (E2) exert a primary role in gonadal development locally, and have profound positive and negative feedback actions at the levels of the brain and pituitary [8]. In addition to these core members of hypothalamo-pituitary-gonadal (HPG) axis, there is complex and interactive regulation of GnRH and LH/FSH by a multitude of neuroendocrine factors including classical neurotransmitters and other neuropeptides [8], [9], [10], [12], [22]. Glutamate, gamma-aminobutyric acid (GABA), and a multitude of neuropeptides are predominantly stimulatory on LH release whereas dopamine is the single most potent inhibitor of LH release in teleosts [8], [12], [22].

It has been extensively described that reproductive seasonality in teleosts is characterized by their seasonal profiles or changes in serum hormone levels [23], [24], [25], brain hormone content [26], brain enzyme activities [23], morphological [27] and histological [28] phenotypes. More recently, there is recognition that seasonal expression of the neuroendocrine genes including neurohypophysial hormones, synthesis enzymes, and related signaling pathway proteins may be utilized to link the external environmental influences and the establishment of fish reproductive cyclicity. It has been observed that in teleosts many neuroendocrine genes including aromatase b [29], GnRH receptors [30], glutamic acid decarboxylase (GAD) [31], cholecystokinin (CCK) [32], preprotachykinin [33] and secretogranin-II [34] have seasonal expression profiles, correlating with seasonal changes in gonadal size and sex steroid serum levels [23], [24], [35], [36]. However, no study at the transcriptomic level has been conducted to investigate the global gene expression changes in fish neuroendocrine systems throughout an annual reproductive cycle. We previously utilized microarrays to identify hormone or endocrine disrupting chemical-regulated genes in the goldfish neuroendocrine system [37], [38], [39], [40]. All experiments were based on goldfish sampled at different seasonal time points; thus, these microarray datasets contain important seasonal gene expression information. In the present study, on the basis of multiple goldfish microarray datasets, we used comprehensive normalizations, differential gene expression identification, multivariate and gene ontology (GO) analyses to characterize the season-related expression patterns in the female goldfish brain. A core gene set was identified whose expression exhibits a seasonal change with fish reproductive cycle. We illustrate that the gene expression change is regulated by photoperiod. Gene ontology (GO) analysis further revealed that neuroendocrine-related genes are significantly enriched in an expression pattern, which follows a high-low-low profile along the spring-summer-winter seasonal axis. This pattern was confirmed by a series of independent seasonal brain sampling and real-time PCR verification on five important neuroendocrine genes. We hypothesize that the seasonal expression patterns may account for dynamic neuroendocrine regulation of reproductive cyclicity.

Results

Extracting Neuroendocrine Brain Gene Expression Information from Multiple Female Goldfish Microarray Datasets

Experimental microarray data from the AURATUS goldfish environmental genomics project were used to define seasonal gene expression information. In order to remove systemic bias and random variation within and between different experiments, we conducted a series of data transformation and normalization procedures. For each experiment, we first subjected the slides to Lowess and Scale normalizations. The raw control sample intensity was extracted and used to represent transcriptomic status of brain samples in different seasons. We further removed the bias between different experiments using Quantile normalization. This normalization procedure was firstly used for the slides (n = 24) with an mRNA source from female hypothalamus (Hyp), and Figure 1 shows data distribution for all slides during normalizations. Datasets can be divided into three seasonal time points (May, August, and December), which correspond to the distinct physiological periods of sexually mature prespawning, sexually regressed, and early redevelopment of goldfish ovary, respectively.

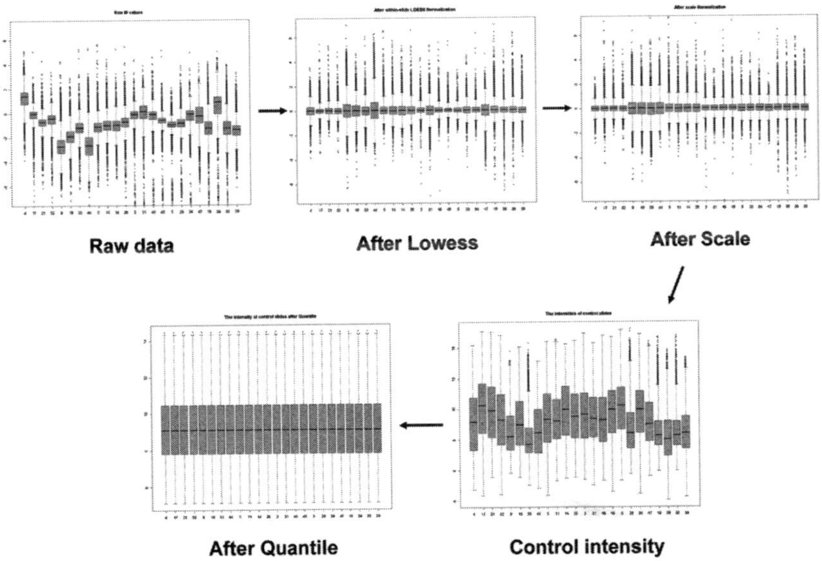

Figure 1. Boxplots of M values showing data distribution during the normalization procedures.

The global transcriptomic relationships among slides were evaluated by principal component analysis (PCA). PCA is a dimension-reducing multivariate statistical technique for simplifying complex datasets [41], [42], [43]. The results of the PCA are visualized in a two-dimensional plot. This sub-space captures the highest amount of total variability. The points in the plots represent the global transcriptomes of the different slides. The transcriptomes that are clustered together are considered overall to be similar, while more dissimilar transcriptomes are further apart. Although from different experiments, global transcriptomes of the Hyp slides from a given season are closely clustered together. This indicates that our normalization procedure was effective in identifying the season-dependent gene expression information in these microarray datasets.

Differentially Expressed Genes in Female Hypothalamus Along the Reproductive Seasonal Cycle

We identified the differentially expressed genes between these three seasonal time points. The optimal discovery procedure (ODP) method [44] implemented in an EDGE program [45] was utilized for this purpose. A total of 873 unique differentially expressed genes were identified with high statistical significance (q value<0.0001). These differential genes constitute about 10% total genes (8448 genes) represented on the arrays and include 662 genes whose function is at least partially characterized and 211 unidentified EST sequences. All of the genes were further subjected to hierarchical cluster analysis (HCA) using Pearson correlation as a distance function. In the HCA, not only the relationships between different samples can be classified, but also the genes with similar expression patterns can be grouped by visual inspection of the hierarchical cluster results. As shown in Figure 2, the identified differentially expressed genes fall into four gene expression clusters comprising H-L-L, H-H-L, L-H-H and L-L-H (L, relatively low expression and H, relatively high expression) patterns along the seasonal cycle (May-August-December). Most of the identified genes belong to H-L-L and L-H-H expression clusters.

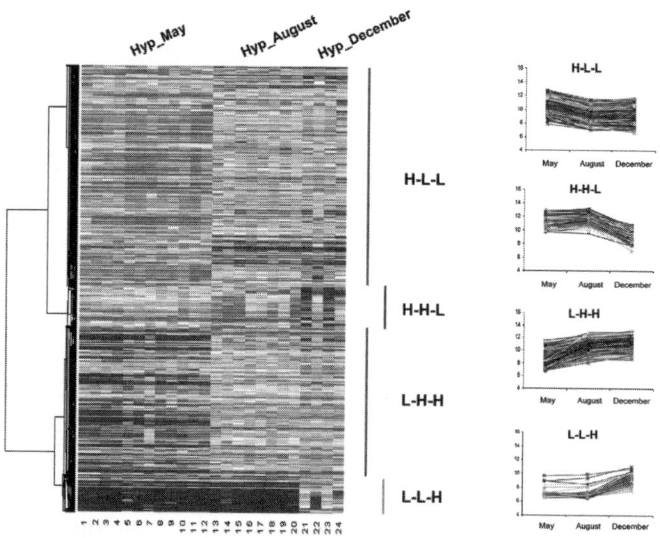

Figure 2. Hierarchical clustering of expression profiles of significantly differentially expressed genes between three reproductive seasonal time points (May, August, and December). Signal represents log ratio intensity of the gene expression. Red indicates relatively high expression and blue indicates relatively low expression.

Telencephalon Exhibits Similar Transcriptomic Patterns as Hypothalamus

We next examined the global transcriptome similarity between Hyp and Tel during the seasonal cycle. Both Hyp and Tel are important brain regions involved in neuroendocrine control of growth and reproduction (Figure 3a). Five expression datasets (n = 20) for female Tel in May and August were available for this study. After the previously applied data normalizations, PCA was performed for both Hyp and Tel datasets. The output of PCA showed that transcriptomes of Hyp and Tel samples in both May and August are overlapped (Figure 3b). This indicates that these tissues have highly similar gene expression profiles in the same season (e.g. in May and August).

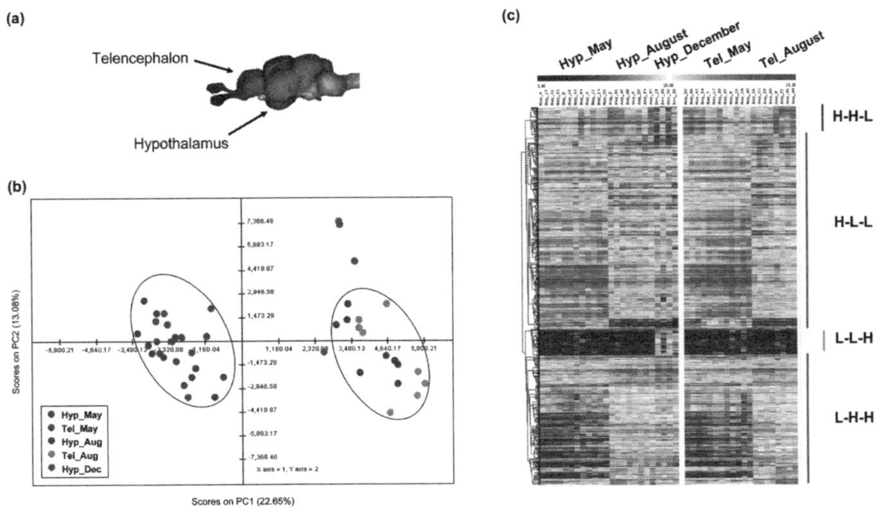

Figure 3. Similar transcriptomic patterns in both hypothalamus and telencephalon.
(a) Location of hypothalamus (Hyp) and telencephalon (Tel) in goldfish brain. (b) Two-dimensional PCA plot for transcriptomes of both Hyp and Tel slides in three seasonal time points (May, August, and December). (c) Hierarchical clustering of the expression profiles of the significantly differentially expressed genes in Hyp and Tel along the seasonal cycle.

We further investigated whether the expression patterns of differentially expressed genes identified in Hyp are similar in the Tel. HCA was carried out on a combined data set including the expression values of all differential genes from Tel slides and those from the Hyp slides. We found that as in Hyp, those genes also exhibit clear differential expression patterns in Tel between May and August (Figure 3c). This was also observed by clustering global transcriptomes of both Hyp and Tel (not shown). Therefore, using multivariate analysis we illustrate that the transcriptomes from two neuroendocrine brain regions (Hyp and Tel) that are controlling reproduction are surprisingly similar within a given season and the differential gene patterns identified in Hyp was also evident in Tel.

The Identified Gene Expression Patterns Can Be Regulated by Photoperiod

Photoperiod is a major environmental factor that governs fish reproductive seasonality [3], [4], [5], [46]. Photoperiodic manipulations have been used to control the egg production and fertilization success of diverse species such as goldfish [47], [48], rainbow trout (Salmo gairdneri) [49], Japanese medaka (Oryzias latipes) [6] and European sea bass (Dicentrarchus labrax L.) [3]. Moreover, some photoperiod-regulated genes have also been described in the ovine pituitary gland [50] and the hamster hypothalamus [51]. Therefore, it can be hypothesized that the identified gene signature which is differentially expressed between seasons is regulated by photoperiod. One dataset provided a unique opportunity to clarify this question in which Hyp brain samples were taken from fish in October that had been acclimated to a long-daylength photoperiod (16 h), typical of the springtime breeding season in May (Figure 4a). The prediction was that if the identified gene expression signature is photo-responsive, the genes in 16 h-October fish should have similar expression patterns to fish in May acclimated for 3 weeks under our seasonally varying daylength protocol and sampled at 15.8 h light-length.

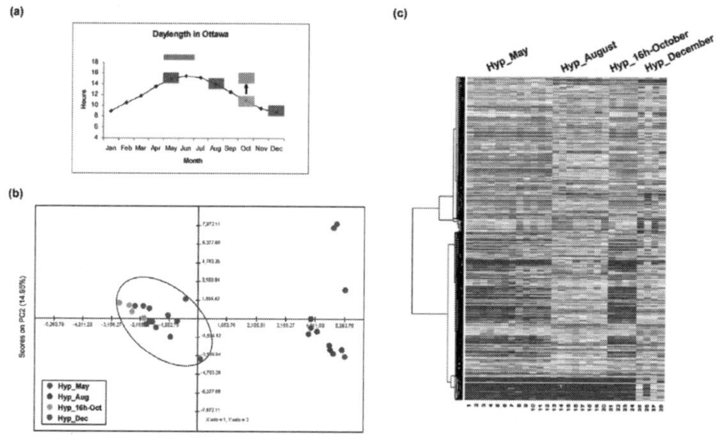

Figure 4. Regulation of brain gene expression patterns by photoperiod.

(a) Seasonal profile of the daylength in Ottawa, Canada (45°27′N, 75°42′W). Pink, blue, grey and red boxes indicate daylength in May, August, October and December, respectively. The light blue box indicates a 16-hour daylength that we used for one fish group in October. Green bar indicates breeding season for goldfish typical in temperate North America. (b) Two-dimensional PCA plot for transcriptomes from Hyp slides in four seasonal time points (May, August, 16 h-October, and December). (c) Hierarchical clustering of expression profiles of significantly differentially expressed genes in Hyp slides along four seasonal time points.

We first examined the relationship of the transcriptomes between the 16 h-October Hyp samples and other three normal Hyp samples using PCA. As shown in Figure 4b, the data points of 16 h-October slides are closely clustered with May slides, indicating similar global transcriptome status between the May and 16 h-October groups. We also compared the gene expression relationship of the identified differentially expressed genes in these four cases. As shown in Figure 4c, the expression of these 873 genes highly resembles that observed in May. Moreover, the gene expression patterns are reversed in 16 h-October compared to other seasons August and December. When we combined this dataset for a re-analysis to identify differential genes between seasons, we obtained a very similar gene list and gene patterns (data not shown). Therefore, we concluded that the identified differential expression patterns of these genes are likely regulated by photoperiod.

Gene Ontology Analysis of Differentially Expressed Genes

We have shown that a core set of genes were differentially expressed between seasons in both Hyp and Tel regions and they are photoperiod-responsive. We hypothesize that the affected genes should be related to neuroendocrine regulation of the fish reproductive cyclicity. We utilized GO term analysis to explore such potential functional significance among these genes. A series of GO categories implicated in neuroendocrine function are significantly over-represented for the genes of the H-L-L pattern. Figure 5 shows the primary GO categories with member genes, some of which can be found in multiple categories. Importantly, many genes have been functionally assigned to hormone activity, including two GnRH forms (salmon GnRH and chicken GnRH-II), growth hormone (GH), isotocin, LH, tachykinin, neuropeptide Y (NPY), pituitary adenlyate cyclase activating polypeptide 1 beta (PACAP 1β). Most of them, especially the GnRHs, are well-known and central to neuroendocrine control of reproduction. Another identified molecular function is $GABA_A$ receptor activity. Four subunits, $GABA_A$ alpha1, beta4, gamma1 and gamma2, were highly expressed in May. The $GABA_A$ receptor is a pentameric protein that is a ligand-gated Cl^- ion channel whose activation by GABA or the specific $GABA_A$ agonist muscimol results in rapid and sustained LH release in goldfish [52]. Many other genes identified are involved in several biological processes including: transmission of nerve impulse, glutamine family amino acid metabolic process, G-protein coupled receptor (GPCR) signaling pathway and neuron differentiation. Some genes are shared between these biological process and molecular function categories, among which transmission of nerve impulse is a central core linking other categories. Successful identification of a series of neuroendocrine function-related genes by GO enrichment analysis supports the biological implication of gene expression change of the H-L-L

(May-August-December) gene pattern in reproductive seasonality. Moreover, since the timing of major expression change corresponds to the transition from pre-spawning/spawning to sexually regressed female goldfish, the high expression status of these genes around May, therefore, should be required for the fish spawning period.

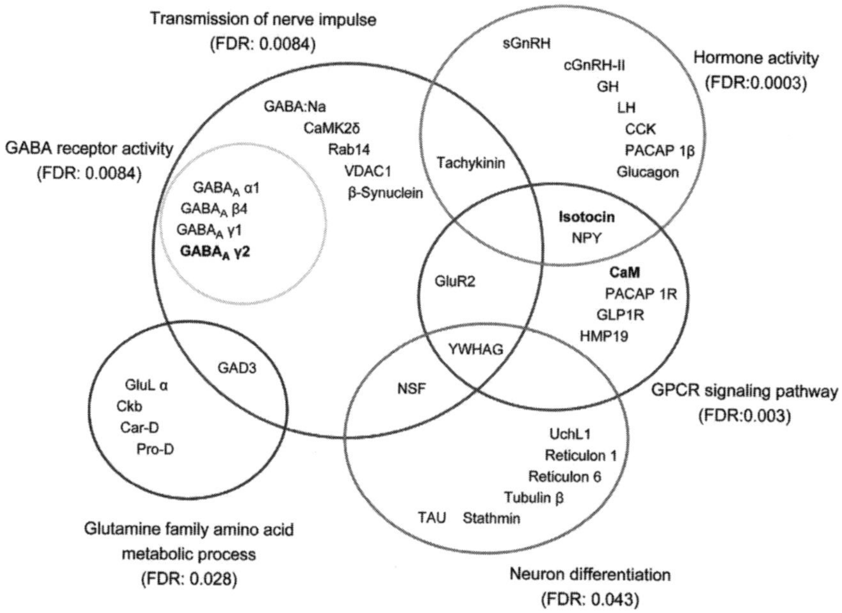

Figure 5. The neuroendocrine genes which are significantly enriched in the H-L-L expression pattern and their associated Gene Ontology categories.

Gene abbreviations: CaM, calmodulin; CaMK2δ, calcium/calmodulin-dependent protein kinase (CaM kinase) II delta; Car-D, carnosine dipeptidase 1; CCK, cholecystokinin; cGnRH-II, chicken gonadotropin-releasing hormone II; Ckb, creatine kinase, brain; GABA$_A$ α1, β4, γ1 and γ2, GABA$_A$ receptor alpha1, beta4, gamma1 and gamma2; GABA:Na, GABA:Na symporter; GAD3, glutamate decarboxylase isoform 3; GH, growth hormone; GLP1R, Glucagon-like peptide-1 receptor; GluL α, glutamate-ammonia ligase alpha; GluR2, glutamate receptor, metabotropic 2; HMP19, neuron specific gene family member 2 (protein p19); LH, luteinizing hormone; NPY, Neuropeptide Y; NSF, n-ethylmaleimide-sensitive factor; PACAP 1β, pituitary adenlyate cyclase activating polypeptide 1 beta; PACAP 1R, PACAP type 1 receptor; Pro-D, Proline dehydrogenase; Rab14, member 14 of RAS oncogene family; sGnRH, salmon gonadotropin-releasing hormone; β-Synuclein, synuclein beta; TAU, microtubule-associated protein tau; UchL1, ubiquitin carboxyl-terminal esterase L1 (also called ubiquitin thiolesterase); VDAC1, voltage-dependent anion channel 1; YWHAG, 3-monooxygenase/tryptophan 5-monooxygenase activation protein, gamma polypeptide 2.

Seasonal Goldfish Brain Sampling and Real-Time RT-PCR Verifications

Following bioinformatics analysis, we collected goldfish brain samples at six seasonal time points for experimental verification of gene expression patterns.

Gonadosomatic index (GSI), an indication of the sexual maturity of the fish, shows variation along the reproductive season and reaches a peak around May (Figure 6a). The observed changes are comparable to previous studies in Japan [35] and Canada [24]. We used real-time RT-PCR to examine the seasonal expression profiles of five genes: isotocin, $GABA_A$ gamma2, calmodulin, ependymin II and aromatase b. Among these genes, isotocin, $GABA_A$ gamma2 and calmodulin are enriched in GO term analysis (Figure 5). Isotocin is a nine amino acid neuropeptide whose function is related to the control of reproductive and social behaviors in teleost fish [53], [54]. The effects on LH release by its mammalian homologue, oxytocin, was also reported in mammalian models [55], [56]. $GABA_A$ gamma2 is one subunit of $GABA_A$ receptor complex that is involved in binding of $GABA_A$ receptor modulators such as benzodiazepines [57]; it is a representative of GABA system genes. Calmodulin is a calcium ion-binding protein and has diverse regulatory roles in L-type calcium channel, signal transduction, synthesis and release of neurotransmitters [58], and transcriptional action of estrogen receptors [59]. Ependymin II is a brain glycoprotein and is implicated in neuronal regeneration [60]. Aromatase b is the brain type aromatase, a P450 family enzyme expressed exclusively in teleost radial glial cells [61], and contributes to catalyzing the aromatization of testosterone (T) to 17beta-estradiol (E2). Although aromatase b is not identified by our meta-analysis, we are interested in it because both T and E2 have regulatory effect on the neuroendocrine system and the seasonal activity of aromatase b in fish brain is known to correlate significantly with seasonal reproductive cycles [23].

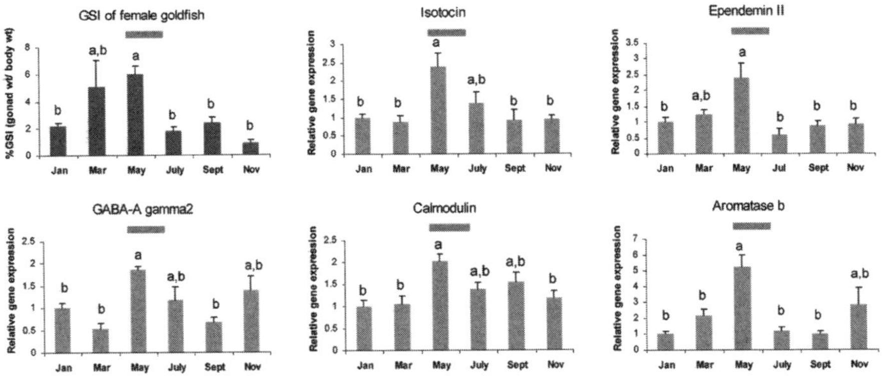

Figure 6. Seasonal profiles in GSI values and brain gene expression in female goldfish. (a) GSI values of female goldfish. (b)–(f) relative gene expression of neuroendocrine genes in female goldfish Hyp along a whole seasonal cycle. The horizontal bar indicates the typical breeding period for goldfish in temperate North America.

As shown in Figure 6b–f, seasonal expression profiles of these five genes in Hyp exhibit a clear increased expression in May. Significant differences ($P<0.05$;

Holm-Sidak multiple test) can also be observed for the expression of most genes between May and other seasons. The range of overall fold changes for those genes (Figure 6) is similar to microarray analysis (Figure 2). Therefore, these results strongly support the patterns found in our meta-analysis of microarray data with an apparent expression peak in May representing the prespawning season of the goldfish.

Discussion

An emerging concept about neuroendocrine regulation of biological rhythms is the recognized importance of the seasonal gene expression profiles in the neuroendocrine brain. Although several elegant studies have attempted to define gene expression change associated with hibernation in whole squirrel brain [62], migration in sparrow telencephalon [63] or photo-response in hamster hypothalamus [51], no system-level study has been conducted to profile the global gene expression changes along a whole reproductive cycle. The purpose of this study was to extract and evaluate such seasonal gene expression characteristics from the existing multiple microarray datasets using the goldfish model. A series of normalization procedures were utilized to remove systematic and experimental biases. The overall similarity among the slides from the same season as determined with PCA indicates that this strategy is effective. On the basis of this meta-type dataset, we identified four clear seasonal gene expression patterns in female goldfish brain. Interestingly, approximately 90% of genes represented on the array exhibit no detectable variation between the examined seasons. This suggests that the fish brain is a relatively stable system. The array does not cover the full transcriptome, but many biological pathways and processes are represented and they did not exhibit any detectable seasonal variation. In contrast, a set of genes were found differentially expressed between seasons and most of them fall into two distinct expression patterns (H-L-L and L-H-H). We observed that both female Hyp and Tel have a similar transcriptome status and same set of genes are differentially expressed between seasons. We further illustrated that artificially increasing daylength in October (16 h light) resulted in an expression profile in Hyp very similar to that of female goldfish in May (15.8 h light). This supports the hypothesis that photoperiod is a major environmental cue for vertebrate reproduction [4] because gene expression profiles in the neuroendocrine brain were photoperiodically-regulated to some degree.

Most differential genes showed expression changes between May and August in H-L-L and L-H-H patterns. May and August correspond to fish reproductive status of sexually mature prespawning and sexual regression respectively, and histological (ovary) and serum hormone changes have been extensively described between these two stages [23], [24], [28]. Thus, expression transition of this core gene set in the neuroendocrine brain may provide the molecular basis for phenotypic changes in the pituitary gland, circulating serum hormone levels and gonad status between seasons. Indeed, a series of GO categories related to neuroendocrine function and process are significantly overrepresented for the genes of H-L-L pattern (Figure 5). Many associated genes have been shown directly involved in regulating brain function and seasonal reproduction, as exemplified by the neuropeptides GnRH, isotocin, NPY and CCK. Moreover, other biological processes have also been identified including glutamate/GABA systems, GPCR signaling pathway and transmission of nerve impulse. Although no GO category is significantly enriched for the genes of L-H-H pattern, we speculate that they may include some inhibitory factors for neuroendocrine function such as the dopamine system related genes [12]. Other potential functions of these L-H-H genes may be involved in the marked seasonal growth phase of goldfish, which is different from reproductive phase [8]. We further verified the seasonal expression profiles for five genes which represent the GABA system (GABAA gamma2), estrogen production (aromatase b), peptide hormones (isotocin), calcium binding proteins (ependymin II and calmodulin) and GPCR signaling pathway (calmodulin). Their seasonal expression profiles all exhibit similar predicted patterns with the apparent peak around May, thereby suggesting that a seasonal expression change is employed by their represented systems (e.g. E2 and GABA) to coincide with fish reproductive development.

Our study may serve as a basis for understanding reproductive seasonality of other fish species. Seasonal patterns in serum levels of reproductive hormones has been documented in many teleost species (e.g., Florida gar (Lepisosteus platyrhincus) [36], masu salmon (Onchorhynchus masou) [64], and European sea bass [25]), including the goldfish [35]. It can be hypothesized that although different fish species have their own reproductive strategies and particular seasonal pattern, they may share one core set of genes with an increased expression prior to the spawning period. Indeed, similar increased expression pattern of several neuroendocrine genes has been observed in other fish species. For example, Okuzawa and colleagues [65] have shown that GnRHs exhibit a similar seasonal expression pattern in the brain of female red seabream (Pagrus major) with an expression peak around April, correlating with high steroid level and mature gonad state. Increased expression of GnRHs can also be found in the turbot (Scophthalmus maximus) [66], in parallel with an increase in oocyte diameter. Aromatase is another example; in the midshipman fish (Porichthys notatus), aromatase b in the magnocellular preoptic area of the brain showed higher expression in the pre-nesting period when compared to nesting and non-reproductive periods [67]. Moreover, in both red-spotted grouper (Epinephelus akaara) [68] and in the European sea bass [25], aromatase b showed enhanced activity during their own prespawning periods, which are corresponding to April and January, respectively.

Overall, using both bioinformatic and experimental strategies, we defined a core gene set whose expression exhibits a seasonal change along a fish reproductive cycle and illustrated that the gene expression change is primarily regulated by photoperiod. Moreover, most genes undergo an expression transition between May and August, which corresponds to the transition from prespawning to sexual regression. Thus the gene expression patterns may account for dynamic neuroendocrine regulation on fish seasonal reproductive development and the increased expression of these neuroendocrine related genes in May is crucial for ensuing spawning period. This is all the more critical in species with a short breeding period or where seasonally timed spawning occurs only once in the year.

Materials and Methods

Ethics Statement

The care and treatment of animals used in this study were in accordance with the guidelines of the Animal Care Committee, University of Ottawa and the Canadian Council on Animal Care.

Experiments and Microarray Datasets

All gene expression data were from version 1.0 Carp-Goldfish arrays which were printed in at University of Liverpool Microarray Facility, UK. The Gene Expression Omnibus (GEO) database accession number of the microarray platform is GPL3735. A total of 48 cDNA microarray slides were used for this study, and each slide was comprised of 6119 unique gene sequences and 2329 EST sequences printed in duplicate on each slide. Microarray datasets of gene expression in female goldfish hypothalamus and telencephalon have been collected to represent four seasonal time points.

Data Normalizations

For each array, spots that had been manually flagged due to poor hybridization and spots in which the estimated fluorescence intensity was below or equal to the estimated background signal intensity in either channel were removed before further analysis. Lowess normalization [69] with span 0.4 was used to decrease the intensity dependent biases within slide, followed by Scale normalization [69].

Thereafter, control sample intensities were recovered and extracted to build a new series of expression data for all the slides. Following this, Quantile normalization [70] was used for across-slide normalization to minimize biases among different experiments.

Identification of Differentially Expressed Genes

An optimal discovery procedure (ODP) method implemented in the Extraction of Differential Gene Expression (EDGE) program [44] was used to assess the significance of differential expression of the genes between different seasonal time points. The false discovery rate for selected genes was monitored and controlled by calculating the q value. Genes with q value<0.0001 were considered significantly differentially expressed genes between comparison groups.

Multivariate Data Analysis

Two multivariate methods including principal component analysis (PCA) and hierarchical clustering analysis (HCA) were utilized in this study. PCA was used to examine the transcriptome similarity relationship among slides that correspond to different seasonal time points. HCA was used to classify/cluster the gene expression patterns. Briefly, PCA combines two or more correlated factors (i.e. transcripts) into one new variable, a principal component (PC) [41], [42], [43]. In PCA the dimensionality of the dataset is reduced by replacing the original variables by a smaller number of newly formed variables that are linear combinations of the original variables and that explain the majority of the information (variability) from the experiment.

A hierarchical clustering tree [71] generates a binary dendrogram representing the association structure of pairs of arrays or genes, which was measured in terms of the Pearson correlation of the standardized intensity measurements. The algorithm identifies the gene pair with the smallest distance and groups them with a link, where distance is defined to be one-minus the correlation. The algorithm proceeds in a recursive manner to build the tree structure step by step. PCA and HCA were conducted by the MultiExperiment Viewer (MeV) [72] and EDGE [44] programs.

Gene Ontology Analysis

The resulting differentially expressed genes from each pattern were submitted to the Blast2GO web server [73]. Enrichment of multiple-level GO functional categories was determined using the GOSSIP/Fisher Exact test package [74], through a comparison of the categories found in our list of genes compared to the list of all genes found on the goldfish cDNA microarray. FDR controlled p-values (FDR<0.05) were used for the assessment of functional significance.

Animal Husbandry and Tissue Sampling

Adult female goldfish (Carassius auratus) were purchased from Aleong's International Inc. (Mississauga, ON, Canada), acclimated for 3 weeks to 18°C under a natural seasonally variable simulated photoperiod except for the animals sampled in October. In this case, animals were acclimated to 18°C and 16 h daylength that is typical of springtime. Fish were fed and maintained on standard flaked goldfish food. For the seasonal profile, 18 female fish were collected every two months starting in January, for a total of six time points (January, March, May, July, September, and November). Goldfish were anesthetized using 3-aminobenzoic acid methylester (MS222; Aquatic Eco-Systems, Apopka, FL) for all handling and dissection procedures. Care was taken to standardize all handling and sample protocols. Goldfish were sacrificed by spinal transection and 3 hypothalami were pooled for a total n = 6. Samples were rapidly dissected and immediately frozen on dry ice. At the time of sampling, body weights and gonadal weights were recorded and the gonadosomatic index (GSI; body weight/gonad weight ×100) was calculated.

RNA Extraction, Quality Assessment and cDNA Synthesis

RNA was isolated using the RNeasy Plus Mini kit (Qiagen) as described in the manufacturer's protocol. Briefly, brain samples were homogenized using the stainless steel beads before using the RNeasy Plus Mini kit. Upon purification, concentration and quality of all samples was assessed using the 2100 Bioanalyzer (Agilent). RIN values of RNA samples ranged from 7 to 9.5, which is above the recommended minimum value of 5 for quantitative real-time RT PCR applications [75]. Total cDNA was prepared from 1 µg total RNA and 200 ng random hexamer primers (Invitrogen) using Superscript II RNase H- reverse transcriptase (SSII) as described by the manufacturer (Invitrogen). Each 20 µl reaction was diluted 10-fold in nuclease-free water and used as the template for the real-time RT-PCR assays. All brain cDNA samples from different seasonal time points were synthesized in parallel.

Real-Time RT-PCR Verification

SYBR green real-time RT-PCR assay was used to validate relative expression of five genes (isotocin, $GABA_A$ gamma2, calmodulin, ependymin II and aromatase b) at all six time points along the reproductive cycle. Primers were designed using Primer3 [76] and synthesized by Invitrogen. The Mx3000 Quantitative PCR System

(Stratagene, La Jolla, CA) was used to amplify and detect the transcripts of interest. Each real-time RT-PCR reaction contained the following final concentrations: 25 ng first strand cDNA template, 1× QPCR buffer, 3.5 mM $MgCl_2$, optimized concentrations (150 nM–300 nM) of gene-specific primers, 0.25× SYBRGreen (Invitrogen), 200 µM dNTPs, 1.25U HotStarTaq (Invitrogen), and 100 nM ROX reference dye, in a 25 µL reaction volume. The thermal cycling parameters were an initial 1 cycle Taq activation at 95°C for 15 min, followed by 40 cycles of 95°C for 15 s, optimized annealing temperature (58–60°C) for 5 s, 72°C for 30 s, and a detection step at 80°C for 8 s. Dilutions of cDNA (1:10 to 1:31,250) from all samples were used to construct a relative standard curve for each primer set. After the reaction was complete, a dissociation curve was produced starting from 55°C (+1°C/30 s) to 95°C. For each PCR reaction, negative controls were also introduced including a no template control (NTC) where RNase-free water was added to the reaction instead of the template (cDNA) and a no reverse transcriptase control (NRT) where RNase-free water was added to the cDNA synthesis reaction (previously described) instead of the SSII enzyme. The SYBR green assay for every target gene was optimized for primer concentration and annealing temperature to obtain a dRn value higher than 0.98, an amplification efficiency between 90%–110% and a single sequence-specific peak in the dissociation curve.

Data were analyzed using the MxPro software package. The relative standard curve method was used to calculate relative mRNA abundance between samples, which was then presented as means+SEM of gene expression from six biological replicates (assayed in duplicate) for each time point. A one-way analysis of variance (ANOVA) followed by a Holm-Sidak post-hoc test in SigmaStat3.5 (SPSS Inc.) was used to evaluate significant changes in gene expression between different time points. All data were first tested for normality and those data with non-normal distribution were subjected to square root transformations prior to statistical analyses.

Acknowledgements

The collaboration of Dr. Donald E. Tillitt (USGS) is acknowledged. We also thank Dr. Margaret Hughes (Liverpool Microarray Facility, UK) for microarray production and assistance.

Author Contributions

Conceived and designed the experiments: DZ HX VLT. Performed the experiments: DZ HX JAM VLM. Analyzed the data: DZ HX JTP. Contributed reagents/materials/analysis tools: JAM JTP CJM KC ARC XX. Wrote the paper: DZ HX VLT.

References

1. Ripperger JA (2007) The rhythms of life. Genome Biol 8: 313.
2. Davis SJ, Millar AJ (2001) Watching the hands of the Arabidopsis biological clock. Genome Biol 2: REVIEWS1008.
3. Bayarri MJ, Rodriguez L, Zanuy S, Madrid JA, Sanchez-Vazquez FJ, et al. (2004) Effect of photoperiod manipulation on the daily rhythms of melatonin and reproductive hormones in caged European sea bass (Dicentrarchus labrax). Gen Comp Endocrinol 136: 72–81.
4. Peter RE, Crim LW (1979) Reproductive endocrinology of fishes: gonadal cycles and gonadotropin in teleosts. Annu Rev Physiol 41: 323–335.
5. Blazquez M, Bosma PT, Fraser EJ, Van Look KJ, Trudeau VL (1998) Fish as models for the neuroendocrine regulation of reproduction and growth. Comp Biochem Physiol C Pharmacol Toxicol Endocrinol 119: 345–364.
6. Koger CS, Teh SJ, Hinton DE (1999) Variations of light and temperature regimes and resulting effects on reproductive parameters in medaka (Oryzias latipes). Biol Reprod 61: 1287–1293.
7. Ruan G, Allen GC, Yamazaki S, McMahon DG (2008) An Autonomous Circadian Clock in the Inner Mouse Retina Regulated by Dopamine and GABA. PLOS Biology 6: 2248–2265.
8. Trudeau VL (1997) Neuroendocrine regulation of gonadotrophin II release and gonadal growth in the goldfish, Carassius auratus. Rev Reprod 2: 55–68.
9. Thiery JC, Chemineau P, Hernandez X, Migaud M, Malpaux B (2002) Neuroendocrine interactions and seasonality. Domest Anim Endocrinol 23: 87–100.
10. Lincoln GA, Richardson M (1998) Photo-neuroendocrine control of seasonal cycles in body weight, pelage growth and reproduction: lessons from the HPD sheep model. Comp Biochem Physiol C Pharmacol Toxicol Endocrinol 119: 283–294.
11. Callard GV, Kruger A, Betka M (1995) The goldfish as a model for studying neuroestrogen synthesis, localization, and action in the brain and visual system. Environ Health Perspect 103: Suppl 751–57.

12. Popesku JT, Martyniuk CJ, Mennigen J, Xiong H, Zhang D, et al. (2008) The goldfish (Carassius auratus) as a model for neuroendocrine signaling. Mol Cell Endocrinol 293: 43–56.

13. Peter RE, Chang JP, Nahorniak CS, Omeljaniuk RJ, Sokolowska M, et al. (1986) Interactions of catecholamines and GnRH in regulation of gonadotropin secretion in teleost fish. Recent Prog Horm Res 42: 513–548.

14. Davies B, Hannah LT, Randall CF, Bromage N, Williams LM (1994) Central melatonin binding sites in rainbow trout (Onchorhynchus mykiss). Gen Comp Endocrinol 96: 19–26.

15. Barrell GK, Thrun LA, Brown ME, Viguie C, Karsch FJ (2000) Importance of photoperiodic signal quality to entrainment of the circannual reproductive rhythm of the ewe. Biol Reprod 63: 769–774.

16. Darrow JM, Goldman BD (1985) Circadian regulation of pineal melatonin and reproduction in the Djungarian hamster. J Biol Rhythms 1: 39–54.

17. Ekstrom P, Vanecek J (1992) Localization of 2-[125I]iodomelatonin binding sites in the brain of the Atlantic salmon, Salmo salar L. Neuroendocrinology 55: 529–537.

18. Mazurais D, Brierley I, Anglade I, Drew J, Randall C, et al. (1999) Central melatonin receptors in the rainbow trout: comparative distribution of ligand binding and gene expression. J Comp Neurol 409: 313–324.

19. Vernadakis AJ, Bemis WE, Bittman EL (1998) Localization and partial characterization of melatonin receptors in amphioxus, hagfish, lamprey, and skate. Gen Comp Endocrinol 110: 67–78.

20. Somoza GM, Miranda LA, Strobl-Mazzulla P, Guilgur LG (2002) Gonadotropin-releasing hormone (GnRH): from fish to mammalian brains. Cell Mol Neurobiol 22: 589–609.

21. Sorensen PW, Christensen TA, Stacey NE (1998) Discrimination of pheromonal cues in fish: emerging parallels with insects. Curr Opin Neurobiol 8: 458–467.

22. Trudeau VL, Spanswick D, Fraser EJ, Lariviere K, Crump D, et al. (2000) The role of amino acid neurotransmitters in the regulation of pituitary gonadotropin release in fish. Biochem Cell Biol 78: 241–259.

23. Pasmanik M, Callard GV (1988) Changes in brain aromatase and 5 alpha-reductase activities correlate significantly with seasonal reproductive cycles in goldfish (Carassius auratus). Endocrinology 122: 1349–1356.

24. Trudeau VL, Peter RE, Sloley BD (1991) Testosterone and estradiol potentiate the serum gonadotropin response to gonadotropin-releasing hormone in goldfish. Biol Reprod 44: 951–960.

25. Gonzalez A, Piferrer F (2003) Aromatase activity in the European sea bass (Dicentrarchus labrax L.) brain. Distribution and changes in relation to age, sex, and the annual reproductive cycle. Gen Comp Endocrinol 132: 223–230.

26. Joy KP, Tharakan B, Goos HJ (1999) Distribution of gamma-aminobutyric acid in catfish (Heteropneustes fossilis) forebrain in relation to season, ovariectomy and E2 replacement, and effects of GABA administration on plasma gonadotropin-II level. Comp Biochem Physiol A Mol Integr Physiol 123: 369–376.

27. McNulty JA (1982) Morphologic evidence for seasonal changes in the pineal organ of the goldfish, Carassius auratus; a quantitative study. Reprod Nutr Dev 22: 1061–1072.

28. Munkittrick KR, Leatherland JF (1983) Seasonal changes in the pituitary- gonad axis of feral goldfish, Carassius auratus L., from Ontario, Canada. Journal of Fish Biology 24: 75–90.

29. Gelinas D, Pitoc GA, Callard GV (1998) Isolation of a goldfish brain cytochrome P450 aromatase cDNA: mRNA expression during the seasonal cycle and after steroid treatment. Mol Cell Endocrinol 138: 81–93.

30. Jodo A, Kitahashi T, Taniyama S, Bhandari RK, Ueda H, et al. (2005) Seasonal variation in the expression of five subtypes of gonadotropin-releasing hormone receptor genes in the brain of masu salmon from immaturity to spawning. Zoolog Sci 22: 1331–1338.

31. Lariviere K, Samia M, Lister A, Van Der Kraak G, Trudeau VL (2005) Sex steroid regulation of brain glutamic acid decarboxylase (GAD) mRNA is season-dependent and sexually dimorphic in the goldfish Carassius auratus. Brain Res Mol Brain Res 141: 1–9.

32. Peyon P, Saied H, Lin X, Peter RE (1999) Postprandial, seasonal and sexual variations in cholecystokinin gene expression in goldfish brain. Brain Res Mol Brain Res 74: 190–196.

33. Peyon P, Saied H, Lin X, Peter RE (2000) Preprotachykinin gene expression in goldfish brain: sexual, seasonal, and postprandial variations. Peptides 21: 225–231.

34. Samia M, Lariviere KE, Rochon MH, Hibbert BM, Basak A, et al. (2004) Seasonal cyclicity of secretogranin-II expression and its modulation by sex steroids and GnRH in the female goldfish pituitary. Gen Comp Endocrinol 139: 198–205.

35. Sohn YC, Yoshiura Y, Kobayashi M, Aida K (1999) Seasonal changes in mRNA levels of gonadotropin and thyrotropin subunits in the goldfish, Carassius auratus. Gen Comp Endocrinol 113: 436–444.
36. Orlando EF, Binczik GA, Denslow ND, Guillette LJ Jr (2007) Reproductive seasonality of the female Florida gar, Lepisosteus platyrhincus. Gen Comp Endocrinol 151: 318–324.
37. Martyniuk CJ, Xiong H, Crump K, Chiu S, Sardana R, et al. (2006) Gene expression profiling in the neuroendocrine brain of male goldfish (Carassius auratus) exposed to 17alpha-ethinylestradiol. Physiol Genomics 27: 328–336.
38. Marlatt VL, Martyniuk CJ, Zhang D, Xiong H, Watt J, et al. (2008) Auto-regulation of estrogen receptor subtypes and gene expression profiling of 17beta-estradiol action in the neuroendocrine axis of male goldfish. Mol Cell Endocrinol 283: 38–48.
39. Mennigen JA, Martyniuk CJ, Crump K, Xiong H, Zhao E, et al. (2008) Effects of fluoxetine on the reproductive axis of female goldfish (Carassius auratus). Physiol Genomics 35: 273–282.
40. Williams DR, Li W, Hughes MA, Gonzalez SF, Vernon C, et al. (2008) Genomic resources and microarrays for the common carp Cyprinus carpio L. Journal of Fish Biology 72: 2095–2117.
41. Basilevsky A (1994) Statistical Factor Analysis and Related Methods: Theory and Applications.
42. Pearson K (1901) On lines and planes of closest fit to systems of points in space. Philosophical Magazine 2: 449–572.
43. Sherlock G (2001) Analysis of large-scale gene expression data. Brief Bioinform 2: 350–362.
44. Storey JD, Dai JY, Leek JT (2007) The optimal discovery procedure for large-scale significance testing, with applications to comparative microarray experiments. Biostatistics 8: 414–432.
45. Leek JT, Monsen E, Dabney AR, Storey JD (2006) EDGE: extraction and analysis of differential gene expression. Bioinformatics 22: 507–508.
46. Duston J, Bromage N (1987) Constant photoperiod regimes and the entrainment of the annual cycle of reproduction in the female rainbow trout (Salmo gairdneri). Gen Comp Endocrinol 65: 373–384.
47. Razani H, Hanyu I, Aida K (1987) Critical daylength and temperature level for photoperiodism in gonadal maturation of goldfish. Exp Biol 47: 89–94.

48. Hontela A, Peter RE (1983) Characteristics and functional significance of daily cycles in serum gonadotropin hormone levels in the goldfish. J Exp Zool 228: 543–550.

49. Bromage NR, Whitehead C, Breton B (1982) Relationships between serum levels of gonadotropin, oestradiol-17 beta, and vitellogenin in the control of ovarian development in the rainbow trout. II. The effects of alterations in environmental photoperiod. Gen Comp Endocrinol 47: 366–376.

50. Dupre SM, Burt DW, Talbot R, Downing A, Mouzaki D, et al. (2008) Identification of melatonin-regulated genes in the ovine pituitary pars tuberalis, a target site for seasonal hormone control. Endocrinology 149: 5527–5539.

51. Prendergast BJ, Mosinger B Jr, Kolattukudy PE, Nelson RJ (2002) Hypothalamic gene expression in reproductively photoresponsive and photorefractory Siberian hamsters. Proc Natl Acad Sci USA 99: 16291–16296.

52. Trudeau VL, Sloley BD, Peter RE (1993) GABA stimulation of gonadotropin-II release in goldfish: involvement of GABAA receptors, dopamine, and sex steroids. Am J Physiol 265: R348–355.

53. Goodson JL, Bass AH (2000) Vasotocin innervation and modulation of vocal-acoustic circuitry in the teleost Porichthys notatus. J Comp Neurol 422: 363–379.

54. Thompson RR, Walton JC (2004) Peptide effects on social behavior: effects of vasotocin and isotocin on social approach behavior in male goldfish (Carassius auratus). Behav Neurosci 118: 620–626.

55. Evans JJ (1996) Oxytocin and the control of LH. J Endocrinol 151: 169–174.

56. Rettori V, Canteros G, Renoso R, Gimeno M, McCann SM (1997) Oxytocin stimulates the release of luteinizing hormone-releasing hormone from medial basal hypothalamic explants by releasing nitric oxide. Proc Natl Acad Sci USA 94: 2741–2744.

57. Sigel E (2002) Mapping of the benzodiazepine recognition site on GABA(A) receptors. Curr Top Med Chem 2: 833–839.

58. Zhang J, Suneja SK, Potashner SJ (2004) Protein kinase A and calcium/calmodulin-dependent protein kinase II regulate glycine and GABA release in auditory brain stem nuclei. J Neurosci Res 75: 361–370.

59. Li L, Li Z, Sacks DB (2003) Calmodulin regulates the transcriptional activity of estrogen receptors. Selective inhibition of calmodulin function in subcellular compartments. J Biol Chem 278: 1195–1200.

60. Shashoua VE (1991) Ependymin, a brain extracellular glycoprotein, and CNS plasticity. Ann N Y Acad Sci 627: 94–114.

61. Forlano PM, Deitcher DL, Myers DA, Bass AH (2001) Anatomical distribution and cellular basis for high levels of aromatase activity in the brain of teleost fish: aromatase enzyme and mRNA expression identify glia as source. J Neurosci 21: 8943–8955.

62. Williams DR, Epperson LE, Li W, Hughes MA, Taylor R, et al. (2005) Seasonally hibernating phenotype assessed through transcript screening. Physiol Genomics 24: 13–22.

63. Jones S, Pfister-Genskow M, Cirelli C, Benca RM (2008) Changes in brain gene expression during migration in the white-crowned sparrow. Brain Res Bull 76: 536–544.

64. Westring CG, Ando H, Kitahashi T, Bhandari RK, Ueda H, et al. (2008) Seasonal changes in CRF-I and urotensin I transcript levels in masu salmon: correlation with cortisol secretion during spawning. Gen Comp Endocrinol 155: 126–140.

65. Okuzawa K, Gen K, Bruysters M, Bogerd J, Gothilf Y, et al. (2003) Seasonal variation of the three native gonadotropin-releasing hormone messenger ribonucleic acids levels in the brain of female red seabream. Gen Comp Endocrinol 130: 324–332.

66. Andersson E, Fjelldal PG, Klenke U, Vikingstad E, Taranger GL, et al. (2001) Three forms of GnRH in the brain and pituitary of the turbot, Scophthalmus maximus: immunological characterization and seasonal variation. Comp Biochem Physiol B Biochem Mol Biol 129: 551–558.

67. Forlano PM, Bass AH (2005) Seasonal plasticity of brain aromatase mRNA expression in glia: divergence across sex and vocal phenotypes. J Neurobiol 65: 37–49.

68. Li GL, Liu XC, Lin HR (2007) Seasonal changes of serum sex steroids concentration and aromatase activity of gonad and brain in red-spotted grouper (Epinephelus akaara). Anim Reprod Sci 99: 156–166.

69. Yang YH, Dudoit S, Luu P, Lin DM, Peng V, et al. (2002) Normalization for cDNA microarray data: a robust composite method addressing single and multiple slide systematic variation. Nucleic Acids Res 30: e15.

70. Bolstad BM, Irizarry RA, Astrand M, Speed TP (2003) A comparison of normalization methods for high density oligonucleotide array data based on variance and bias. Bioinformatics 19: 185–193.

71. Heyer LJ, Kruglyak S, Yooseph S (1999) Exploring expression data: identification and analysis of coexpressed genes. Genome Res 9: 1106–1115.

72. Saeed AI, Sharov V, White J, Li J, Liang W, et al. (2003) TM4: a free, open-source system for microarray data management and analysis. Biotechniques 34: 374–378.

73. Conesa A, Gotz S, Garcia-Gomez JM, Terol J, Talon M, et al. (2005) Blast2GO: a universal tool for annotation, visualization and analysis in functional genomics research. Bioinformatics 21: 3674–3676.

74. Bluthgen N, Brand K, Cajavec B, Swat M, Herzel H, et al. (2005) Biological profiling of gene groups utilizing Gene Ontology. Genome Inform 16: 106–115.

75. Fleige S, Pfaffl MW (2006) RNA integrity and the effect on the real-time qRT-PCR performance. Mol Aspects Med 27: 126–139.

76. You FM, Huo N, Gu YQ, Luo MC, Ma Y, et al. (2008) BatchPrimer3: a high throughput web application for PCR and sequencing primer design. BMC Bioinformatics 9: 253.

Assortative Mating Among Lake Malawi Cichlid Fish Populations is not Simply Predictable from Male Nuptial Color

Jonatan Blais, Martin Plenderleith, Ciro Rico, Martin I. Taylor, Ole Seehausen, Cock van Oosterhout and George F. Turner

ABSTRACT

Background

Research on the evolution of reproductive isolation in African cichlid fishes has largely focussed on the role of male colors and female mate choice. Here, we tested predictions from the hypothesis that allopatric divergence in male color is associated with corresponding divergence in preference.

Methods

We studied four populations of the Lake Malawi Pseudotropheus zebra complex. We predicted that more distantly-related populations that independently evolved similar colors would interbreed freely while more closely-related populations with different colors mate assortatively. We used microsatellite genotypes or mesh false-floors to assign paternity. Fisher's exact tests as well as Binomial and Wilcoxon tests were used to detect if mating departed from random expectations.

Results

Surprisingly, laboratory mate choice experiments revealed significant assortative mating not only between population pairs with differently colored males, but between population pairs with similarly-colored males too. This suggested that assortative mating could be based on non-visual cues, so we further examined the sensory basis of assortative mating between two populations with different male color. Conducting trials under monochromatic (orange) light, intended to mask the distinctive male dorsal fin hues (blue v orange) of these populations, did not significantly affect the assortative mating by female P. emmiltos observed under control conditions. By contrast, assortative mating broke down when direct contact between female and male was prevented.

Conclusion

We suggest that non-visual cues, such as olfactory signals, may play an important role in mate choice and behavioral isolation in these and perhaps other African cichlid fish. Future speciation models aimed at explaining African cichlid radiations may therefore consider incorporating such mating cues in mate choice scenarios.

Background

One of the most significant recent development in speciation theory has been the increased attention given to sexual selection as an evolutionary force capable of rapidly inducing reproductive isolation among populations [1]. Sexual selection is thought to have played an important role in major adaptive radiations, such as those of Hawaiian Drosophila [2,3] and East African cichlid fishes [4-6]. Because many closely related species differ in sexually dimorphic male breeding color, many studies on cichlid fishes from Lakes Malawi and Victoria have emphasized the possible role of female choice of male nuptial color as a driving force for speciation [5,7-10]. The observations that individual females can have preferences for different male color patterns and that these preferences lead to reproductive isolation between incipient species have been used to model speciation in cichlid radiations [e.g. [11-14]]. The breakdown of assortative mating between a pair of sympatric Lake Victoria cichlid species in turbid waters [15], the breakdown of assortative mating during laboratory experiments under monochromatic light [9], and the observation that non-hybrid females prefer hybrid males that have the colors of conspecifics [16] provided evidence that female mating preferences for male courtship hue are important in reproductive isolation in Lake Victoria cichlids.

Cichlid fishes specialised to live on rocky shores are known to be philopatric and poor dispersers, with significant genetic structure among populations isolated by habitat discontinuities [17,18]. It has been proposed that differentiation of mating traits under divergent sexual selection among allopatric populations may contribute to the high diversity of these lineages [4-6,19]. Allopatric color variants are common among the rock-dwelling haplochromine species of Lake Malawi [20], and a substantial proportion of the estimated species richness of these fish is based on the allocation of species status to allopatric color variants [21,22]. Laboratory mate choice trials with closely related populations have indicated substantial, albeit incomplete, assortative mating between populations that differ substantially in male color and random mating between populations with more similar color [6], suggesting that color differentiation may play a pivotal role in prezygotic isolation.

Similar adult male coloration can be found in populations from widely separated localities within Lake Malawi [23]. Molecular phylogenetic analysis using microsatellites [24] and genome-wide surveys of Amplified Fragment Length Polymorphisms (AFLPs) [25] indicate that some of these populations have evolved similar colors independently. A powerful way of testing the role of male color-female preference coevolution in producing prezygotic isolation among these species is to test whether assortative mating of populations can be predicted based on similarity of independently derived color patterns. Indeed, if male color tightly coevolved with female preference, we would expect to find cases of parallel speciation whereby unrelated populations that independently evolved similar male coloration would freely interbreed. We would furthermore predict that females discriminate against males from more closely-related populations with different color patterns [6,23,25]. That parallel speciation may occur in nature has been shown for sympatric benthic and limnetic populations of sticklebacks (Gasterosteus spp.) in North American lakes that have adapted to different ecological conditions [26,27]. Although sympatric species of these fish are largely reproductively isolated by mate choice, allopatric species of similar ecotype that adapted independently to similar conditions in different lakes (e.g. benthics from different lakes) lack behavioral isolation. These findings raise the possibility that under similar ecological conditions, populations in different places can evolve in parallel and upon secondary contact form one polyphyletic species [28].

However, recent laboratory studies [29] have shown that the strong tendency of females of the Lake Malawi cichlid Pseudotropheus emmiltos to choose conspecific males broke down when only visual cues were available, but was restored when the experimental design allowed olfactory cues while still preventing direct physical contact. Males of the sympatric species used had distinct dorsal fin markings and background hue that would have been visible under the experimental conditions. This suggests that male color and female preference for male color may not have coevolved closely with each other in this species.

On this background, we examine the hypothesis that parallel evolution of color in Lake Malawi cichlids resulted in parallel speciation. We took advantage of naturally occurring color diversity by using four allopatric populations of the rock-dwelling Pseudotropheus zebra species complex from Lake Malawi. Territorial males of all populations are blue with black vertical bars, but males of two populations have orange dorsal fins, while those from the other two sites have blue dorsal fins. According to Schluter and Nagel [27], at least three criteria must be met to demonstrate parallel speciation. First, populations exposed to the same conditions must be phylogenetically independent so that shared traits responsible for reproductive isolation evolved separately. Second, populations exposed to different environmental conditions must be reproductively isolated. Third, populations exposed to the same environmental conditions must not be reproductively isolated. Here, we selected a pair of orange/blue populations from the northwestern part of the lake, P. emmiltos and P. zebra, and another orange/blue pair from the southeastern area, P. thapsinogen and P. zebra. The northwestern populations are closely related and belong to a clade of northwestern and central eastern populations according to AFLP data [25]. The southeastern orange population (P. thapsiongen) was shown to have evolved independently from the northwestern orange P. emmiltos population and to be more closely related to the southeastern blue P. zebra population than from the northwestern populations according to microsatellite analysis [25]. These populations therefore satisfy the first criterion of parallel speciation established by Schluter and Nagel [27].

We tested whether partial assortative mating among allopatric populations is the result of co-evolution between female preference and male color, and investigated the roles of different sensory modalities in female mate choice experiments following two different approaches. First, we specifically address Schluter and Nagel's [27] second and third criteria. Under the scenario of parallel speciation through color-preference co-evolution, we predict random mating between distantly-related populations with similar male colors and assortative mating between more closely-related populations with clearly different male colors. Second, in order to further evaluate the role of color divergence as the trait hypothesized to cause reproductive isolation in the system, we artificially manipulated color and other cues in mate choice trials involving one of these population pairs (P. emmiltos and northern P. zebra from Nkhata Bay). We tested the importance of visual cues by comparing mate choice under normal (white) light and monochromatic light [9], while other cues were manipulated by comparing female choice of males under white light behind transparent partitions [29].

Methods

a) Experimental Animals

The study populations were members of the Pseudotropheus zebra species complex, cichlid fish endemic to Lake Malawi. The taxonomy of this group is confused, and these fish are sometimes assigned to the genus or subgenus Maylandia or the genus Metriaclima. Many allopatric color variants have been assigned species names, while others have not. Territorial males of all four study populations are pale blue with strongly contrasting black vertical bars, but populations differ in the color of the dorsal fin. Male P. zebra from Nkhata Bay (11°36'S, 34°18'E) on the northwestern shore and Chiofu Bay (13°13'S, 34°52'E) on the southeastern shore have blue dorsal fins (henceforth referred to as 'northern blue' and 'southern blue', respectively). Male P. thapsinogen from Chimwalani (Eccles) Reef (13°46'S, 34°58'E) off the southeastern shore and male P. emmiltos from Mpanga Rocks (10° 25'S 34°16'E) near the northwestern shore have orange dorsal fins and are referred to as 'southern orange' and 'northern orange,' respectively (Figure 1). P. thapsinogen (southern orange) males have yellow throat membranes, whereas those of the northern orange males (P. emmiltos) are dark grey. These throat membranes are normally not visible unless protruded beneath the gill-cover as part of a lateral display. However, these displays are typically part of antagonistic rather than courtship interactions, and therefore it is unlikely that females use this trait in mate choice. The taxonomy, morphology and ecology of these populations were discussed by Stauffer et al. [30] and the biology of these and related 'mbuna' species reviewed by Genner & Turner [31]. Fish used were wild-caught, although some first-generation laboratory stock bred from wild-caught parents were used in experiment 2. Housing and maintenance conditions have been described previously [6,29].

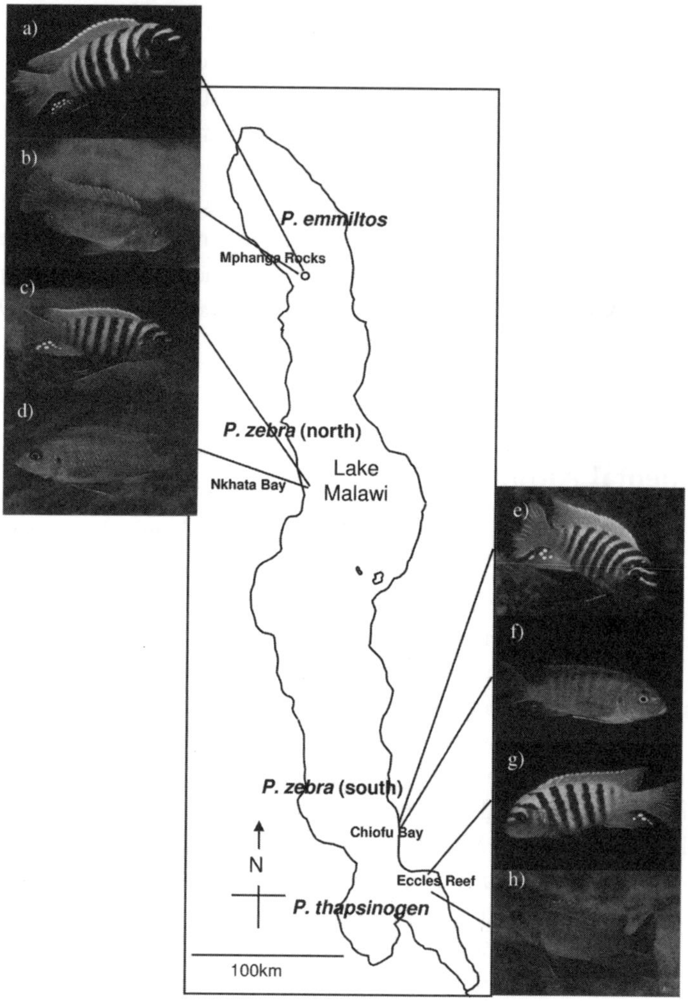

Figure 1. Map of Lake Malawi showing sampling sites of study populations. Male (a) and female (b) Pseudotropheus emmiltos (northern orange) from Mphanga Rocks; male (c) and female (d) Pseudotropheus zebra (northern blue) from Nkhata Bay; male (e) and female (f) Pseudotropheus zebra (southern blue) fromChiofu Bay and male (g) and female (h) Pseudotropheus thapsinogen (southern orange) from Eccles Reef.

b) Experimental Procedures

Experiment 1: Test of Parallel Speciation Through Color-Preference Coevolution

Levels of behavioral reproductive isolation were measured in glass aquaria 6 m long × 0.7 m wide × 0.4 m deep under a 'partial partition' design [6,32], where males were confined to territories of floor dimensions 60 × 70 cm using plastic grids adjusted to allow the smaller females to pass through and spawn with any preferred male. Standard fluorescent lamps were used throughout the study. Although Pseudotropheus species express a UV-sensitive cone opsin, examinations of photographs of these and related species taken under UV light did not reveal any UV-specific markings, simply a greater degree of luminosity of the bright blue parts of their color, which does not vary noticeably between the populations. It is therefore unlikely that different results could have been obtained with UV-enhanced tubes. Terracotta flower pots were provided as spawning caves. Four males (size-matched within 5 g) and 10–15 females of each of two populations were present throughout the experiment. Males' positions in the tank were randomized, except that two males of the same population were not allowed to be neighbours. After at least one clutch had been produced by females of both populations, one or two males of each population were removed, and replaced by size-matched males of the same population, and all males were moved to new territories. For each population comparison, 5–7 rearrangements were carried out, and 8–11 males per population used to minimise the influence of individual variation in male attractiveness. Courtship behavior is known to be highly conserved in haplochromine cichlids across broad taxonomic scales [33] and we did not observe any differences in courtship behavior among populations. It was estimated that 63% of males used sired part or all of one or more clutches. Four population combinations were tested. Of these, two reciprocal trials compared more distantly-related populations with similar male colors: 1) P. emmiltos females choosing between males of the same population and P. thapsinogen males and vice versa (northern v southern orange) and 2) P. zebra females from Nkhata Bay choosing between males of the same population and P. zebra males from Chiofu Bay and vice versa (northern v southern blue). The other two reciprocal trials compared more closely-related and geographically proximal populations with distinct male colors: 1) P. emmiltos females choosing between males of the same population and P. zebra males from Nkhata Bay and vice versa (northern orange v blue), and 2) P. thapsinogen females choosing between males of the same population and P. zebra males from Chiofu Bay and vice versa (southern orange v blue).

Before introduction into the experimental aquarium, all males and females were PIT-tagged for individual identification, and fin-clipped for DNA testing. A small piece of tissue was cut from the soft-rayed part of the dorsal fin and placed in 100% ethanol. Mouthbrooding females were removed from the aquarium ca. 5–10 days after spawning and their offspring removed by gently opening the female's mouth and allowing the fry to drop into a tray of water containing an anaesthetic overdose (MS-222). Fry were then preserved in 100% ethanol. Paternity was assigned by genotyping all mothers, potential fathers and 3–15 offspring per clutch at between two and four microsatellite loci: UNH130 [34],

UNH002 [35], UME002 [36], and Pzeb2 [37]. DNA was extracted from tissue samples using the HOTSHOT method [38] and amplified using a FAM/HEX/TET-labelled forward primer. PCR conditions consisted of 30 cycles of 30 sec at 95°, 60 sec at 55°, 60 sec at 72° in a 20 μl reaction volume containing 0.25 mM of each dNTP, 2 μl of 10× reaction buffer (Bioline, London, UK) 1.5 mM MgCl2, 0.5 U of taq DNA polymerase (Bioline, London, UK), 25 ρmol of each primer and approximately 1 μg of template genomic DNA in a TGradient thermocycler (Biometra, Göttingen, Germany). PCR products were resolved on an ABI 310 (Applied Biosystems, Foster City, U.S.A.), with ROX-labelled internal size standards. Alleles were sized using Genemapper (Applied Biosystems).

Experiment 2: Test for Role of Color in Behavioral Isolation

Experiments were run in aquaria of dimensions 180 × 45 × 33 cm or 200 × 50 × 31 cm in which mate choice was determined from a group of 10–20 females of a single species at a time (northern orange P. emmiltos or northern blue P. zebra from Nkhata Bay). In all trials, the males were size-matched for weight (± 5 g) and standard length (± 5 mm). Female choice was assayed by counting eggs that had collected underneath mesh false-floors, where they were inaccessible to the females that would normally collect them in their mouths [29]. The male that received the largest number of the eggs was taken to be preferred. Twelve males from each population (northern orange P. emmiltos/northern blue P. zebra from Nkhata Bay) were used in each experimental treatment. Mate choice was tested under three treatments: (i) White light with full contact, in which males were kept in compartments with false floors, separated from a female-only compartment by plastic grids that allowed females to pass through and spawn inside the male compartment, the tank being illuminated with standard fluorescent lights (figure 2a); (ii) Monochromatic light with full contact, in which the apparatus was the same as the previous treatment, but the experiment was conducted in a darkroom in which the aquarium lights were covered with a red optical filter (primary red, Lee Filters 106, http:/ / www.leefilters.com/ lighting/ products/ finder/ act:colordetails/ colorRef:C4630710C51307/). This effectively masked the color difference between males of the two populations by transmitting light from only the orange and red portion of the white spectrum (500–700 nm), thereby eliminating the difference between the orange and blue dorsal fin color; (iii) White light, visual communication only, in which the tank, illuminated by standard fluorescent lights, was separated into three sections by transparent 5 mm thick acrylic partitions, with the females confined in a central section which had a tray covered with plastic mesh adjacent to each male compartment, which contained an artificial cave used as a spawning site next to the female compartment (figure 2b).

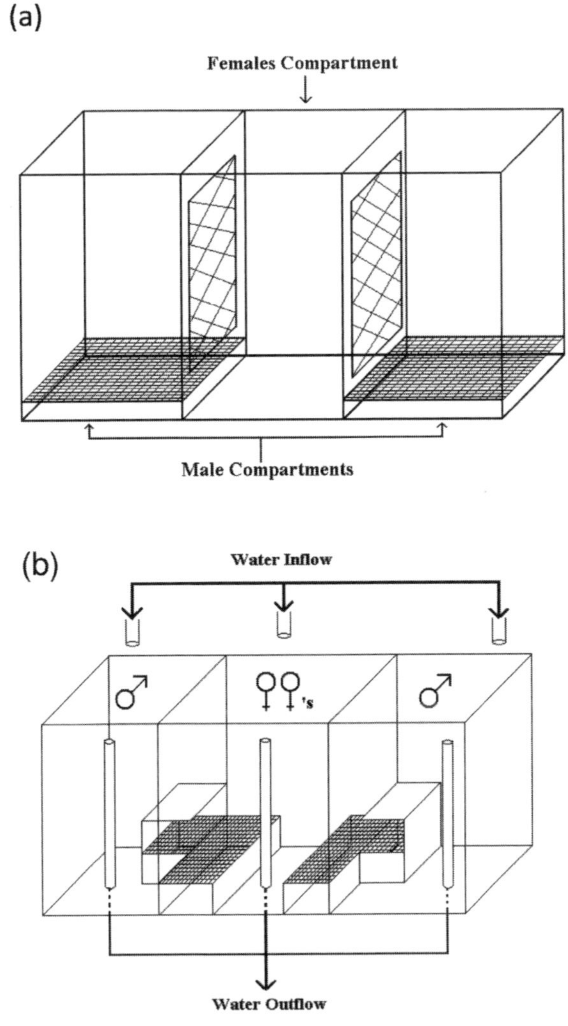

Figure 2. Experimental set up for experiment 2. Control and monochromatic light experiments (a), and visual-only experiment (b). Males in the two end compartments courted females from the central compartment.

c) Data Analysis

Simulation of parentage assignment of 10,000 randomly generated offspring genotypes in Cervus v. 3.0.3 [39] showed that typing between two to four (average of three) of the four loci used resulted in an 82% successful assignment rate to one of the eight potential fathers given known mother genotype when using a delta LOD criterion with 95% confidence interval and allowing for 1% of mistyped loci. We first typed two loci for all individuals, and progressively increased to three and then four loci whenever unambiguous paternal assignment to one of the eight possible males could not be obtained. At the end of this process, we were able to unambiguously assign paternity to a single male for virtually all fry of all clutches. In most cases, clutches were sired by a single male or by several males of the same population. We found only a few cases of multiple paternity involving males of different populations. Multiple paternity was resolved by a majority consensus rule where the sire of the majority of the first five eggs sampled at random was considered as the preferred male, thus providing an unbiased statistical sample at the population level. To minimise the impact of variation in clutch sizes between broods, final analyses were based on the first 5 eggs typed, or the total, if smaller.

The significance of assortative mating of population pairs was assessed by one-tailed Fisher Exact Tests, which tests the hypothesis of assortative mating at the level of the tested species pair as a whole (at the level of the whole 2 × 2 table). It is thus more powerful than testing individual reciprocal crosses. We also report, for individual reciprocal crossings, one-tailed Binomial Tests on numbers of broods and Wilcoxon tests on the number of eggs typed for each clutch assigned to sires of the two populations. The binomial test is based on the binomial distribution, while the Wilcoxon signed-rank test does not assume anything about the shape of the distribution. The full data set of experiment 1 included broods where analysis of maternal genotypes suggested that they were produced by females that had already produced one brood earlier in the experiment, or broods where males had sired the majority of eggs sampled from a number of broods produced by different females. To ensure that these non-independent points did not impact our conclusions, we analysed a reduced dataset, keeping only the first brood produced by each female and eliminating second or subsequent broods with the same majority sire. This eliminated possible influences of over-representation of particular individual female preferences, particularly successful males, or mate choice copying, although mate choice copying by Malawi cichlids has never been reported, and many of the clutches were spawned weeks or months apart, often after a male had been moved to a new territory.

Results

Experiment 1: Test of Parallel Speciation Through Color-Preference Co-Evolution

Eighty-nine clutches were produced and 3–15 fry (mean ± S.E. = 6.6 ± 1.9) from each of these tested for paternity. Of these, 12 (13% of total) were sired by males of different populations. Forty-nine broods with dam or majority sire involved in earlier broods were removed from the original dataset before analysis. Once these broods were eliminated from the data set, only seven clutches sired by males of different populations remained.

Test of Random Mating Among Similarly Colored Populations

Contrary to the prediction of random mating among geographically distant populations of the same color, Fisher's exact tests indicated significant assortative mating among both pairs of distantly-related populations with similar-colored males (Table 1). In two populations (southern blue P. zebra and northern orange P. emmiltos), Binomial and Wilcoxon signed-rank test indicated that males from the female's population were significantly more likely to sire the majority of the clutches (Table 1). Northern blue P. zebra and southern orange P. thapsinogen females also spawned more often with males from their own population, although not significantly.

Table 1. Mate choice trials among populations of Pseudotropheus zebra and related species in large arena tanks.

Dam Population	Majority Sire Population		Binomial Test one-tailed *P*	Wilcoxon Test one-tailed *P*	Fisher's Exact Test one-tailed *P*
	NB	SB			
NB	4	0	0.063	*Z* = 1.826, *P* = 0.068	
SB	0	5	**0.031**	*Z* = 2.023, *P* = **0.043**	
					< 0.01
	NO	SO			
NO	6	0	**0.016**	*Z* = 2.201, *P* = **0.028**	
SO	2	5	0.227	*Z* = 0.507, *P* = 0.612	
					0.016
	SB	SO			
SB	2	2	0.500	*Z* = 0.000, *P* = 1.000	
SO	0	4	0.063	*Z* = 1.826, *P* = 0.068	
					0.214
	NB	NO			
NB	4	0	0.063	*Z* = 1.618, *P* = 0.106	
NO	2	5	0.227	*Z* = 0.676, *P* = 0.499	
					0.045

Forty-nine broods from dams or sires involved in previous spawnings were eliminated from the original dataset to ensure data independence. The results were inconsistent with the predictions of the hypothesis of parallel speciation by divergence of male colour and associated female preference, with significant assortative mating between geographically and phylogenetically more distant populations with similarly-coloured males [northern orange *P. emmiltos* (NO) v southern orange *P. thapsinogen* (SO); northern blue *P. zebra* from Nkhata Bay (NB) v southern blue *P. zebra* from Chiofu Bay (SB)] but not between more closely related and geographically proximal populations with differently-coloured males (SO v SB; NO v NB). Shown are the frequencies with which males were estimated to have sired the majority of the offspring genotyped in a clutch. The Binomial Test and Fisher Exact Tests were based on these figures, while the Wilcoxon Test was based on the actual numbers of eggs typed for each clutch assigned to sires of the two populations. Significant p-values (α = 0.05) are in bold.

Test of Assortative Mating Among Differently-Colored Populations

When comparing geographically proximate populations with different colors, Fisher's exact test indicated significant assortative mating only between P. emmiltos and P. zebra from Nkhata Bay (Table 1). Although female P. emmiltos, P. zebra from Nkhata Bay and P. thapsinogen spawned more often with males of their own populations, assortative mating was not significant in any of the individual reciprocal crosses (Binomial and Wilcoxon tests $p > 0.05$; Table 1). Females of the southern P. zebra population of Chiofu Bay did not differ from random expectation and spawned half of their clutches with P. thapsinogen males (Table 1).

Experiment 2: Test for Role of Color in Behavioral Isolation

Mate choice of female P. emmiltos (northern orange): Under control conditions of full sensory contact under white light, female P. emmiltos spawned exclusively with males of their own (northern orange) population, rather than northern blue P. zebra males (Table 2; Binomial $p < 0.0001$). This preference persisted in monochromatic light, as females spawned with same-population males in 11 out of 12 trials (Binomial $p < 0.005$), which does not differ significantly from the control (Fisher's exact test, $p \sim 1$). When free contact was blocked, but color differences were visible under white light, they showed no significant preference, spawning with the same-population male only in 5 of the 12 trials (Binomial $p = 0.386$). Female choice with visual contact only differed significantly from that in the experiment with full sensory communication under white light (Fisher's exact test, $p = 0.04$).

Table 2. Mate choice trials under treatments to control sensory cues in northern blue P. zebra from Nkhata Bay and northern orange P. emmiltos

Female	Treatment	Male: *P. emmiltos*	Male: *P. zebra*	Cumulative Binomial Probability (one-tailed)
P. emmiltos (NO)	Control: free contact; white light	12	0	**< 0.001**
	Treatment 1: free contact, monochromatic light	11	1	**0.005**
	Treatment 2: visual only, white light	7	5	0.386
P. zebra (NB)	Control: free contact, white light	5	7	0.386
	Treatment 1: free contact, monochromatic light	7	5	0.386

Female preference for conspecific *P. emmiltos* males was maintained under monochromatic light, but lost when non-visual cues were blocked (response variable is the number of replicates in which the majority of the clutch was laid next to a male of a particular species). Significant p-values ($\alpha = 0.05$) are in bold.

Mate choice of female P. zebra (northern blue): With full contact under white light, northern blue females' preference for males of their own population was not significant (Table 2; Binomial $p = 0.386$), spawning with the same-population male in 7 out of 12 trials. Under monochromatic light, same-population males were chosen in 5 out of the 12 trials (Binomial p = 0.386). As northern blue females had not shown significant assortative mating in the control treatment of experiment 2, they were not tested with free contact blocked.

Discussion

The suggestion that sexual selection may have played a major role in the diversification of the cichlid fishes of the African Great Lakes has been largely inspired by the enormous diversity in color pattern among the Malawian and Victorian haplochromine cichlids. One of the main assumptions of speciation models by sexual selection so far was that evolutionary divergence in female color preference was driving the emergence of behavioral isolation [40]. The hypothesis of parallel speciation of populations having independently evolved the same color patterns examined in this paper derives from this idea. The importance of male color hue for reproductive isolation has been shown in a pair of Lake Victoria cichlid species where assortative mating broke down under monochromatic light [9] and where non-hybrid females choosing among hybrid males prefer those that have the color of conspecific males [10]. These species are fully sympatric, differ in visual pigment sensitivities [41] and in their behavioral response to blue and red light [42]. They appear to hybridise more often in habitats with low water transparency [15]. Strong assortative mating based on visual cues (possibly melanin pattern rather than hue) have also been shown in the polymorphic Lake Victoria cichlid Neochromis omnicaeruleus [43,44] and the polymorphic Lake Malawi cichlid Pseudotropheus "zebra gold" [45].

Between-population assortative mating based on color has yet to be convincingly demonstrated with Lake Malawi cichlid species. Experiments with some Lake Malawi species have shown that assortative mating can persist when transparent partitions prevented direct contact but not visual communication [40,46,47]. These results suggest a significant role for visual cues in mating preference. However, these cues are not necessarily confined to the male's hue, but could also include male melanin pattern, eggspot size and number, body shape, and male responsiveness to visual stimuli from the female. Indeed, the pair examined by Kidd et al. [40] differ in melanin pattern rather than hue, and Jordan et al. [47] found that assortative mating was maintained even under monochromatic light masking the species' hue differences, implying that some visual features other than hue were used by females. Even when it has been demonstrated that a single cue may be sufficient for assortative mating, other cues may also be influential. For example, olfactory cues may be important in the later stages of courtship, while females are initially attracted by visual signals. If a pair of species had attained reproductive isolation based on the preference for divergent signals in one signal modality, there could subsequently be strong selection to recognise and avoid courting with heterospecifics using other sensory modalities for signalling. In such cases, visual signals could be sufficient to lead to assortative mating in experiments where access to olfactory signals was prevented, even if divergence in olfactory signals had been the initial cause of reproductive isolation or vice versa.

Here, laboratory mate choice tests with allopatric populations of Pseudotropheus zebra and related nominal species were inconsistent with the predictions we derived from the hypothesis that parallel correlated divergence in male color and female preferences had resulted in parallel speciation among these fishes. Firstly, we demonstrated that allopatric populations differing in dorsal fin color did not show complete assortative mating, and probably should not be considered as biological species. Unexpectedly, allopatric populations with similar colors showed relatively high levels of (partial) assortative mating. This suggests that color differentiation is not necessarily a prerequisite for the evolution of significant prezygotic isolation.

We predicted that geographically proximal and phylogenetically related populations showing clear differences in male breeding color would mate assortatively. This prediction was supported in one of the two population pairs with different male colors. By contrast, the second prediction that geographically and phylogenetically distant populations with similar male color should mate randomly was not verified. Instead, we found assortative mating in both pairs of similarly colored populations from distant locations. Of course, it is still possible that further comparisons may show evidence for parallel speciation, perhaps a weaker form of incipient parallel speciation, where pairs of independently derived similar-colored populations may show lower levels of assortative mating than among pairs of more closely-related differently-colored populations.

Like most studies of behavioral evolution, we are required to make inferences about past events based on knowledge of current populations and their past relationships, and this is often aided by having a robust phylogeny of the study taxa and an estimate of divergence times. Recently, it has been shown that Lake Malawi suffered a major lake level fall of around 580 m until about 70,000 years ago, with present lake levels only being restored within the past 50,000 years [48,49]. At such low lake levels, the nearest refuge for the southern populations used in the present study would have been around Nkhata Bay- the site of one of the northern populations. Given that the northern and southern populations do not yet seem to mate completely assortatively and probably occurred sympatrically 70,000 years ago, it is likely the study populations are descended from a single common ancestor which was split into multiple populations as new habitat patches became available (and were lost or fused) during the rapid expansion of the lake around 60–70,000 years ago. There is no evidence that any of these study populations have ever co-existed in sympatry and maintained complete reproductive isolation.

The two population pairs used in this study were carefully chosen for the similarity of their color pattern apart from the color of the dorsal fin (and the yellow throat membrane of P. thapsinogen). We were unable to find any other differences among populations in hue or melanin pattern. However, subtle differences cannot be entirely ruled out, although we consider it unlikely.

Another potential limitation concerns the possible influence of color differences only visible under ultraviolet (U.V.) light. Although we could not find any such differences when examining photographs of the specimens under U.V. light and we found significant assortative mating in many populations without enhanced U.V. lighting, further tests might reveal an impact of U.V. illumination in mate choice in these populations.

Experimentally controlling the effect of visual and non-visual cues revealed that the strong tendency of northern orange (P. emmiltos) females to mate with same-population males over northern blue males (P. zebra) persisted when male color differences were masked under monochromatic light. However, assortative mating broke down when direct contact was prevented, even when male color differences and other visual features should have been apparent under white light. These results are inconsistent with color being a necessary requirement for the assortative mating observed between northern orange females and northern blue males, and with visual cues being sufficient for reproductive isolation of this pair. They are, however, in agreement with Plenderleith et al.'s [29] results demonstrating the role of olfactory cues for mate choice when P. emmiltos females were presented with conspecific males and those of the sympatric P. fainzilberi. This is also consistent with the findings of experiment 1 where assortative mating was observed in pairs where color differences were absent or minimal. The lack of significant assortative mating of P. zebra females in experiment 2 compared with experiment 1 is possibly simply due to limited statistical power, but perhaps it may also be because experiment 2 did not allow for mate choice among males of the same population, as only one male of each population was used at a time. Thus, assortative mating might have been reduced when an otherwise attractive male of another population might be chosen in preference to a male of the female's own population that was considered less attractive, perhaps in traits that did not differ among populations.

Future studies might aim at testing the generality of the importance of non-visual cues in other haplochromine species. Among the cues cichlids may use in mate choice, those involving olfactory, acoustic and other non-visual modalities remain little investigated. Plenderleith et al. [29] proposed that olfactory signals may be particularly promising for explaining some cases of assortative mating. Recently, Blais et al. [50] showed that there was evidence of adaptive divergence at MHC class II genes between P. emmiltos (northern orange) and the sympatric P. fainzilberi and suggested a mechanism by which this divergence may have contributed to reproductive isolation of the pair. MHC genes are known to influence mating behavior and kin recognition in many fishes through olfaction [e.g. [51-53]] and to lead to female rejection of mates with highly dissimilar MHC multilocus genotypes in humans, Malagasy giant jumping rats, Hypogeomys antimena, and threespine sticklebacks [52,54,55]. There is therefore one potential candidate gene for odour-based mate choice known to have diverged between a pair of the P. zebra complex.

Haplochromine cichlids are usually strongly sexually dimorphic in color. We do not yet know whether olfactory signals involved in mate choice are expressed by both sexes. If so, it would greatly increase the likelihood that mate preferences may be learned by imprinting on the mother during mouthbrooding, a situation that may enhance the probability of speciation [56]. Sexual imprinting on maternal traits, perhaps olfactory signals, has recently been shown in a pair of Lake Victoria cichlid fish [57]. By contrast, assortative mating based on male breeding dress is less likely to be aided by imprinting, as male haplochromine cichlids play no role in parental care and male nuptial colors are not or only weakly expressed in females [57]. Moreover, because of their chemical nature, olfactory signals might be more likely than others to vary with small differences in ecological traits such as diet, microhabitat or parasite infections [50,58,59], perhaps augmenting the likelihood of reproductive isolation among populations.

Conclusion

The three principal implications of the findings presented here are that: (i) behavioral mating preferences in cichlid fish depend, in some populations, not on color, but on other forms of sensory communication, which in some cases is not visual; (ii) between-population divergence in male color does not necessarily imply divergent female preferences, so parallel evolution of male color is not necessarily coupled with parallel evolution of female mating preference; (iii) sensory cues involved in behavioral isolation among haplochromine cichlids may vary among species and lakes and so speciation models based on one particular mate choice scenario may not apply to whole or all radiations of cichlids. This last point is worth bearing in mind because the type of trait involved in species isolation might provide important clues about which evolutionary force has driven population divergence and speciation.

Author Contributions

JB performed microsatellite genotyping, statistical analyses, and drafted the manuscript. MP performed behavioral experiments. CR provided critical review and discussion of draft versions. MIT performed microsatellite genotyping and contributed to preparation of the final manuscript. OS provided critical review and discussion of draft versions and supervised the data collection for experiment 1. CVO contributed with critical discussion of draft versions. GFT coordinated the study and was involved throughout the design, analysis, interpretation and writing of the work. All authors have read and approved the final manuscript.

Acknowledgements

This research was supported by a NSERC and FQRNT scholarship to J.B., a NERC scholarship to M.P. and a NERC research grant to G.F.T, C.R. and O.S. and a Junta de Andalusia research grant to C.R. We are grateful to Andy Gould for carrying out much of the practical work for experiment 1, Ad Konings for the photographs in figure 1 and Louis Bernatchez for helpful comments on earlier drafts of this manuscript.

References

1. Coyne JA, Orr HA: Speciation. Sunderland, MA: Sinauer Associates Inc; 2004.
2. Coyne JA, Kim SY, Chang AS, Lachaise D, Elwyn S: Sexual isolation between two sibling species with overlapping ranges: Drosophila santomea and Drosophila yakuba. Evolution 2002, 56(12):2424–2434.
3. Boake CRB, Andreadis DK, Witzel A: Behavioral isolation between two closely related Hawaiian Drosophila species: the role of courtship. Anim Behav 2000, 60:495–501.
4. Dominey WJ: Effects of sexual selection and life history on speciation: Species flocks in African cichlids and Hawaiian Drosophila. In Evolution of fish species flocks. Edited by: Echelle AA, Kornfield I. Orono, Maine: University of Maine at Orono Press; 1984:231–249.
5. Kocher TD: Adaptive evolution and explosive speciation: The cichlid fish model. Nat Rev Genet 2004, 5(4):288–298.
6. Knight ME, Turner GF: Laboratory mating trials indicate incipient speciation by sexual selection among populations of the cichlid fish Pseudotropheus zebra from Lake Malawi. Proc R Soc Lond B Biol Sci 2004, 271(1540):675–680.

7. Maan ME, Seehausen O, Söderberg L, Johnson L, Ripmeester EAP, Mrosso HDJ, Taylor MI, van Dooren TJM, van Alphen JJM: Intraspecific sexual selection on a speciation trait, male coloration, in the Lake Victoria cichlid Pundamilia nyererei.Proc R Soc Lond B Biol Sci 2004, 271:2445–2452.
8. Pauers MJ, McKinnon JS, Ehlinger TJ: Directional sexual selection on chroma and within-pattern color contrast in Labeotropheus fuelleborni. Proc R Soc Lond B Biol Sci 2004, 271:S444–S447.
9. Seehausen O, van Alphen JJM: The effect of male coloration on female mate choice in closely related Lake Victoria cichlids (Haplochromis nyererei complex). Behav Ecol Sociobiol 1998, 42:1–8.
10. Stelkens RB, Pierotti MER, Joyce DA, Smith AM, Sluijs I, Seehausen O: Disruptive sexual selection on male nuptial coloration in an experimental hybrid population of cichlid fish. Philos Trans R Soc Lond B Biol Sci 2008, 363:2861–2870.
11. Dieckmann U, Doebeli M: On the origin of species by sympatric speciation. Nature 1999, 400:354–357.
12. Higashi M, Takimoto G, Yamamura N: Sympatric speciation by sexual selection. Nature 1999, 402:523–526.
13. Kawata M, Shoji A, Kawamura S, O S: A genetically explicit model of speciation by sensory drive within a continuous population in aquatic environments. BMC Evol Biol 2007, 7:99.
14. Turner GF, Burrows MT: A model of sympatric speciation by sexual selection. Proc R Soc Lond B Biol Sci 1995, 260:287–292.
15. Seehausen O, van Alphen JJM, Witte F: Cichlid fish diversity threatened by eutrophication that curbs sexual selection. Science 1997, 277:1808–1811.
16. Stelkens RB, Pierotti MER, Joyce DA, Smith AM, Sluijs I, Seehausen O: Disruptive sexual selection on male nuptial coloration in an experimental hybrid population of cichlid fish. Philos Trans R Soc Lond B Biol Sci 2008, 363(1505):2861–70.
17. Taylor MI, Ruber L, Verheyen E: Microsatellites reveal high levels of population substructuring in the species-poor eretmodine cichlid lineage from Lake Tanganyika. Proc R Soc Lond B Biol Sci 2001, 268:803–808.
18. van Oppen MJH, Turner GF, Rico C, Deutsch JC, Ibrahim KM, Robinson RL, Hewitt GM: Unusually fine-scale genetic structuring found in rapidly speciating Malawi cichlid fishes. Proc R Soc Lond B Biol Sci 1997, 264:1803–1812.
19. Kornfield I, Smith PF: African cichlid fishes: Model systems for evolutionary biology. Annu Rev Ecol Syst 2000, 31:163–196.

20. Ribbink AJ, Marsh BA, Marsh AC, Ribbink AC, Sharp BJ: A preliminary survey of the cichlid fishes of rocky habitats in Lake Malawi. S Afr J Zool 1983, 18:147–310.

21. Genner MJ, Cleary DFR, Knight ME, Michel E, Seehausen O, Turner GF: How does the taxonomic status of allopatric populations influence species richness within African cichlid fish assemblages? J Biogeogr 2004, 31:93–102.

22. Stauffer JJR, Bowers NJ, McKaye KR, Kocher TD: Evolutionary significant units among cichlid fishes: The role of behavioral studies. In Evolution and the aquatic ecosystem: Defining unique units in population conservation. Edited by: Nielsen JL. Bethesda, Maryland: American Fisheries Society; 1995:227–244.

23. Rico C, Bouteillon P, Van Oppen MJH, Knight ME, Hewitt GM, Turner GF: No evidence for parallel sympatric speciation in cichlid species of the genus Pseudotropheus from north-western Lake Malawi. J Evol Biol 2003, 16:37–46.

24. Smith PF, Kornfield I: Phylogeography of Lake Malawi cichlids of the genus Pseudotropheus: significance of allopatric color variation. Proc R Soc Lond B Biol Sci 2002, 269:2495–2502.

25. Allender CJ, Seehausen O, Knight ME, Turner GF, Maclean N: Divergent selection during speciation of Lake Malawi cichlid fishes inferred from parallel radiations in nuptial coloration. Proc Natl Acad Sci USA 2003, 100:14074–14079.

26. Rundle HD, Nagel L, Wenrick Boughman J, Schluter D: Natural selection and parallel speciation in sympatric sticklebacks. Science 2000, 287:306–308.

27. Schluter D, Nagel L: Parallel speciation by natural selection. Am Nat 1995, 146:292–301.

28. McKinnon JS, Mori S, Blackman BK, David L, Kingsley DM, Jamieson L, Chou J, Schluter D: Evidence for ecology's role in speciation. Nature 2004, 429:294–298.

29. Plenderleith M, van Oosterhout C, Robinson RL, Turner GF: Female preference for conspecific males based on olfactory cues in a Lake Malawi cichlid fish. Biology Letters 2005, 1:411–414.

30. Stauffer JJR, Bowers NJ, Kellog KA, McKaye KR: A revision of the blue-black Pseudotropheus zebra (Teleostei: Cichlidae) complex from Lake Malawi, Africa, with a description of a new genus and ten new species. Proc Acad Nat Sci Phila 1997, 148:189–230.

31. Genner MJ, Turner GF: The mbuna cichlids of Lake Malawi: a model for rapid speciation and adaptive radiation. Fish Fish 2005, 6:1–34.

32. Turner GF, Seehausen O, Knight ME, Allender CJ, Robinson RL: How many species of cichlid fishes are there in African lakes? Mol Ecol 2001, 10:793–806.

33. McElroy DM, Kornfield I: Sexual selection, reproductive behavior, and speciation in the mbuna species flock of Lake Malawi (Pisces: Cichlidae). Environ Biol Fishes 1990, 28:273–284.

34. Lee W-J, Kocher TD: Microsatellite DNA markers for genetic mapping in Oreochromis niloticus. J Fish Biol 1996, 49:169–171.

35. Kellogg KA, Markert JA, Stauffer JR JR, Kocher TD: Microsatellite variation demonstrates multiple paternity in lekking cichlid fishes from Lake Malawi, Africa. Proc R Soc Lond B Biol Sci 1995, 260:79–84.

36. Parker A, Kornfield I: Evolution of the mitochondrial DNA control region in the mbuna (cichlidae) species flock of Lake Malawi, East Africa. J Mol Evol 1997, 45:70–83.

37. van Oppen MJH, Rico C, Deutsch JC, Turner GF, Hewitt GM: Isolation and characterization of microsatellite loci in the cichlid fish Pseudotropheus zebra. Mol Ecol 1997, 6:387–388.

38. Truett GE, Heeger P, Mynatt RL, Truett AA, Walker JA, Warman ML: Preparation of PCR-quality mouse genomic DNA with Hot Sodium Hydroxide and Tris (HotSHOT). BioTechniques 2000, 29:52–54.

39. Kalinowski ST, Taper ML, Marshall TC: Revising how the computer program cervus accommodates genotyping error increases success in paternity assignment. Mol Ecol 2007, 16:1099–1106.

40. Kidd MR, Danley PD, Kocher TD: A direct assay of female choice in cichlids: all the eggs in one basket. J Fish Biol 2006, 68:373–384.

41. Carleton KL, Parry JWL, Bowmaker JK, Hunt DM, Seehausen O: Color vision and speciation in Lake Victoria cichlids of the genus Pundamilia. Mol Ecol 2005, 14:4341–4353.

42. Mann ME, Hofker KD, van Alphen JJM, Seehausen O: Sensory drive in cichlid speciation. The American Naturalist 2006, 167:947–954.

43. Pierotti MER, Seehausen O: Male mating preferences predate the origin of a female trait in an incipient species complex of Lake Victoria cichlids. J Evol Biol 2007, 20:240–248.

44. Seehausen O, van Alphen JJM, Lande R: Color polymorphism and sex ratio distortion in a cichlid fish as an incipient stage in sympatric speciation by sexual selection. Ecol Lett 1999, 2:367–378.

45. Pierotti MER, Knight ME, Immler S, Barson NJ, Turner GF, Seehausen O: Individual variation in male mating preferences for female coloration in a polymorphic cichlid fish. Behav Ecol 2008, 19:483–488.

46. Couldridge VCK, Alexander GJ: Color patterns and species recognition in four closely related species of Lake Malawi cichlid. Behav Ecol 2002, 13:59–64.

47. Jordan R, Kellogg K, Juanes F, Stauffer JJ: Evaluation of Female Mate Choice Cues in a Group of Lake Malawi Mbuna (Cichlidae). Copeia 2003, 2003(1):181–186.

48. Cohen AS, Stone JR, Beuning KRM, Park LE, Reinthal PN, Dettman D, Scholz CA, Johnson TC, King JW, Talbot MR, et al.: Ecological consequences of early Late Pleistocene megadroughts in tropical Africa. Proc Natl Acad Sci USA 2007, 104:16422–16427.

49. Brown ET, Kalindekafe L, Amoako PYO, Lyons RP, Shanahan TM, Castañeda IS, Heil CW, Forman SL, McHargue LR, Beuning KRM, et al.: East African megadroughts between 135 and 75 thousand years ago and bearing on early-modern human origins. Proc Natl Acad Sci USA 2007, 104:16416–16421.

50. Blais J, Rico C, van Oosterhout C, Cable J, Turner GF, Bernatchez L: MHC adaptive divergence between closely related and sympatric African cichlids. PLoS One 2007, 2(8):e734.

51. Landry C, Garant D, Duchesne P, Bernatchez L: "Good genes as heterozygosity": MHC and mate choice in Atlantic salmon (Salmo salar). Proc R Soc Lond B Biol Sci 2001, 268:1279–1285.

52. Milinski M, Griffiths S, Wegner KM, Reusch TBH, Haas-Assenbaum A, Boehm T: Mate choice decisions of stickleback females predictably modified by MHC peptide ligands. Proc Natl Acad Sci USA 2005, 102(12):4414–4418.

53. Olsén KH, Grahn M, Lohm J, Langefors Å: MHC and kin discrimination in juvenile Arctic charr, Salvelinus alpinus (L.). Anim Behav 1998, 56:319–327.

54. Jacob S, McClintock MK, Zelano B, Ober C: Paternally inherited HLA alleles are associated with women's choice of male odor. Nat Genet 2002, 30:175–179.

55. Sommer S: Major histocompatibility complex and mate choice in a monogamous rodent. Behav Ecol Sociobiol 2005, 58:181–189.

56. Verzijden MN, Lachlan RF, Servedio MR: Female mate-choice behavior and sympatric speciation. Evolution 2005, 59:2097–2108.

57. Verzijden MN, ten Cate C: Early learning influences species assortative mating preferences in Lake Victoria cichlid fish. Biology Letters 2007, 3:134–136.

58. Rundle HD, Chenoweth SF, Doughty P, Blows MW: Divergent selection and the evolution of signal traits and mating preferences. PLoS Biol 2005, 3:e368.

59. Maan ME, van Rooijen AMC, van Alphen JJM, Seehausen O: Parasite-mediated sexual selection and species divergence in Lake Victoria cichlid fish. Biol J Linn Soc 2007, 94:53–60.

Hibernation in an Antarctic Fish: On Ice for Winter

Hamish A. Campbell, Keiron P. P. Fraser, Charles M. Bishop, Lloyd S. Peck and Stuart Egginton

ABSTRACT

Active metabolic suppression in anticipation of winter conditions has been demonstrated in species of mammals, birds, reptiles and amphibians, but not fish. This is because the reduction in metabolic rate in fish is directly proportional to the decrease in water temperature and they appear to be incapable of further suppressing their metabolic rate independently of temperature. However, the Antarctic fish (Notothenia coriiceps) is unusual because it undergoes winter metabolic suppression irrespective of water temperature. We assessed the seasonal ecological strategy by monitoring swimming activity, growth, feeding and heart rate (f_H) in N. coriiceps as they free-ranged within sub-zero waters. The metabolic rate of wild fish was extrapolated from f_H recordings, from oxygen consumption calibrations established in the laboratory prior to fish release. Throughout the summer months N. coriiceps spent a considerable proportion of its time foraging, resulting in a growth rate (G_w)

of 0.18±0.2% day^{-1}. In contrast, during winter much of the time was spent sedentary within a refuge and fish showed a net loss in G_w (-0.05±0.05% day^{-1}). Whilst inactive during winter, N. coriiceps displayed a very low fH, reduced sensory and motor capabilities, and standard metabolic rate was one third lower than in summer. In a similar manner to other hibernating species, dormancy was interrupted with periodic arousals. These arousals, which lasted a few hours, occurred every 4–12 days. During arousal activity, f_H and metabolism increased to summer levels. This endogenous suppression and activation of metabolic processes, independent of body temperature, demonstrates that N. coriiceps were effectively 'putting themselves on ice' during winter months until food resources improved. This study demonstrates that at least some fish species can enter a dormant state similar to hibernation that is not temperature driven and presumably provides seasonal energetic benefits.

Introduction

A number of temperate fish species become dormant during winter months [1]. During this time the fish remain inactive, cease feeding, and reduce protein synthesis and growth [2], [3]. However, dormancy in fish is thought to significantly differ from obligatory hibernating vertebrates [1], [2], [3], [4], [5], [6]. This is because the reduction in metabolism during winter correlates with the declining water temperature and can be overturned by temperature reversal [4], [5], [6]. Recent studies have found that otoliths from Antarctic Notothenioid fish display distinct growth annuli [7], [8], demonstrating that they too have seasonal variations in growth. The cessation of growth by Notothenioids during winter months appears paradoxical, because the Antarctic marine environment is considered one of the most thermally stable regimes on the planet [9] and these fish are often demersal omnivores living in shallow productive waters, where suitable prey are available all year round [10], [11], [12].

The Notothenioids include many species that remain in the inshore waters of the Antarctic continent and sub-Antarctic islands year round. This fish group has been overwhelmingly successful in the Southern Ocean and no other oceanic ecosystem is so dominated by a single taxonomic group of fish [13]. Previous studies that have tagged and recaptured Notothenioid fish, have demonstrated a 5-fold decline in growth rates from summer to winter months [12], [14], [15], [16]. It is very unlikely that the small seasonal change in near-shore Antarctic sea water temperatures is the major factor driving the observed large seasonal change in growth rates, as this would imply an unrealistic thermal sensitivity (a 5-fold change in growth for a seasonal 2°C temperature change). The virtual absence of solar radiation during winter, coupled with expansion of the sea ice, is responsible for producing amongst the lowest phytoplankton standing stocks anywhere on earth and, therefore, the lowest energy transfer through the marine food web [9]. Clarke [17] proposed that this temporal variation in food supply formed the basis of seasonal growth patterns in polar invertebrates. However, there is no evidence to suggest that the food supply of Antarctic fish is restricted during winter [10], [11], [12]. Moreover, the inshore fishes Notothenia coriiceps and Harpagifer antarcticus exhibit a reduction in feeding activity and a mobilisation of lipid reserves when acclimated to a winter photoperiod in the laboratory [12], [16], [18]. This decline in feeding occurred even when food was available in excess of that eaten; suggesting that decrease in growth rate during winter is a product of appetite suppression rather than food limitation per se. We hypothesised that the reduction in growth observed in Nototheniod fish during the Antarctic winter occurs due to a seasonal switch in ecological strategy, from one which maximises the procurement of food to another which minimises the energetic cost of living.

To test this hypothesis we have examined the behavioral and metabolic strategy of the Antarctic fish Notothenia coriiceps, a widespread omnivorous predator of the Antarctic and sub Antarctic inshore waters. Swimming activity, heart rate (fH) and metabolism were recorded by miniaturised electronic devices over a full annual cycle, as the fish responded to the annual physical and biotic fluxes of the Southern Ocean. From this extensive dataset, together with seasonal measurements of growth and feeding from wild fish, we show that N. coriiceps employs a hibernation–like ecological strategy during winter months. This finding is of profound interest, firstly because this type of ecological survival strategy has typically been considered to only occur in higher terrestrial vertebrates, and secondly, because these fish already live at the extreme lower thermal limit for physiological and metabolic processes.

Results

Growth and Feeding

A total of 118 immature adult N. coriiceps were caught by either fyke net or rod and line between Jan 2004 and Feb 2005. Of these, 21 were recaptured a second time within the year and 4 fish were recaptured more than once (Fig. 1). The highest specific growth rates (G_w) were recorded in fish captured and recaptured between January–April (the austral summer) than at other times of year (Tukey HSD modified for unequal groups, $F = 8.6$, $P<0.05$, $n = 25$). The maximum G_w (0.21% bd. wt. d^{-1}) was measured in a fish that was at liberty for 62 days between January and March. Nine fish that were tagged in the autumn and were recaptured in the early spring, thus the days of liberty only included winter months, exhibited close to zero or negative G_w. Growth rates in these animals can be considered as representative of winter animals. The Gw of two fish that were at liberty for almost a full year was 0.052 and 0.041% bd. wt. $d^{-1}$and are probably fairly representative of yearly field growth rates of N. coriiceps at Rothera.

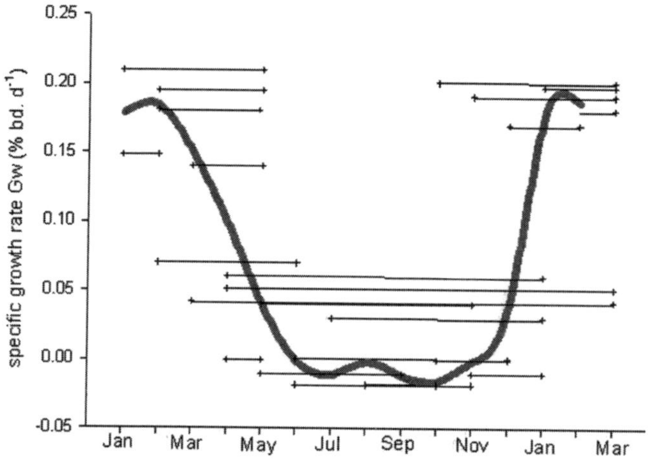

Figure 1. The mass specific growth rate (Gw) of free-ranging and sea-caged N. coriiceps. A total of 21 immature adult N. coriiceps (4 recaptured twice) were tagged, released and then recaptured between 2nd Jan 2004 and 12th March 2005. The black crosses indicate the date of first and subsequent capture and the average Gw for an individual fish during its days at liberty indicated by the connecting black line. The red line connects calculations of Gw (n = 6) measured from sea-caged fish every 8 weeks. For clarity the standard error is not shown on the graph, which from May to Oct was <0.003% bd. wt. d–1, and from Nov to Apr <0.031% bd. wt. d–1).

It was necessary to keep the fish fitted with fH dataloggers in sea-cages to enable serial sampling throughout the year. Gw was also determined in these fish on a bimonthly basis and showed a seasonal pattern similar to that seen in free-ranging fish (Fig. 1). That is, much higher growth rates in summer compared to winter. In February (summer) the Gw rate was 0.178% bd. wt. d–1. It declined rapidly after March (autumn) and reached a minimum of –0.02% bd. wt. d–1 in June (winter). During winter growth rates were negative, hence fish lost body mass. The decline in body mass continued until October (spring), after which the Gw rapidly increased, peaking at 0.172% bd. wt. d–1 in January 2004. The Gw from June to August was significantly lower than the Gw between December and February. (Tukey HSD, F2, 10 = 10.8, P<0.01). The within group variance was very low during winter months, when most fish showed negligible or negative growth rates, but had increased 10-fold by the summer when growth rate was high. This indicates that net energy gain differed between sea-caged fish; nevertheless, the large summer increase in Gw was still significant (ANOVA, F 2, 10 = 19.1, P<0.01).

Capture rates during winter months were very low, but 5 fish were captured in July so allowing gut content analysis. These fish showed significantly (Tukey HSD test with modification for unequal n numbers F = 30.12, P<0.01, n = 5 & 9) less food in the gut (8.1±1.9 mg.g bd. wt.-1) than 9 fish captured in January (27.2±3.4 mg.g bd. wt.-1). The food in the gut of winter caught fish also consisted mainly of digested matter (61.42±5.4%), which composed a much smaller portion (18.17±3.2%) of the gut content of fish caught during summer (F = 24.3, P<0.01, n = 5 & 9).

Activity

Fish movement was tracked continuously within a 1 km^2 area using a static hydrophone array between March 2004 and March 2005. Only seven of the original 20 fish implanted with acoustic transmitters confined their complete daily and annual behavioral repertoire within the boundaries of the tracking zone. A sweep search every 100 m, up to 1 km outside the study area with a boat-mounted hydrophone, found 6 individuals 28–543 m outside of the study area. Other tagged individuals may have migrated further from the study site, but loss of study animals by transmitter fault or predation cannot be discounted. Even when fish were located within the tracking zone the software could not always determine location. This occurred because the acoustic pulse from the transmitter was not received by all 3 hydrophones in every instance and hence the position could not be triangulated. An in ability to fix individual fish occurred far more frequently during winter months. Investigation by SCUBA divers suggested fish that could not be fixed were located in crevices or under rocks, thereby blocking the transmitter signal.

For each of the 4 designated seasons the fish showed a restricted home range, and movements within this area were non-random as tested by the Moran statistic (Table 1; X2>0.25, P<0.05, n = 7). All fish showed high site fidelity to a central area of approximately 5–10 m2. However, the relative size of the home range, and the variance in probability distribution for the number of fixes made between cells, showed significant differences between seasons (Fig. 2, Table 1). Fish occupied a relatively large home range during summer months. However, ranges reduced in size during autumn and by winter occupied a 6-fold smaller area than during the summer. The mean daily swimming distance was reduced 20-fold between summer and winter, and the spatial distribution of N. coriiceps within its home range was concentrated within a much smaller core area during winter (Fig. 2, Table 1). In spring (Sept–Nov) the home range increased in size but the fish still concentrated a large portion of their activity within a small core area (5–10 m2). By December this had changed significantly (Table 1), and the fish were now spending equal proportions of their time at more locations throughout their home range. The maximum distance a fish traveled within an hour was also significantly greater (ANOVA, F 4, 5664 = 54.05, P<0.01) in summer than in other seasons (Table 1).

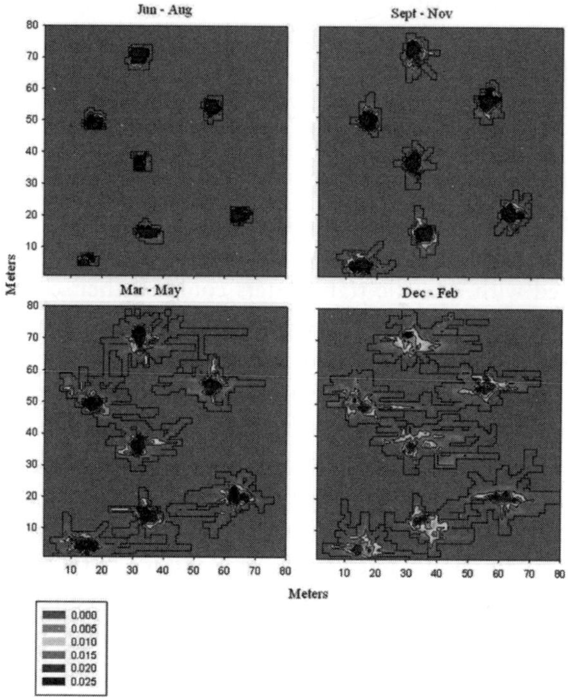

Figure 2. Seasonal home ranges of N. coriiceps. The precise locations of free-ranging fish were determined continually by acoustic telemetry throughout the year (n = 7). The 3-D spatial distribution plot was created by assigning positional data within 1 m2 cells over the range of the fish. The number of fixes per cell was calculated as a probability distribution of the total positional fixes for each 3 month period. The probability scale interval is 0.005, and purple areas indicate cells where the probability distribution was >0.03, the black outline indicates the outer boundary of the home range.

Table 1. Activity parameters for free-ranging N. coriiceps.

	Mar–May *(autumn)*	**Jun–Aug** *(winter)*	**Sep–Nov** *(spring)*	**Dec–Feb** *(summer)*
% total of position fixes	31.2 ± 12.2^a	12.2 ± 5.4^b	23.4 ± 8.8^c	41.8 ± 13.5^a
Size of range (m^2)	180.4 ± 9.8^a	60.8 ± 3.4^b	80.5 ± 6.7^c	233.0 ± 14.2^d
Area spend >5% of time	105 ± 14.2^a	27 ± 2.4^b	58 ± 7.5^c	137 ± 18.4^a
Variance of Prob. Dist.	4887 ± 330^a	18143 ± 114^b 3	14453 ± 90^c 1	4973 ± 246^a
Total No tracking days (N=7)	546	598	532	560
Days of no activity	147	511	378	77
Mean swimming speed ($m\ h^{-1}$)	12.1 ± 3.2^a	0.86 ± 0.12^b	5.3 ± 2.3^c	18.1 ± 2.2^d
Max swimming speed ($m\ h^{-1}$)	24.3 ± 5.2^a	23.1 ± 4.6^a	21.3 ± 3.1^a	38.4 ± 3.2^b

Fish were tracked within a 1 km^2 area by acoustic telemetry and a static hydrophone array over a 365 day period (mean±S.E, N=7).
[a,b,c,d]indicates seasonal mean data that is significantly different as calculated by multivariate analysis (*P*<0.05).

Heart Rate and Metabolism

Prior to the fish fitted with fH dataloggers being released into sea-cages, their f_H and oxygen consumption (MO_2) were measured in the laboratory using a static respirometry chamber. There was a strong linear relationship (r^2 = 0.86, n = 559) between f_H and MO_2 for N. coriiceps (Fig. 3). Both f_H and MO_2 showed similar ranges from minimum to maximum values in January and July. However, the whole range was displaced to lower values in both measures in winter. Therefore, the ratio of MO_2 with f_H did not change significantly between measurements taken in January and July (F = 2.54, P = 1.05, number of fish = 6, number of observations = 559), and a single overall regression equation (0.148 X–0.14, r^2 = 0.86) was therefore used to estimate field metabolic rate from the f_H of sea-caged fish throughout the year.

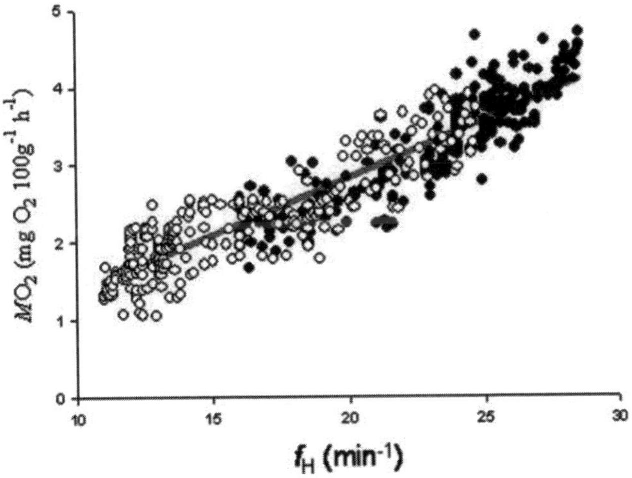

Figure 3. The seasonal relationship between heart rate (fH) and oxygen consumption (MO2) in N. coriiceps. Data were recorded in June (winter, open circles), and January (summer, closed circles). Fish were collected from sea-cages and immediately placed into a closed circuit respirometer, oxygen consumption and fH were recorded simultaneously. ANOVA of fitted regression lines demonstrated that the slopes were not significantly different between summer and winter, and therefore a single regression line was fitted to explain the relationship between fH and MO2 throughout the year (0.148 X–0.14, r2 = 0.86, N = 6 fish, n = 559 observations).

The mean fH recorded from fish inhabiting sea-cages during February was 25.2±1.2 min−1 and the estimated field MO2 for this month was 3.59±0.78 mg O2 100 g−1 h−1 (Fig. 4). Between February and April there was a 23% decline in fH and therefore MO2. The fall in sea water temperature to the yearly low (−1.8±0.02°C) occurred in mid-April, and thus succeeded the decline in N. coriiceps metabolism. Between June and October fH showed little variability and remained around 11±0.8 min−1 with a calculated metabolic rate of 1.48 mg O2 100 g−1 h−1. Consequently, summer and winter metabolism differed by 58%. During November and December fH and water temperature steadily increased, however by mid-December water temperatures reached the summer maximum of 0.7±0.02°C whilst fH and MO2 continued to rise through December until February.

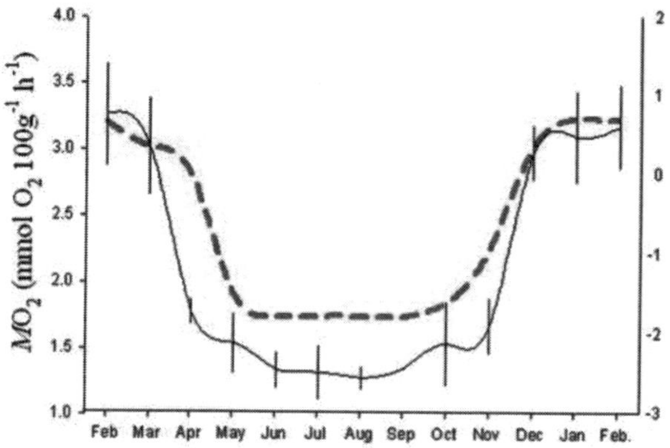

Figure 4. The free-ranging metabolic rate of N. coriiceps. The mean monthly MO2 of wild N. coriiceps (black line, n = 6) was extrapolated from continual field recordings of fH using the equation given in Fig. 3. Water temperature was measured by an onboard temperature sensor (dotted).

In a separate, controlled laboratory experiment we measured standard metabolic rate (SMR) and resting fH in starved non-swimming N. coriiceps. The difference between summer and winter standard metabolism in these fish was 29% (T 2, 12 = 3.6, P<0.01,). Reversing the small difference between summer and winter water temperatures produced no significant change (T 2, 12 = −0.4, P = 0.64) in fH or MO2 of laboratory fish (Table 2).

Table 2. The effect of altering winter and summer water temperature on f_H and MO_2 in N. coriiceps.

	January		June	
	MO_2 100 g fish (mg O_2 h^{-1})	f_H (min^{-1})	MO_2 100 g fish (mg O_2 h^{-1})	f_H (min^{-1})
Ambient temperature	2.72 ± 0.21^a	19.3 ± 2.4^b	1.95 ± 0.20^c	14.1 ± 2.6^d
Switched seasonal temp	2.64 ± 0.25^a	18.8 ± 2.1^b	2.05 ± 0.24^c	14.8 ± 2.1^d

^{a,b,c,d}are used to denote significantly different means as tested between rows (Students paired t-test, $P<0.05$).

Periodic Arousals

The activity of free-ranging fish during summer months varied between 0 and 38.4 m h^{-1}, with no evident circadian rhythm. This contrasted to winter months when the fish were sedentary for much of the time, but exhibited short bouts (1–3 hours) of activity where rates were similar to summer. These short bouts of activity in winter occurred between every 4 to 12 days (Fig. 5A).

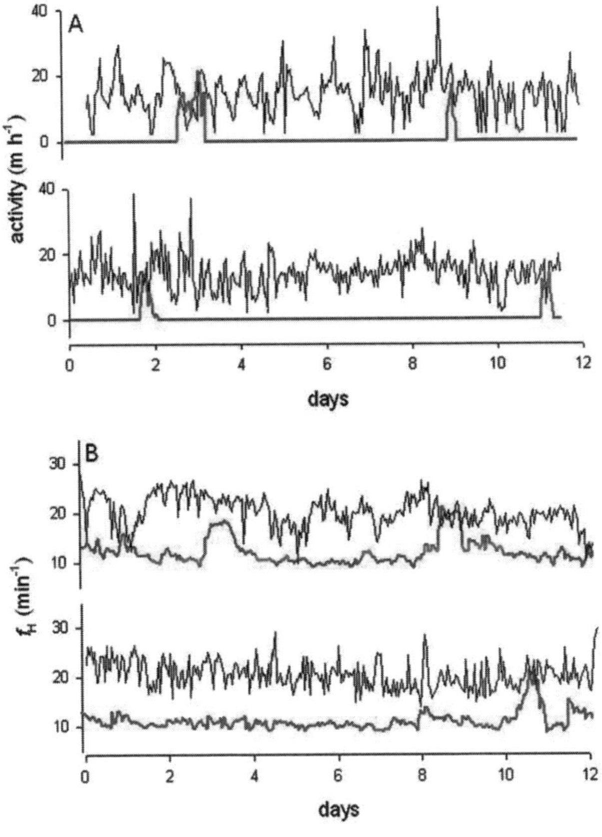

Figure 5. Summer and winter comparison of hourly activity and f_H profiles in N. coriiceps. Panel A shows activity profiles from two separate fish remotely recorded by acoustic telemetry during summer months (black), and the activity profile for the same fish is also shown during winter months (red). The flat line indicates when the fish were sedentary. Panel B shows the mean hourly f_H from two sea-caged N. coriiceps during summer (black) and f_H for the same fish during winter (red).

A similar seasonal profile was seen in the fH of fish held in sea-cages. In summer fH was very variable and ranged between 12 and 26 beats min−1, and there was no evidence of circadian rhythmicity (Fig. 5B). During winter months the fH was less variable and >2-fold slower than during the summer months (ANOVA, F 2, 55447 = 584, P<0.01).The low winter fH was intermittently elevated every 4–12 days for periods lasting a few hours. During these periods fH was elevated to rates similar to summer months.

SCUBA divers visiting winter refuges of fish in which movement could not be detected by acoustic telemetry, found the fish to be initially unresponsive to handling (Fig. 6). After 20 to 60 sec of handling the fish would become active and swim off, albeit sluggishly. In summer SCUBA divers could not handle wild N. coriiceps in this way.

Figure 6. SCUBA diver holding a non-responsive hibernating N. coriiceps. The fish was collected from under the Antarctic sea-ice in August from 18 m depth, temperature –1.8°C.

Discussion

The Antarctic marine environment is characterised by continual near-freezing temperatures, and highly seasonal primary productivity driven by the lack of light in winter [17]. In these conditions the Antarctic Notothenioid fish have flourished, although the ecological strategy adopted by these fish to endure the Antarctic winter is poorly understood. Data reported here from field observations over an annual cycle illustrate, for the first time, a tactic of metabolic suppression that displays many parallels with hibernation-like responses from higher vertebrate classes.

Notothenia coriiceps in this study were exposed to natural conditions typical of the Southern Ocean: large seasonal differences in photoperiod and primary productivity, but little variation in temperature. Free-ranging and sea-caged N. coriiceps in our study exhibited a significant winter decrease in activity, fH, metabolism and growth. Previous studies on Antarctic notothenioids have also demonstrated a winter reduction in growth rates [10], [11], [12], [14], [15], [16]. N. coriiceps is likely to have exhibited a loss of mass during winter due to reduced feeding and a reliance on endogenous lipid reserves [18]. Prey capture is likely to be associated with foraging activity. There could be many reasons for the reduction in foraging activity, including a reduced ability in a visual predator to capture food in low light conditions in winter. Irrespective of the underlying cause, the sedentary behavior of the fish during winter months partially explains the reduction in weight. This winter switch in ecological strategy, from one which capitalizes on energy gain through foraging to another which curtails the energetic cost of metabolism, is a common theme for hibernating organisms [19].

The current study estimated MO2 from field recordings of fH, and for the first time determined field metabolism in a fish from polar waters. This was possible because cardiac output (CO) is a robust indicator of metabolic rate [20] and Antarctic fish modulate CO chiefly by changes in fH [21]. In N. coriiceps the oxygen pulse (MO2: fH) did not vary between summer and winter seasons, and therefore, confirmed that for this species fH was a highly reliable indicator of metabolism. Our results demonstrated a 58% suppression in the total metabolic rate (TMR) from wild N. coriiceps between summer to winter months, and the seasonal difference in water temperature was not responsible for the seasonal shift in fH or MO2. The active suppression of MO2 and fH, irrespective of temperature, is a novel observation in fish and may correspond to the metabolic rate depression at the onset of hibernation of other vertebrates, where entry is also independent of body temperature [22].

The disparity in standard metabolic rate between the winter dormant period and the summer active period was relatively small (29%) compared to that of hibernating endotherms [22]. Nevertheless, the difference between summer and winter metabolism occurred in the absence of thermoregulatory requirements, and suggests an active suppression of physiological processes. Indeed, fH was reduced 2-fold and food processing may have been down-regulated in dormant N. coriiceps (as implied by the 3-fold increase in digested matter found in the gut of wintering compared to summer fish). Laboratory studies on seasonally acclimated N. coriiceps have also reported a winter depression in muscle and liver enzyme activity [23], and the non-responsiveness of N. coriiceps to human handling when taken from a winter refuge was reflective of the reduced sensory and motor capabilities documented in hibernating species, another energy saving physiological response [25].

In this study, although a significant seasonal difference was found in the metabolic rates of laboratory held fish, the change from summer to winter standard metabolic rates was much less than that recorded in wild N. coriiceps. A possible reason for the lack of dormancy behavior in the laboratory fish may be due to a disturbance effect, which is well documented to interrupt hibernation and delay the entry into dormancy of many hibernating animals [23]. This may also explain why previous studies that transported N. coriiceps outside Antarctica, where little or no seasonal signal was evident, and then acclimated those fish to seasonal conditions in the laboratory observed negligible metabolic suppression [18], [24].

The suppression of standard metabolism and fH during winter correlated with the cessation of activity. Interestingly, wintering N. coriiceps showed short periodic episodes when both fH and behavioral activity were elevated to summer rates. This demonstrated that during the dormant period N. coriiceps maintained the ability to up- or down-regulate metabolic and physiological processes and undertake swimming activity. The reason for such arousals is at present unclear, but the necessity of N. coriiceps to partake in these energetically expensive arousals draws further parallels with other hibernating groups. Why hibernating animals undergo expensive arousals is a subject of much speculation, and along with the proximal signals remain a long-standing, unresolved question of hibernate research [25].

It is considered that the 'Zeitgeber' for most animals to enter seasonal hibernation is either environmental temperature or photoperiod [4], [5]. Fish have an overt sensitivity to light and the seasonal extremes of photoperiod at high latitudes in the absence of thermal change make it an obvious environmental cue for N. coriiceps. Polar fish certainly have the ability to anticipate winter conditions, e.g. Atlantic wolf fish held in constant darkness upregulate antifreeze proteins and blood electrolytes to protect themselves from expected winter ice crystal formation [26]. Most hibernators seek constant darkness to assist with entry into hibernation and this has recently been shown to stimulate the expression of genes facilitating hibernation in mammals [27].

Conclusions

Hibernation is a complex subject, with animals continually being described as hibernators that challenge traditional views [28], [29].The winter dormancy we have documented in the Antarctic Notothenia coriiceps is distinct from the facultative dormancy observed in temperate fish species by the levels, and duration of the reduced physiological state, and is instead similar in many ways to the state achieved by truly hibernating species. There may be a continuum between mild levels of metabolic depression seen in many species over short timescales to full hibernation where dormancy in N. coriiceps forms a significant link. The seasonal hibernation strategy observed in N. coriiceps may be common to many, if not all Notothenioid fishes as illustrated by otolith growth annuli [7], [8]. This, and the ability to anticipate and compensate for marked seasonal effects may have contributed to the success of the Notothenioids in the Southern Ocean. Finally, the ability of N. coriiceps to actively suppress physiological processes beyond what is generally viewed as the extreme lower thermal threshold highlights that the 'over-wintering' Antarctic fish may be harbouring further cellular secrets.

Methods

Study Site and Fish

The study was conducted in a 1 km^2 inshore area off Rothera Research Station (British Antarctic Survey), Adelaide Island (67° 34′S 68° 08′W), Antarctica. Notothenia coriiceps (457±28 g, n = 166) were caught by fyke net and only immature adults were selected for the study, in order to avoid factors associated with reproductive cycles. Seasonal variation in growth has been shown to have no gender dependence and therefore this was not taken into consideration when sampling [10], [14]. After capture fish were taken immediately to an aquarium at ambient sea water temperature and photoperiod. All surgical procedures for tag attachment were undertaken within an air-cooled room (0°C), and the fish were anaesthetised in MS222 (0.3 g l^{-1}), before being placed on an operating table where their gills were irrigated with aerated seawater containing MS222 (0.1 g l^{-1}).

Growth and Feeding of Free-Ranging Fish

Between Jan 2004–Mar 2005 118 N. coriiceps (418±13 g) were caught, weighed, their lengths determined, and a numbered T-bar anchor tag (FD-64, Floy tag, Seattle, U.S.A.) fitted between the dorsal rays. The procedure took less than 2 min and the fish were returned at point of capture. Fish which were subsequently recaptured were measured, weighed and released, following weekly net-laying. Specific growth rate (G_w in % body weight day^{-1}) was calculated using the following equation [14]:

$$G_w = \frac{\log Wt_2 - \log Wt_1}{(t_2 - t_1)} * 100$$

Where: $W = body\ weight$
$t = time.$

Gut content analysis was undertaken on 5 N. coriiceps (425±28 g) captured by fyke nets in winter months and 9 (489±22 g) captured during summer. The fish were weighed and total gut content expressed per mg of total fresh body mass.

Acoustic Tracking

In March 2004, 20 N. coriiceps (436±28 g), were implanted with an acoustic transmitter (VS8, Vemco Ltd, Nova Scotia, Canada), which emitted a pulse every 15–30 s and had an acoustic range of 250m. The transmitter, which had a battery life of 400 d, was surgically implanted into the peritoneal cavity via a 2 cm incision made behind the left pectoral fin. The muscle was closed using an interrupted stitch with a 5/0 catgut suture and the skin closed using a non-interrupted stitch with 5/0 nylon monofilament; surgery took <12 min. The fish were recovered and kept under observation for 24 h before being released at the point of capture. Fish position was determined in the field using a purpose built static hydrophone array, consisting of three fixed acoustic receivers (VR2 receiver, Vemco Ltd) with a detection radius of 210 m. These were anchored to the seabed using climbing pitons. Receivers were placed 130 m apart and floated 2–9 m above the sea-bed, so that each was at the same depth (depth was accounted for when fish position was determined). The exact location in longitude and latitude coordinates of each hydrophone was determined by a hand held GPS system (GPS 76, Garmin, U.S.A.), taking the average latitude and longitude measurements over a 5 min period at the water surface directly above the hydrophone. ARCGIS software used these coordinates to plot position onto a scaled, orthorectified aerial photograph of the study area (MAGIC, British Antarctic Survey). The position of each fish that carried an acoustic transmitter was calculated by triangulation based on the relative position of each hydrophone, the speed of sound through the water and the difference in timing of the pulse arriving at each specific hydrophone. Pilot studies found this method to be accurate to <0.3 m.

Recording Heart Rate From Fish Held in Sea-Cages

In February 2004, miniature electronic micro-controlled data loggers (DL), capable of making high resolution recordings of ECG during free-ranging activity were attached to 6 N. coriiceps (589±34 g). The DL and housing was neutrally buoyant in water. A rubber saddle was permanently secured through the dorsal rays of the fish with nylon T-bar tags (FD-64, Floy tag Seattle, U.S.A), and the ECG electrodes (0.2 mm, Teflon coated 7-strand stainless steel wire, A–M systems, Connecticut, U.S.A.) were placed subcutaneously using a hypodermic needle a few mm through the septum behind the 4th left gill arch. The saddle and recording electrodes could be quickly attached or detached from the DL. The ECG was analysed in situ by a microprocessor using proprietary software that performed waveform analysis to generate inter-beat intervals and, thus, instantaneous f_H. The logger was primarily used in inter-beat mode but was also programmed to record two complete ECG waves every 4000 beats, to validate the quality of the ECG signal and the veracity of the calculated f_H (details in 30). The DL also recorded ambient temperature every minute to a resolution of ±0.3°C. Fish fitted with the DL were housed individually within 4 m^3 cages secured to the sea bed.

The DL needed to be exchanged every 60 d for battery replacement and data download. To enable DL exchange with minimal disturbance to the fish a cylindrical tube 20 cm D×50 cm L with double threaded end caps was placed in each pen, in which the fish would naturally seek refuge. The fish would always retreat inside the refuge when a SCUBA diver was in close proximity. The diver would open the pen, attach threaded end caps to both ends of the refuge, and remove it from the pen. On arrival at the water surface the refuge was immediately submerged into a large insulated holding tank containing sea water at ambient temperature then transported to an air temperature controlled aquarium approximately 100 m away. In the aquarium, each refuge was submerged in light anaesthetic (MS222 0.1 g−1) for 5 min. Once anaesthetized the fish was removed and placed on an operating table where its gills where flushed with fresh aerated sea water, to commence recovery. The housing cap was removed and the datalogger quickly exchanged (<2 min). The caps were secured onto the refuge and submerged in a large holding tank. The fish was returned to its original sea cage using the reverse of the collection procedure.

The cages had a large mesh size so that smaller prey items could enter the sea-cage, but to assist in attracting small invertebrate prey items a bait ball of chopped fish was inserted in the cage every month.

Extrapolation of Field MO_2 from f_H

Prior to release into the sea-cages, in January and June 2004 fish were first placed into a cylindrical respirometry chamber (8 cm D×30 cm L; vol 5 l) for 72 hours. All chambers were immersed into an aerated water bath that was continually flushed with fresh sea water at ambient sea temperature. Each chamber was fitted with two submersible pumps (100 l h^{-1}, Interpet, UK), one circulated the water around the chamber whilst the other flushed the chamber with aerated water from the water bath. This created a flow rate of 0.3 cm s^{-1} within the chamber. Water was extracted automatically from each chamber in series, via a rotor valve (Omnifit, Birmingham, UK), and injected into a purpose made flow cell containing a polarographic oxygen electrode (Strathkelvin, Glasgow, UK). The oxygen concentration from each chamber was sampled at 1 Hz for 15 minutes every two hours and during this period fH was recorded simultaneously by the attached data logger (see above). The relative oxygen depletion was calculated as a rate constant (only depletion traces with an r^2>0.90 were used in calculations). Adjustments were made for the dissolved oxygen in seawater at ambient temperature, the volume of the chamber, and the mass of the fish and subsequent water displacement. MO_2 values were then converted by the mass exponent for summer (0.82) and winter (0.76) immature adult N. coriiceps [22].

In a separate respirometry study, N. coriiceps (343±23g, n = 6) caught by fyke net in January and July were held in a large circular tank held at ambient sea temperature under local photoperiod. The instantaneous ECG and MO2 were recorded in fish after 72 h undisturbed rest in a respirometry chamber (as above). After the recordings were made on resting fish the seasonal temperature was reversed so that the June external water bath (–1.8°C) was warmed to 1°C, and in January lowered to –1.8°C. Recordings of fH and MO2 were made 48 h later. Photoperiod was kept as per local conditions i.e. constant light in January and darkness in June.

Statistical Analysis

Variance in activity parameters was tested using the F-test statistic and the means compared by multivariate analysis. To test that fish movement within the home range was not random, a weighting or connectivity matrix was generated to represent the spatial arrangement of each cell. The Moran statistic was employed to determine if data values within a cell were influenced by data values of other nearby cells [31]. Seasonality effects or the effects of treatment on mean physiological parameters were tested using the Students-paired t-test or Durban Watson statistic for serial autocorrelation. To test for seasonal shifts in specific growth rate the Tukey HSD test was employed and fish were grouped depending on the season which the initial and subsequent measurements were recorded. Sometimes this statistic needed to be adjusted for unequal size of groups. All statistics were applied using Statgraphics 5.1 or Minitab 12.0 software and were deemed significant at $P<0.05$.

Acknowledgements

The resolute support of the Rothera Research Station SCUBA team, and station staff is gratefully acknowledged. The help of Andrew Davies in datalogger construction, Julian Klepacki and Hamish Ross in designing statistical software, and the work of Andrew Miller and Karen Webb in the field is greatly appreciated.

Author Contributions

Conceived and designed the experiments: HC SE. Performed the experiments: HC. Analyzed the data: HC. Contributed reagents/materials/analysis tools: KF CB LP SE. Wrote the paper: HC SE.

References

1. Crawshaw LI (1980) Low-temperature dormancy in fish. Rev. Physiol. 42: 473–491.
2. Sayer MDJ, Reader JP (1996) Exposure of goldskinny, rock cook and corkwing wrasse to low temperature and low salinity: survival, blood physiology and seasonal variation. J. Fish Biol. 49: 41–63.
3. Cooke SJ, Grant EC, Schreer JF, Philipp DP, Devries AL (2003) Low temperature cardiac response to exhaustive exercise in fish with different levels of winter quiescence. Comp. Biochem. Physiol. A: 157–165.
4. Lyman CP, Willis JS, Malan A (1982) Hibernation and Torpor in Mammals and Birds. New York: Academic Press.
5. Gieser F, Hulbert AJ, Nicol SC (1996) Adaptations to the Cold. University of New England Press.
6. Brett JR, Groves DD (1979) Physiological energetics. In: Hoar WS, Randall DJ, Brett JR, editors. Fish Physiology vol VIII. New York: Academic Press.
7. North AW (1988) Age of Antarctic fish: Validation of the timing of annuli formation in otoliths and scales. Cybium 12: 91–115.
8. White MG (1991) Age determination in Antarctic fish. In: di Prisco G, Maresca B, Tota B, editors. Biology of Antarctic fish. Berlin: Springer-Verlag. pp. 87–100.
9. Brockington S, Clarke A, Chapman AA (2001) Seasonality of feeding and nutritional status during the austral winter in the Antarctic sea urchin (Sterechinus neumayeri) Mar. Biol. 139:: 127–138.
10. Casaux C, Mazzotta AS, Barrera-Oro S (1990) Seasonal aspects of the biology and diet of nearshore nototheniid fish at Potter Cove, South Shetland Islands, Antarctica Polar Biol 11: 63–72.
11. Everson I (1970) The population dynamics and energy budget of Notothenia neglecta at Signey Island, South Orkney Islands. Br. Antarct. Surv. Bul. 23: 25.
12. Targett TE (1990) Feeding, digestion and growth in Antarctic fishes: ecological factors affecting rates and efficiencies. Second International Symposium on the biology of Antarctic fishes. Naples: IIGB Press. pp. 37–39.
13. Clarke A, Johnston IA (1996) Evolution and adaptive radiation of Antarctic fishes. T.R.E.E. 11, (5): 212–2128.
14. Coggan R (1997) Seasonal and annual growth rates in the Antarctic fish Notothenia coriiceps R. J. Exp. Mar. Biol. Ecol., 213: 215–229.

15. Kawaguchi K, Ishikawa S, Matsude O, Naito Y (1989) Tagging experiments of nototheniid fish, Trematomus bernachii B. under the coastal fast ice in Lutzow-Holm Bay Antarctica. Pol. Biol., 2: 111–116.

16. Coggan R (1996) Growth:Ration relationships in the Antarctic fish Notothenia coriiceps R. maintained under different conditions of temperature and photoperiod. J. E.M.B.E., 210: 23–35.

17. Clarke A (1988) Seasonality in the Antarctic Marine-Environment. Comp. Biochem. Physiol. B 90: 461–473.

18. Johnston IA, Battram J (1993) Feeding energetics and metabolism in demersal fish species from Antarctic, temperate and tropical environments. J Mar Biol 115: 7–14.

19. McNab BK (2002) The physiological ecology of vertebrates. A view from energetics. Cornell: Cornell University.

20. Farrell AP, Jones DR (1992) The heart. In Fish Physiology, Vol XIIA, In: Hoar WS, Randall DJ, editors. New York: Academic Press.

21. Axelsson M, Davison W, Forster ME, Farrell AP (1992) Cardiovascular responses of the red-blooded Antarctic fishes Pagothenia bernacchii and P. borchgrevinki J Exp Biol 167: 179–201.

22. Elvert R, Heldmaier G (2005) Cardio-respiratory and respiratory reactions during entry into torpor in dormice, Glis glis. J Exp Biol 208: 1373–1383.

23. Grigg GG, Beard L (1996) Hibernation in the echidna: not an adaptation to cold? In: Geiser F, Hulbert AJ, Nicol SC, editors. Adaptations to the Cold. NSW, Australia: University of New England Press. pp. 13–21.

24. Johnston IA, Clarke A, Ward N (1991) Temperature and metabolic-rate in sedentary fish from the Antarctic, North-Sea and Indo-West Pacific-Ocean. Mar. Biol., 109: 191–195.

25. Choi IH, Cho Y, Oh YK, Jung NP, Shin HC (1998) Behavior and muscle performance in heterothermic bats. Physiol. Biochem. Zool. 71: 257–266.

26. Desjardins M, Le Francois NR, Fletcher GL, Bleir PU (2006) Seasonal modulation of plasma antifreeze protein levels in Atlantic (Anarhichas lupus) and spotted wolfish (A.minor). J.E.M.B.E 335: 142–150.

27. Zhang J, Kaasik K, Blackburn MR, Lee CC (2006) Constant darkness is a circadian metabolic signal in mammals. Nature, 439: 340–343.

28. Cossins AR, Barnes BM (1996) Southern discomfort. Nature 382: 582.

29. Dausmann KH, Glos J, Ganzhorn JU, Heldmaier G (2004) Hibernation in a tropical primate. Nature 429: 825–826.

30. Campbell HA, Bishop CM, Davies DA, Egginton S (2005) Recording long-term heart rate in the Black cod (Paranotothenia angustata) using an electronic datalogger. J. Fish Biol. 66: 1–7.

31. Cliff AD, Ord JK (1973) Spatial autocorrelation. London: Pion.

Evolutionary History of the Fish Genus *Astyanax* Baird & Girard (1854) (Actinopterygii, Characidae) in Mesoamerica Reveals Multiple Morphological Homoplasies

Claudia Patricia Ornelas-García,
Omar Domínguez-Domínguez and Ignacio Doadrio

ABSTRACT

Background

Mesoamerica is one of the world's most complex biogeographical regions, mostly due to its complex geological history. This complexity has led to interesting biogeographical processes that have resulted in the current diversity and distribution of fauna in the region. The fish genus Astyanax represents a useful model to assess biogeographical hypotheses due to it being one of the most diverse and widely distributed freshwater fish species in the New World. We used mitochondrial and nuclear DNA to evaluate phylogenetic relationships within the genus in Mesoamerica, and to develop historical biogeographical hypotheses to explain its current distribution.

Results

Analysis of the entire mitochondrial cytochrome b (Cytb) gene in 208 individuals from 147 localities and of a subset of individuals for three mitochondrial genes (Cytb, 16 S, and COI) and a single nuclear gene (RAG1) yielded similar topologies, recovering six major groups with significant phylogeographic structure. Populations from North America and Upper Central America formed a monophyletic group, while Middle Central America showed evidence of rapid radiation with incompletely resolved relationships. Lower Central America lineages showed a fragmented structure, with geographically restricted taxa showing high levels of molecular divergence. All Bramocharax samples grouped with their sympatric Astyanax lineages (in some cases even with allopatric Astyanax populations), with less than 1% divergence between them. These results suggest a homoplasic nature to the trophic specializations associated with Bramocharax ecomorphs, which seem to have arisen independently in different Astyanax lineages. We observed higher taxonomic diversity compared to previous phylogenetic studies of the Astyanax genus. Colonization of Mesoamerica by Astyanax before the final closure of the Isthmus of Panama (3.3 Mya) explains the deep level of divergence detected in Lower Central America. The colonization of Upper Mesoamerica apparently occurred by two independent routes, with lineage turnover over a large part of the region.

Conclusion

Our results support multiple, independent origins of morphological traits in Astyanax, whereby the morphotype associated with Bramocharax represents a recurrent trophic adaptation. Molecular clock estimates indicate that Astyanax was present in Mesoamerica during the Miocene (~8 Mya), which implies the existence of an incipient land-bridge connecting South America and Central America before the final closure of the Isthmus of Panama (~3.3 Mya).

Background

Mesoamerica is one of the most complex biogeographical areas in the world [1-5]. This complexity reflects the confluence of Neotropical and Nearctic biotas and a long history of geological activity, stretching from the Miocene to the present, during which movements of the Cocos, North American, Pacific and Caribbean Plates [6,7] created barriers and land-bridges that have affected the distribution of freshwater fishes [8-13]. For example, the Pliocene (~3.3 Mya) closure of the Panama Strait has been postulated to be one of the most important causes of faunal interchange between Neartic and Neotropical regions [14]. Climatic changes have also been invoked to explain the distribution of Mesoamerican fish fauna [15]. Distinguishing between climatic and geological effects requires information on phylogeny and species boundaries in a diversity of taxa [8,9,16-19].

Special attention has been devoted to understanding the number and timing of colonizations of Mesoamerica by freshwater fishes from South America: a topic which remains somewhat controversial [8,13,20-22]. The most widely accepted theories support two waves of colonization: 1) an ancient episode (70 – 80 Mya) through a proto-Antillean arc and 2) a more recent, Cenozoic episode via the Antillean islands and/or a continental corridor [14,23,24]. Molecular data suggest colonization of Mesoamerica by primary freshwater fishes about 4–7 Mya [8,12,13,22]. This is incongruent with the geological data, which does not support the existence of a continental land-bridge before the closure of the Panama Strait (3.3 Mya). Older colonization events have been postulated for secondary freshwater fishes: for example, Early to Mid Miocene (12.7–23 Mya) colonization for Synbranchidae [13], 14–24 Mya for heroinid cichlids (10 Mya for Mesoamerican lineages) and 18.4–20 Mya for rivulids [22].

The absence of primary freshwater genera (e.g., Hypopomus, Pimelodella, Rhamdia and Roeboides) from Mesoamerica and the Antillean islands argues against an ancient colonization route through a proto-Antillean arc. Instead, it supports a colonization route through an incipient land-bridge formed during the gradual uplifting of the Panama Isthmus [8,12], over a time span of 3–20 Mya, combined with changes in sea level [25]. This is supported by molecular studies of arthropods [26], amphibians [27], and marine geminate species pairs on either side of the Panamanian Isthmus [28].

Support for a more recent colonization of Mesoamerica by primary freshwater fish through the Panama Strait comes from phylogeographical studies of Characids (e.g., Brycon, Bryconamericus, Eretmobrycon, and Cyphocharax). These studies indicate multiple waves of rapid expansion from South America during the Pliocene ~3.3 Mya. [15].

The genus Astyanax provides an ideal model to investigate the relative importance of vicariance and dispersal on biogeographical patterns. This is partly because it is widely distributed across the region [29], and because its dispersal is confined to freshwater routes and dependent, therefore, on the formation of land-bridges.

Characiforms are generally assumed to have a Gondwanan (South American) origin [30-32], as supported by the fossil record [33], so the presence of Characidae in Northern America is viewed as a consequence of dispersal.

Astyanax comprises more than 107 recognized species and is, together with Hyphesobrycon (105 species), the largest and most diverse characiform genus [34,35]. Moreover, Astyanax has the widest distribution of American characids, being found from the Nearctic (Colorado River in Texas and New Mexico) to the Neotropics (Negro River in Patagonia) [4].

Previous phylogenetic studies [36,37] of the biogeography of Astyanax used a small number of samples from Mexico, Belize and Guatemala, and did not find geographical congruence for some of the groups recovered (i.e. Yucatan and Belizean populations were not the most closely related despite their geographical proximity). Furthermore, conspecific cave and nearest surface populations formed two separate lineages, in agreement with an earlier study of the genus [38]. This was attributed to at least two separate colonizations of Mesoamerica from South America during the Pleistocene. Estimated colonization times based on the cytochrome b gene were 1.8 and 4.5 Mya (3.1 Mya), with an estimated divergence rate of 1.5% per pairwise comparison per million years, which coincides with the closure of the Panama Strait (3.3 Mya) [37]. However, incomplete sampling (only few samples were included from upper Central America and Mexico) could lead to erroneous interpretations. This study provides a phylogeographical analysis based on a comprehensive distribution-wide sampling regime and more extensive sampling, and thus should provide new insights into the evolutionary history of the genus.

The genus Astyanax is characterised by high phenotypic plasticity and a capacity to adapt to diverse habitats [36,38-41]. There is clear evidence of extremely rapid adaptations of fish to new habitats and environments, with ecological specialization and morphological differentiation, generally in accordance with genetic divergence [15,42-44]. Considerable attention has been given to the evolution of developmental mechanisms and adaptation to cave environments [36,38,39], but less attention has been given to other habitat associated morphological plasticity. In this regard, Bramocharax, which is sympatric with Astyanax, is characterized by conspicuous trophic specializations, including differences in the number of premaxillary teeth, the presence of diastemas on the maxillary teeth, as well as differences in the shape and number of cuspids on the premaxillary, maxillary and dentary teeth, with some species (B. caballeroi and B.baileyi) having intermediate states between the morphotypes of Astyanax and Bramocharax species [45-47].

In this study we used mitochondrial and nuclear DNA sequences to develop a robust phylogenetic hypothesis for Astyanax and Bramocharax. This allows us to test biogeographical hypotheses for the Mesoamerican fish fauna, including the relative importance of historical geology and climatic factors.

Results

Three mitochondrial (Cytb, COI and 16 S) genes and one nuclear gene (RAG-1) were sequenced, giving a total of 3862 characters (2350 mitochondrial and 1512 nuclear).

RAG-1 was the most conservative of the genes analyzed (Table 1). Among the mitochondrial DNA genes, Cytb was the most variable, with COI exhibiting similarly high levels of variability and 16 s being the least variable. For the joint mitochondrial and nuclear analysis, 931 sites were variable, with 448 (~11%) being parsimony informative.

Table 1. Primers and PCR conditions.

Gene	Primers	Sequence (5'-3')	Tm (°C)	Size (pb)	Reference
RAG1	RAG1 (a, s)	AGCTGTAGTCAGTAYCACAARATG			
	RAG5 (s)	TRGAGTCACACAGACTGCAGA	58*	1512	[67].
	RAG9 (a, s)	GTGTAGAGCCAGTGRTGYTT			
Cytochrome oxidase I (*COI*)	FISHF1 (a, s)	TCAACCAACCACAAAGACATTGGCAC			
	FISHR1 (a)	TAGACTTCTGGGTGGCCAAAGAATCA	54	655	[83]
Cytochrome *b* (*Cytb*)	Glu- F (a, s)	GAAGAACCACCGTTGTTATTCAA			
	Thr- R (a, s)	ACCTCCRATCTYCGGATTACA	48		[84]
	CbCHR (a, s)	TTARTCCGGCTGGGWTNTTTG	48	1140	This study
16 S	16SAR (a, s)	CGCCTGTTTATCAAAAACAT			
	16SBR (a)	CCGGTCTGAACTCAGATCACGT	46	552	[85].

Amplification (a) and sequencing (s) primers used for the *RAG1*, 16S, *COI*, and *Cytb* genes. *Touchdown PCR was performed for the first 10 cycles: from 58°C to 53°C (-0.5°C each cycle), followed by 25 cycles at 53°C.

The topologies recovered by Maximum Parsimony (MP) and Bayesian Inference (BI) for the Cytb data set (Figure 1) and combined data set were similar. In addition, Cytb and the combined data set analyses were also concordant, with discrepancies restricted to the tree topologies in Clades III and IV from Group I (the Maquinas population was grouped with Montebello in the Cytb topology but with Polochic-Grijalva-Usumacinta with the combined data set). The combined data matrix was useful to resolve the deeper nodes and recovered mostly higher support values, providing greater phylogenetic resolution. For this reason, description of the higher-level groups identified was based on the topology obtained with the combined data set.

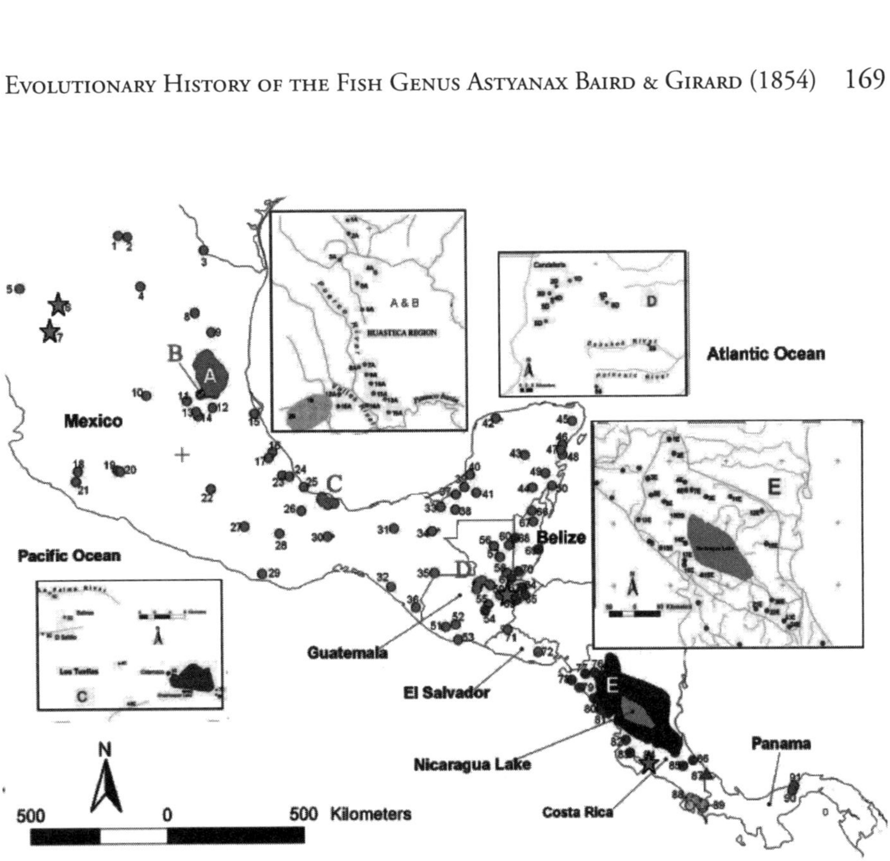

Figure 1. Sampling sites. Map of the sampled localities in Mesoamerica. The six major groups obtained in our phylogenetic analyses are represented by different colors. Stars represent the localities or basins where we found different lineages in sympatry.

All analyses supported the polyphyly of Bramocharax, with species of Bramocharax being sister groups to different clades of Astyanax (Figure 1), making Astyanax paraphyletic.

We identified six major phylogenetic groupings with high bootstrap support and significant posterior probabilities (Figures 1 and 2). Percentage divergences between groupings are given in Table 2. Groups V and VI correspond to the Chagres region (Panama) and Lagarto-Puntarenas basins of Costa Rica in Lower Central America, respectively. Groups II to IV are from Middle Central America and Group I is from Upper Central America and Mexico. These groups are non-overlapping geographically except for I with II and II with III. Their inter-relationships were not resolved with either the Cytb or the combined data set (3.8 Kbp).

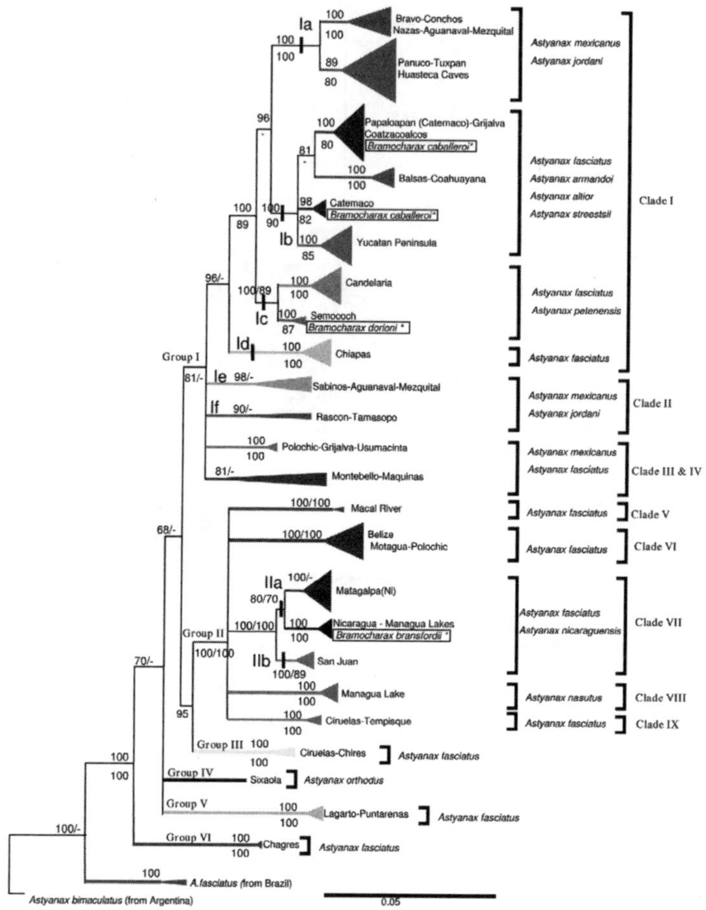

Figure 2. Summarized Phylogenetic tree estimated with Bayesian Inference and maximum-parsimony methods using the Cytb gene. Bayesian and Maximum-Parsimony inference tree for Astyanax and Bramocharax genera (the latter denoted by squares) based on the Cytb gene. The posterior probabilities and bootstrap values are shown. The species considered as valid according to Lima et al. [35] for each linage are also shown. Definition of the Clades was based on the Combined data set tree.

Table 2. Phylogenetic Performance of each gene.

Gene	Size (pb)	Variable sites	PI	% PI	CI	RI
Cytb	1140	505	298	26.14	0.58	0.85
COI	655	137	93	14.19	0.46	0.64
16S	555	74	19	3.42	0.52	0.61
RAG1	1512	215	38	2.51	0.56	0.07
All	3862	931	448	11.09	0.73	0.61

Molecular characterization for the data set for each gene. PI = parsimony informative characters, CI = Consistency index, RI = retention index.

Geographic Structuring Within the Major Phylogenetic Groupings

GROUP I (Mexico and Upper Central America)

Four main clades were recovered from Group I. Clade I comprised most of the Mexican and Upper Central American (Guatemala and Belize) populations. Clades II-IV represented fewer populations with a patchy distribution over the range of Clade I.

Within Clade I we found geographical structure corresponding to the following four lineages: Lineage Id, from the Chiapas region of Mexico, was sister to a clade comprising Lineage Ie from the Candelaria region and a pair of sister lineages (1a and 1b) that occupy a wider region from the Yucatan Peninsula to the Bravo-Conchos basin.

Lineage Ia contained individuals of A. mexicanus and the troglobitic species A. jordani from northern Mexico. This lineage was subdivided in two sublineages. The first included samples from Bravo-Conchos basin in the northern-most part of the range of Astyanax, and the basins of Mezquital and Nazas – Aguanaval. The second sublineage grouped populations from the Panuco, Tuxpan, Nautla and San Fernando-Soto La Marina basins, including most troglobitic populations from the Huasteca region (see region A in the Figure 3). Therefore, this latter sublineage included A. mexicanus and the cave-dwelling nominal species, A. jordani.

Lineage Ib contains Astyanax and Bramocharax from southern TMVB and Belize. It is subdivided into four sublineages. It ranges from the Media Luna Lagoon (Panuco basin, Mexico) to the Mopan Basin (Belize) on the Atlantic slope, and from the Armeria – Coahuayana Basin to the Balsas Basin (both in Mexico) on the Pacific slope. This group is a good example of morphological plasticity with low levels of genetic differentiation.

The first sublineage included Astyanax fasciatus from Puente Nacional to Grijalva – Usumacinta basins (including Papaloapan and Coatzacoalcos basins) on the Atlantic slope, and populations from the type localities of A. armandoi and A. altior (Palenque in the Grijalva – Usumacinta basin and Cenote of Noc-Ac in the Yucatan Peninsula, respectively). Both of these are junior synonyms of A. fasciatus [35,48]. We also found shared haplotypes or low divergence between Bramocharax caballeroi and sympatric A. fasciatus from Lake Catemaco (see region C, Los Tuxtlas, Figure 4). Additionally these distances were lower than those observed within Astyanax populations.

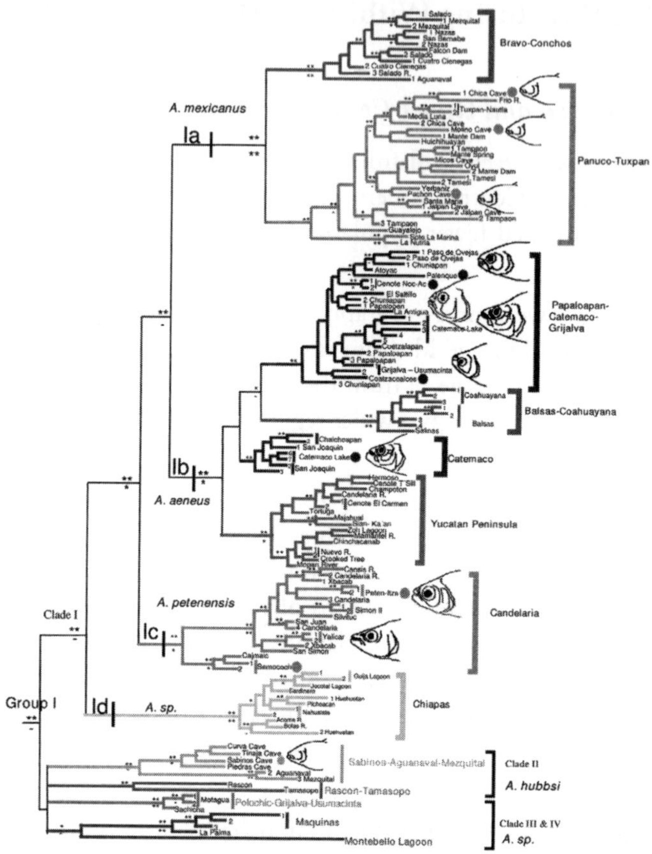

Figure 3. Subtree of Group I based on Cytb gene. Subtree of Group I for the Bayesian Inference and Maximum-Parsimony methods for Astyanax and Bramocharax based on the Cytb gene. Double asterisks indicate Bayesian posterior probabilities ≥ 0.95 or MP bootstrap values ≥ 90; single asterisks identify values between 0.89 and 0.80 or 89 and 80. Circles represent type localities. Definition of the Clades was based on the Combined data set tree.

The second sublineage (Catemaco) was a shallow clade restricted to the Tuxtlas region (lakes Chalchoapan and Catemaco, Figure 3), and comprised morphotypes of A. fasciatus and B. caballeroi, with very low genetic distinctiveness with haplotypes shared in some cases (Figures 1 and 2). The third sublineage included A. fasciatus, mostly from the Pacific slope (Coahuayana to Balsas basins), and one single population of A. mexicanus from the Media Luna Lagoon basin on the Atlantic slope (Figure 3). The fourth sublineage ('Yucatan Peninsula', Figure 1) included A. fasciatus from Grijalva-Usumacinta (Candelaria River), Cenotes from Yucatan Peninsula, Belize and Nuevo basins (Belize) and Mopan basin (Guatemala).

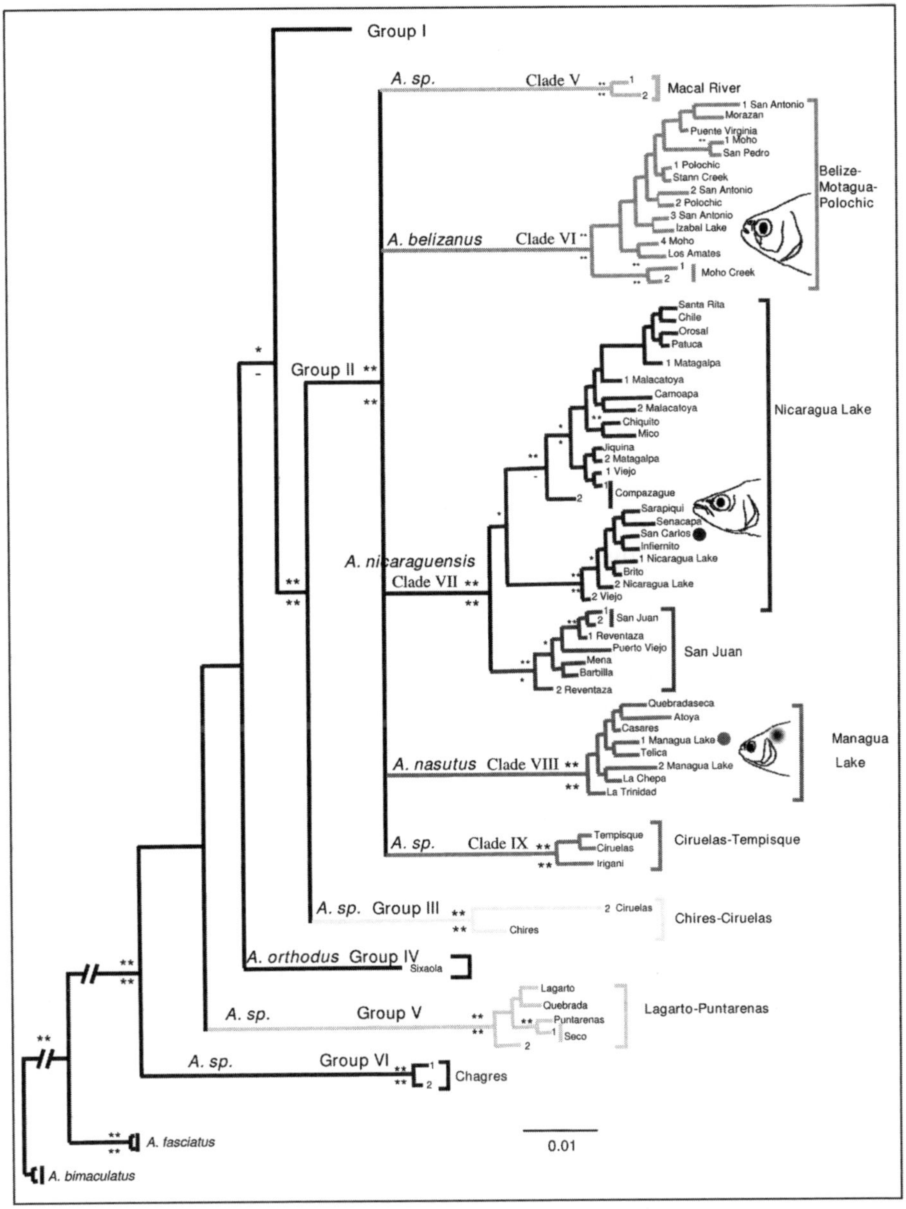

Figure 4. Subtree of Groups II to VI based on Cytb gene. Subtree of Group II – VI for the Bayesian Inference and maximum-parsimony methods for Astyanax and Bramocharax based on the Cytb gene. Double asterisks indicate Bayesian posterior probabilities ≥ 0.95 or MP bootstrap values ≥ 90; single asterisks identify values between 0.89 and 0.80 or 89 and 80. Circles represent type localities. Definition of the Clades was based on the Combined data set tree.

Lineage Ic, (region D, Figure 3) comprised populations from the Atlantic slope: Astyanax fasciatus (a synonym of A. mexicanus sensu Lima et al. [35]) from the type locality of A. petenensis (Peten – Itza Lake, Yucatan Peninsula) and A. fasciatus from the Candelaria karst region in Guatemala. This lineage included Bramocharax dorioni from its type locality (Semococh river). Similar to other instances of sympatry between Bramocharax and Astyanax, these two morphotypes showed low levels of genetic differentiation (less 0.5%).

Lineage Id (Chiapas) grouped Pacific slope populations of A. fasciatus (Figure 1) from the Pichoacan basin (Oaxaca, Mexico) to the El Jococal Lagoon (El Salvador), including the Guatemalan coast, with very low genetic divergence within the clade.

Clade II comprised two lineages. (Ie and If; mean divergence of $\bar{D}_{K81uf} = 2.13\% \pm 1.21$). Lineage Ie (referred to as "Sabinos-Aguanaval-Mezquital") grouped troglobitic morphotypes (A. jordani) from the Piedras, Tinaja, La Curva, and Sabinos caves (type locality of A. hubbsi, synonym of A. jordani sensu Lima et al. [35]) with the surface-dwelling populations of A. mexicanus from the Mezquital and Nazas – Aguanaval basins, with a low level of differentiation between surface and cave populations (mean of $\bar{D}_{K81uf} = 1.4\% \pm 0.9$).

Lineage If included populations of A. mexicanus from the Rascon valley and the Panuco basin in Tamposa. This lineage was highly divergent with regard to the rest of the lineages in this group, and those populations closest geographically.

Clade III grouped Atlantic slope populations of A. fasciatus from the La Palma and Maquinas basins (see Los Tuxtlas region, Figure 3) with those from the upper Polochic basin (Cahabon river) and the Grijalva-Usumacinta basin.

Clade IV contained populations from the Montebello Lagoons (Montebello) in south-eastern Mexico. This group clustered with the Maquinas and La Palma populations in the analysis of Cytb alone, but this clustering was not supported by the combined nuclear and mitochondrial analysis (Figures 2 and 4).

GROUP II (Middle Central America)

Our analyses did not resolve relationships among the five main clades recovered within Group II. This group occurs widely over Middle Central America, ranging from Belize to Nicaragua and Costa Rica (E, Figure 3).

Clade V included A. fasciatus from the Macal Basin (Belize). It is highly divergent with respect to the other clades from Group II ($\bar{D}_{K81uf} = 3.6\% \pm 1.39$).

Clade VI ("Belize-Polochic-Motagua") includes Atlantic slope A. fasciatus from the Moho basin (Belize), and from Guatemala, downstream of the Polochic Basin to the Puente Virginia Basin (Figure 5).

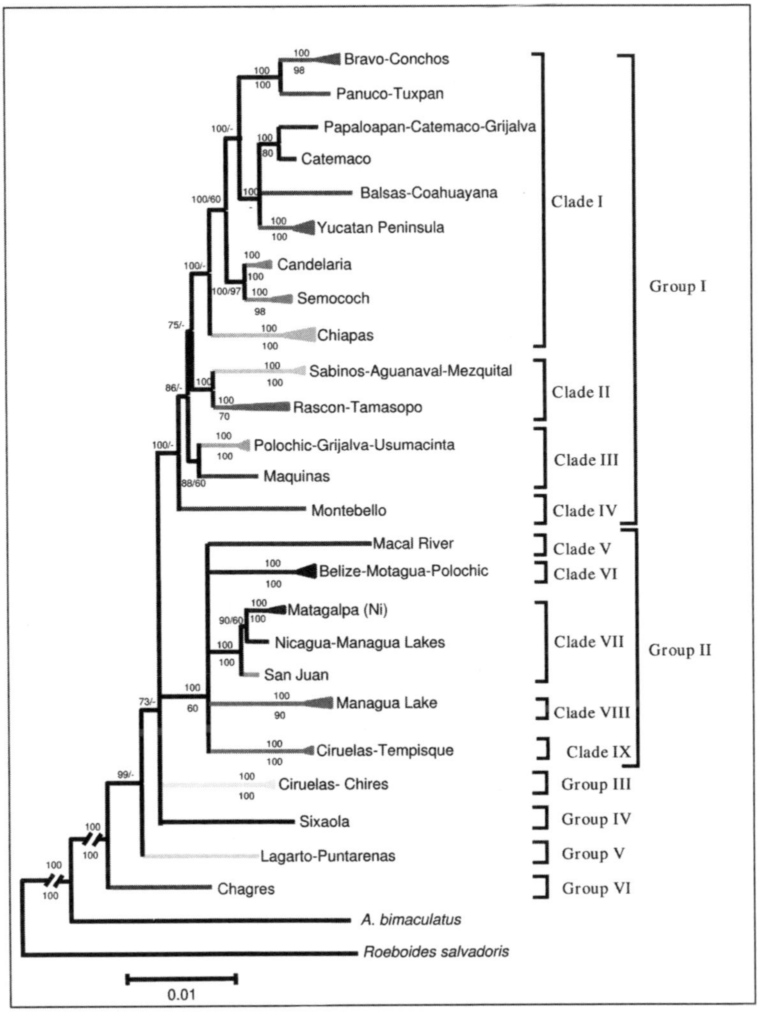

Figure 5. Combined data set tree. Summarized Phylogenetic tree for Astyanax and Bramocharax estimated with Bayesian Inference and Maximum-Parsimony methods using the subset of data (Cytb+COI+16S and RAG1). Posterior probabilities and bootstrap values are shown.

Clade VII comprised two lineages (IIa and IIb). Lineage IIa included Astyanax nicaraguensis and Bramocharax bransfordii. As in other clades containing both morphotypes there were low levels of genetic differentiation ($\bar{D}_{K81uf} = 0.9\% \pm 0.61$). This lineage occurs from the Jigüina Basin to the Sarapiqui basin (Nicaragua) on the Atlantic slope, and included populations from the great Lakes of Nicaragua (Managua and Nicaragua, region E, figures 1 and 4) and the Senacapa and Brito basins (Nicaragua) on the Pacific slope. Lineage IIb included A. fasciatus from

Lake Nicaragua and the San Juan basin (Nicaragua), as well as the Barbilla and Reventaza basins on the Atlantic slope of Costa Rica.

Clade VIII included A. fasciatus from the Pacific tributaries of Nicaragua and A. nasutus from the Managua Lake Basin.

Clade IX included A. fasciatus from a very restricted area comprising the Ciruelas and Tempisque basins on the Pacific slope of Costa Rica ("Ciruelas-Tempisque," Figure 5).

Lower Central America

Four major groups (groups III-VI) in Lower Central America were characterized by more restricted ranges relative to Groups I and II. Group III comprised A. fasciatus populations from the Ciruelas and Chires basins on the Pacific Slope of Costa Rica (figures 1 and 2). Two highly divergent ($\bar{D}_{K81uf} = 2.5\%$ haplotypes (corresponding to Groups II and III) were found in sympatry in the Ciruelas basin (figures 1 and 2).

Group IV included a single and well-differentiated population of A. orthodus from the Sixaola basin on the Atlantic slope of Costa Rica. Group V contained Pacific slope A. fasciatus populations from the Puntarenas basin (Costa Rica) to the Lagarto basin on the Panama-Costa Rica border. Group VI included A. fasciatus from the Chagres region on the Atlantic slope of Panama.

Discussion and Conclusion

Systematics of the Genera Astyanax and Bramocharax

Our analyses do not support the previously proposed monophyly of Bramocharax based on morphological analyses [45-47]. Moreover, Bramocharax specimens were present in two of the seven major Astyanax clades, with low levels of genetic differentiation when both morphotypes were found in sympatry (such as in lake Catemaco where it was possible to find haplotypes shared between individuals from both genera). Differentiation was equally reduced between allopatric populations of Bramocharax and Astyanax.

The genus Astyanax has been considered to be monophyletic in Mesoamerica [37], but polyphyletic in South America [30] on the basis of molecular analyses. Our results support the monophyly of Mesoamerican Astyanax only if we consider Bramocharax species to be morphotypes of Astyanax within the range of its phenotypic plasticity. This hypothesis is supported by the low genetic divergence between specimens of Bramocharax and Astyanax, and the evidence of recurrent evolution of the Bramocharax morphotype within Astyanax (Figure 3). This morphotype is associated with lacustrine habitats, suggesting that its recurrent evolution is a result of morphological convergence to similar ecological factors; similar patterns have been shown in other freshwater fishes [37,38,41,45,49]. If the "recurrent convergence" hypothesis is considered to be correct, then the taxonomy of Bramocharax needs to be revised and the evolutionary mechanisms giving rise to these morphological homoplasies need further investigation. Our analyses question the taxonomic utility of trophic characters (e.g., teeth shape or jaw modification), as previously done by Rosen [47] on the basis of intermediate morphological states between Astyanax fasciatus and Bramocharax baileyi.

Further incidences of morphological convergence were found in troglobitic morphotypes of Astyanax jordani (this has been noted by previous authors [37,38,50]), providing further evidence of independent (at least two different times, see Figure 6) adaptation to troglobitic habitats. The presence of recurrent morphological convergence in Astyanax [50,51] makes the delimitation of species and genera difficult. Thus the absence of congruence between phylogenetic relationships uncovered in this study and previous taxonomic classifications for Astyanax and Bramocharax from Mesoamerica [35,37,52] is not surprising. In addition, our results are not in agreement with the idea that Astyanax (including samples from Mexico and Upper Central America) is a single species (i.e., A. fasciatus) as has previously proposed [37].

Although not a main goal of this study, we propose a provisional taxonomic nomenclature for Astyanax populations from Mesoamerica. The nomenclature proposed is based on well-defined monophyletic groups, high genetic divergences with Cytb (>2% K81uf), and in agreement with geographical distributions. In ascribing species names we gave priority to previous species descriptions and diagnostic morphological traits. Where monophyletic lineages could not be assigned to a valid species name, they were assigned to their own monophyletic group as Astyanax sp.

Genetic and Time Divergences

The penalized likelihood analyses performed in r8s for Cytb sequences was calibrated using the following events: 1) the Merida-Perija uplift about 8–12 Mya, 2) the presence of fossils of Colossoma macropomum in the Magdalena basin (from at least 15 Mya), and 3) the formation of the TMVB about 3–6 Mya (Figure 6) [53]. The analysis gave an average divergence rate of 0.8% per million years with our in-group and the K81uf model of evolution (Figure 6). While this is similar to divergence rates reported for other fishes [Cichlidae (0.7%), [22], Cobitidae (0.68%) [54] (Table 3) and slightly lower than in cyprinid fishes (1.05%) [38], it is much lower than previous molecular clock rates (using fragments from the same gene) proposed for Astyanax (1.5% K2P divergences) [37]. This difference in estimated divergence rate is partially the cause of discrepancies between our study and previous historical biogeographical interpretations for Astyanax [37].

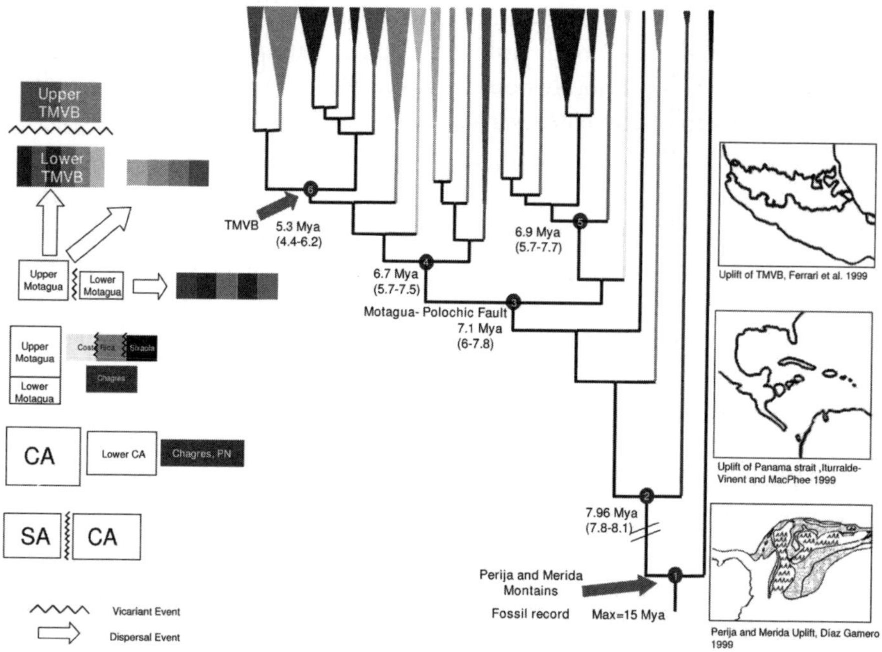

Figure 6. Ultrametric tree based on Cytb topology using semi-parametric penalized likelihood. Ultrametric tree based on the topology obtained with the mitochondrial Cytb gene using semi-parametric penalized likelihood. The calibration points are indicated by arrows, the first, node 1, corresponds with the rising of the Sierra of Perija and Merida Mountains Díaz de Gamero [79], and the second, node 6, corresponds to the final closure of the Trans-Mexican Volcanic Belt Ferrari et al. [82]. The main vicariant events are shown in the diagram as are the dispersal events.

Table 3. Genetic distances in percentage among major groups (below diagonal, uncorrected p sequence divergences; above diagonal, ML distances K81uf)

	Between						Within	
	1	2	3	4	5	6	Uncorrected	GTR
Group I	-	5.74	5.68	4.77	6.44	6.37	2.84	3.25
Group II	4.65	-	5.4	4.73	6.1	5.89	2.51	2.86
Group III	4.64	4.4	-	4.58	6.79	5.58	0.79	0.82
Group IV	4.05	4.03	3.91	-	5.45	5.27	0	0
Group V	5.19	4.91	5.41	4.51	-	7.05	0.23	0.23
GroupVI	5.08	4.77	4.57	4.41	5.6	-	0	0

Biogeographical Implications

We found a pattern of north-south phylogeographical structuring. The major phylogenetic groups were mostly non-overlapping, with the exception of Groups I and II, which overlap in the upper part of the Polochic basin of Guatemala, and Groups II and III, which overlap in the Ciruelas basin of Costa Rica. This north-south pattern is similar to that reported for other freshwater fishes [12,13,22]. We explain the observation of sympatric lineages of Astyanax in terms of niche overlap and lineage turnover, similar to that proposed in biogeographical models for other characins in Mesoamerica [15].

The lack of phylogenetic structuring in Astyanax in Middle Central America (even with the subset of data 3.8 Kbp) can be explained by a more recent colonization and rapid radiation about 6.9 Mya (Table 4); this pattern has been previously observed in other freshwater fishes of the region [12,13,15,22,55].

Table 4. Summary of the timings of the main geological events in Mesoamerica based on freshwater fauna.

Genera	*Gene*	*Calibration %/Mya*	*Event*	*Event Time*
Roeboides, Hypopomus and *Pimelodella** [8]	ATP6&8	1.3% K2P		4–7 Mya Mio-Pliocene
*Astyanax** [37]	*Cytb*	1.5% *K2P*		3 Mya Pliocene
*Rhamdia** [12]	*Cytb +*	1.3% *K2P*		2.5–2.9 Mya guatemalensis clade
	ATP6&8	1.5% (*Cytb*) *HKY85*		6.5–5.6 Mya laticauda clade Mio-Pliocene
*Synbranchus and Ophisternon*** [13]	*Cytb +*	1.3% (*ATP6&8*) *K2P*		12.7 – 23 Mya Miocene
	ATP6&8	1.5% (*Cytb*) *TrN+G*	*Colonization of the Mesoamerica from South America*	
*Cichlidae*** [22]	*Cytb*	0.7% uncorrected		10 Mya Miocene
*Rivulus***[86], recalculated	*Cytb*	1% uncorrected		18–20 Mya Miocene
*Brycon, Bryconamericus, Eretmobrycon and Cyphocharax ** [15]	*ATP6&8*	3.6% Ks		< 3.1 Mya Plio-Pleistocene
Poeciliidae [57]	MtDNA and nuclear genes	Mateos (2002) calibration		Before to Panama closure Cretaceous-Miocene
*Poeciliopsis*** [11]	*Cytb*	1–2% *K2P*	TMVB	8–16 Mya to 2.8–6.4 Mya Mio-Pliocene

*Primary freshwater fishes, ** Secondary freshwater fishes.

Dispersal Hypothesis on the Origin of Genus Astyanax in Mesoamerica

We accept the widely held hypothesis of a South American origin for Astyanax and other Central American characids [15,30,33]. This is supported by the observation that Lower Central America lineages were most closely related to South American samples from Brazil and Argentina. Considering the widely used Cytb calibration rate for fish (1.09%/my HKY distances) [22,38,54] and our mean rate of 0.8%/my K81uf distances from R8s (Figure 6 and Table 4), levels of divergence for populations of South and Central America imply a period of Mesoamerican colonization/expansion of Astyanax from South America about 7.8–8.1 Mya, before the final uplift of the Isthmus of Panama ~3.3 Mya [7,20,56]. The inclusion of more Astyanax samples from both sides of the Sierra de Perija and Merida Andes in further studies could improve this dating scenario.

The colonization of Central America prior to Late Cenozoic closure of the Panama Strait is incongruent with the geological data, and with other studies of characid genera (Brycon, Bryconamericus, Eretmobrycon and Cyphocharax) [15], including a previous study of Astyanax [37], all of which propose that closure of the strait ~3.3 Mya provided the first opportunity for colonization of Central America from South America.

An earlier colonization of Mesoamerica has been proposed for other freshwater fishes (Table 3) [[8,12], [13], [22]]. For example, ancient colonization events have been proposed for the family Poeciliidae (Cretaceous – Rosen model [23]) [57] and the Cichlidae, Rivulidae and Synbranchidae families (Miocene – GAARlandia model as proposed by Iturralde-Vinent and MacPhee [7]) [8,12,13,22]. However in contrast with Astyanax, these families also occur in the Caribbean islands, and as secondary fishes could have crossed through a shallow passage between South America and Central America during the Miocene [13,20,22].

Divergence times similar to those found in this study have been reported in molecular studies of primary freshwater fauna [8,12]. For example, the Bermingham and Martin model [8] proposes a colonization of Mesoamerica between 4 and 7 Mya, prior to the final closure of the Panama Isthmus (~3.3 Mya), based on comparative phylogeography of three genera [Roeboides (Characidae), Pimelodella (Pimelodidae) and Hypopomus (Hypopomidae)] Moreover, Rhamdia seems to have colonized Mesoamerica in two different waves, one prior to the closure of the Panama Strait (6.5 to 5.6 Mya, R. laticauda group) [12]. These estimates are in agreement with our study, and coincident with other primary and secondary freshwater fish fauna [12,13,22], as well as divergence times for invertebrates (9 Mya in pseudoscorpions) [26] and benthic foraminifera fossils (8 Mya) [58].

We found evidence of a unique biogeographical pattern involving multiple waves of expansion of Group I (Clades II – V) Astyanax in the upper part of the Polochic-Motagua fault (Mexico and Chiapas region). These clades have a restricted distribution overlapping that of Clade I, and in general occupy relatively stable ecological environments (springs or lakes), which can be less affected by climate change. Niche overlap and lineage turnover could explain this pattern, except in stable habitats where two lineages are found in sympatry (lineages Ia of Clade I and the lineage of Clade II in the Huasteca region, and in the Mezquital and Nazas-Aguanaval basin).

Main Vicariant Events in Astyanax Populations from Mesoamerica

The vicariance events involving Astyanax in Mesoamerica occurred during the Plio-Miocene (4–8 Mya), occurring earlier in Lower Central America (Panama and mainly Costa Rica) than in Central and Upper Mesoamerica.

A pattern of restricted geographic ranges in Lower Central America (Groups IV-VI) supports pronounced geographical fragmentation as a consequence of tectonics movements [7,20], which eventually resulted in closure of the Panama Strait ~3.3 Mya. Bermingham and Martin [8] have implicated multiple range fragmentation during the Miocene in patterns of diversity in other taxa of the primary freshwater fauna. With our data, five main vicariant events were identified for Lower Central America (Figure 6). These are related to changes in eustatic sea level (5–8 Mya) [25] and to the formation of inter-oceanic biogeographical barriers [58] during the Middle-Late Miocene (8 Mya).

In Upper Central America, the main volcanic activity was produced by the Trans-Mexican Volcanic Belt (TMVB). This region was affected by periods of intense geological activity between 3 and 12 Mya, with some volcanic activity still occurring today [53,59,60]. The geographic structuring evident in Clade I of Astyanax indicates that the TMVB formed an effective geographic barrier during its development during the late Miocene 4 – 6 Mya (Figure 6). This date is in agreement with the geology of the region and previous studies of several groups of vertebrates [1,5,10,11,61,62].

Other Biogeographical Patterns

Other biogeographical patterns were obtained in Lower and Middle Central America. These cannot be explained by geological barriers, but are in accordance with the main biogeographical regions proposed for other freshwater fishes [15,22,63].

While our results contrast somewhat with the Mesoamerican faunal regions recognized by Bussing [14], they do support his Isthmian Region, but with a fragmented pattern similar to that reported for other primary freshwater fishes [8,12,15]. Furthermore, we found that some Belize and Guatemala Astyanax clades (Clades VII and VIII in Figure 5) were joined to the Central America group (group II), a pattern shared with other Mesoamerican Cichlids [22]. Our Chiapas-Nicaraguan lineages (Lineage Id in Figure 4) did not reach the Pacific cost of Costa Rica, but instead have their southern limits in El Salvador. The distribution of lineages from San Juan Region differs from Bussing's proposal [14], occurring from the Barbilla basin (Costa Rica) to Belize (Atlantic slope) and on the Pacific slope from the Rio Grande Basin to the Ciruelas basin in the Nicoya Gulf (Costa Rica). This pattern has been previously reported by other authors [22,63].

Finally, we dated the separation of Groups I and II to about 6 and 7.8 Mya (Table 4). These coincide geographically with Polochic-Motagua Fault [6], reported as a transition region for other freshwater fish groups [22,13]. In addition, we observed the presence of two well differentiated Lineages of Astyanax (Clade IV in Group I and Clade V in Group II) in sympatry in the Polochic basin, a finding that has not been reported for other characids [15]. We explain this pattern in terms of river capture whereby the Cahabon tributary was diverted to the Polochic river as a consequence of tectonic activity (Sierra de Chiapas [64]), while separating the Cahabon river from the Grijalva-Usumacinta (Clade IV in Group I). This has been proposed for Rhamdia [12].

Methods

Tissue Collection and DNA Extractions

A total of 208 specimens of the Astyanax and Bramocharax from 141 localities from Panama to the Mexican-USA border (Figure 3) were analyzed, corresponding to 10 species of Astyanax: A. aeneus, A. altior, A. armandoi, A. fasciatus, A. jordani, A. mexicanus, A. nasutus, A. nicaraguensis, A. orthodus and A. petenensis; and three species of Bramocharax: B. caballeroi, B. dorioni and B. bransfordii. Samples of Roeboides bouchellei (from El Salvador), Astyanax bimaculatus (from Argentina) and Astyanax fasciatus (GenBank sequence from Brazil) were used as outgroups. We collected individuals from the type localities of most of the species (10 nominal species) described in Mesoamerica and considered valid by Lima et al [35]. Specimens were sampled by electro-fishing and netting, individually tagged, and preserved in DMSO/EDTA buffer [65] or 95% ethanol. DNA voucher specimens and their associated lots were subsequently preserved in 10% buffered formalin and deposited in the Museo Nacional de Ciencias Naturales of Madrid, Spain (MNCN), and the Universidad Michoacana de San Nicolas de Hidalgo, Michoacan, Mexico (UMSNH).

DNA Extraction and Sequencing

Genomic DNA was isolated by standard proteinase K and phenol/chloroform extraction methods [66] and stored at 4°C. The entire cytochrome b (Cytb) gene (1140 bp) and fragments of 16 S rRNA (552 bp) and cytochrome oxidase I (COI) (655 bp) genes were amplified. We also amplified exon 3 of the nuclear Recombinant Activating Gene 1 (RAG1) (1512 bp). Polymerase chain reactions (PCRs) were performed in 25-μL reactions containing 0.4 μM of each primer, 0.2 μM of each dNTP, 2 mM MgCl2, and 1.5 units of Taq DNA polymerase (Biotools). PCRs were conducted under the following conditions: 94°C (2 min), 35 cycles of 94°C (45 s), region specific Tm°C (1 min), 72°C (90 s), and 72°C (5 min), for most amplifications (see Table 1), with the exception of the RAG1 gene for which we followed the PCR conditions described in [67]. PCR products were run on 1.0% agarose gels to confirm amplification and purified with the EXOSAP-IT PCR Product Clean – Up (Usb) kit or by ethanol precipitation. Both strands were sequenced (see Table 5 for primers) and run on an ABI 3700 DNA automated sequencer (SECUGEN sequencing service).

Table 5. Estimated dates derived from the r8s molecular dating analyses, along with standard deviations of the node ages derived from Penalized Likelihood bootstrap analyses.

		Estimated ages		
Node	*Calibration point*	Strict Molecular Clock 0.8%/Mya	1.05%/Mya	Relaxed Molecular Clock NPRS with r8s Mya 95% of confidence
1	Merida-Perija Mountains **8–12 Mya**	24.6 ML (18 – 23)	32.3 ML (23.7–42.8)	15
	Colossoma macropomum **15 Mya**	11 *p* (9.8 – 12.4)	14.6 *p* (12.8–16.3)	
2	Mesoamerica Colonization	7.7 ML (6.36–9.32)	10.05 ML (8.35–12.23)	7.9 SD 0.165 (7.8–8.1)
		5.5 *p* (4.9–6.4)	7.3 *p* (6.4–8.5)	
3	Motagua-Polochic fault	4.6 ML (2.9–7.9)	6 ML (3.8–10)	7.1* SD 0.34 (6.0–7.8)
		3.7 *p* (2.5–5.7)	4.9 *p* (3.3–7.5)	
4	Radiation in North Mesoamerica Lineages	3.8 ML (2–6.2)	4.83 ML (2.7–8.2)	6.7* SD 0.37 (5.7–7.5)
		3.1 *p* (1.9–4.8)	4.1 *p* (2.5–6.3)	
5	Central America Radiation	3.7 ML (2–6.2)	5 ML(0.2–7.42)	6.9* SD 0.39 (5.9–7.7)
		3.1 *p* (1.9–4.8)	4.2 *p* (0.2–5.5)	
6	TMVB **3–6 Mya**	3.6 ML (2.3–5.4)	4.7 ML (3.1–7)	5.25* SD 0.37 (4.4–6.2)
		3.07 *p* (2.2–4.3)	4.03 *p* (2.8–5.7)	

The fixed ages are shown in Bold. *P* = Uncorrected distances ML = *K81uf* distances. Minimum and maximum values are given in parenthesis. Asterisks identify values with a normal distribution $p \leq 0.05$.

Data Analysis

Chromatograms and alignments were visually checked and verified. Saturation for transition and transversion substitutions was checked by plotting the absolute number of changes at each codon position against patristic distances for coding genes only.

Phylogenetic reconstruction was performed for Bayesian Inference (BI) using MrBayes version 3.1.2 [68]. We used Modeltest 3.07 [69] to find the best-fit model of evolution for each gene fragment using the Bayesian Information Criterion (BIC) [69]. BI was performed on two data sets as follows: (1) Cytb gene only with separate best-fit models for each codon position and (2) using the three mtDNA and RAG1 (separate best-fit models for each codon position were used) genes with a separate best-fit model of evolution for each gene fragment (partition). Analyses of best-fit models of evolution and BI were performed for a subset of the data.

Bayesian analyses were performed using two independent runs of four Metropolis-coupled chains of 10 million generations each to estimate the posterior probability distribution. The first 10,000 trees were discarded as burn-in. The program Tracer v1.4 [70] was used to assess run convergence and determine burn-in.

Sequence data were also analysed using maximum parsimony (MP) as implemented in PAUP* 4.0 b10 [71], NONA version 2.0 [72] and WINCLADA version 1.00.08 [73]. MP analyses in PAUP* and NONA/WINCLADA were done under the same heuristic search strategy. Statistical support for recovered clades was assessed using bootstrap (1000 pseudo-replications). We applied different weights for transversions and transitions according to the empiric criterion obtained in PAUP* 4.0 b10 [71]. The two datasets run for BI were also run for MP.

Analysis 1 (for which most species and populations of the Astyanax and Bramocharax genera from Mesoamerica were represented) was used to infer phylogenetic relationships among populations. Analysis 2 (for which only a subset of the species/populations were available) was used to infer relationships among the main lineages identified in analysis 1.

Molecular Clock and Divergence Time

Since we have a more complete data matrix, and in order to make our data comparable with previous studies in the region, we calibrated the molecular clock based on Cytb data alone rather than the combined data matrix (which included two more mitochondrial genes and one nuclear DNA gene).

Rate heterogeneity within the dataset was assessed using the likelihood-ratio test (LRT) [74,75]. LRTs were determined by comparing the log likelihood of the optimal topology recovered by maximum likelihood analysis (using appropriate models identified by Modeltest), while enforcing the molecular clock to the log likelihood of the optimal topology recovered by one that did not. The likelihood ratio statistic is twice the difference between the two log likelihoods. This statistic is compared to a $\chi2$ distribution with degrees of freedom equal to the number of terminals minus two following [76].

Because the molecular clock hypothesis was rejected, we conducted a nonparametric rate smoothing approach (NPRS) of divergence time estimation with the r8s package [75] to estimate divergence between the taxa. NPRS relaxes the molecular clock assumption by applying a least squares smoothing of estimates of substitutions rates.

Standard errors of divergence dates were estimated using the boot strapping procedure outlined in, and implemented by, Perl scripts in the r8s bootkit provided by Torsten Eriksson [77]. The first 100 bootstrapped datasets were created from the original Cytb dataset with the program Mesquite v. 2.01 [78]. Branch lengths were then re-estimated for each bootstrapped dataset in PAUP* using the original ML parameters. The resulting trees, with branch lengths, were then imported into r8s. The TN (Truncated Newton) algorithm was implemented.

We constrained the molecular clock considering two main points in the tree topology (see figure 6). First, our ingroup (node 1, Figure 6) was calibrated with the isolation of the Maracaibo basin as a result of the rise of Sierra de Perija and Merida Andes 8–12 Mya: [7,79,80]. Additionally for this node, we also used the oldest fossil record for a Characid (Colossoma macropomum, 15 Mya) in the Magdalena river system [81]. This allowed us to determine the minimum and maximum values for this node (8 and 15 Mya, respectively). We also constrained the molecular clock using the final closure of the Trans-Mexican Volcanic Belt (TMVB) 3–6 Mya (node 6 in the Figure 6) [53,60].

Author Contributions

CPOG collected material, compiled data, performed analyses and wrote the manuscript. ID designed the research project, collected material and wrote the manuscript. ODD collected material. The final draft was read and approved by all the authors.

Acknowledgements

We thank Jerry Johnson and Petter Unmack for providing material. Thanks also to Lourdes Alcaraz for her assistance with the laboratory work, and Carlos Pedraza, Paul Bloor, Alejandro Zaldivar and Guy Reeves for helpful suggestions on an early version of the manuscript. We also thank an anonymous referee for his helpful and meticulous work in revising of this article. CPOG was supported by CONACyT – Fundación Carolina grant 196833.

References

1. Contreras-Balderas SHO, Lozano-Vilano ML: Punta de Morro, an interesting barrier for distributional patterns of Continental fishes in North and Central

Veracruz, Mexico. Publ Biol Fac Cienc Biol Univ Autóm Nuevo Leon México 1996, 16:37–42.

2. Domínguez-Domínguez O, Doadrio I, Pérez-Ponce de León G: Historical biogeography of some river basins in Central Mexico evidenced by their goodeine freshwater fishes: A preliminary hypothesis using secondary Brooks Parsimony Analysis (BPA). J Biogeogr 2006, (33):1437–1447.
3. Huidobro L, Morrone JJ, Villalobos JL, Alvarez F: Distributional patterns of freshwater taxa (fishes, crustaceans and plants) from the Mexican Transition Zone. J Biogeogr 2006, 33(4):731–741.
4. Morrone JJ: Biogeographical regions under track and cladistic scrutiny. J Biogeogr 2002, 29(2):149–152.
5. Zaldivar-Riveron A, Leon-Regagnon V, de Oca ANM: Phylogeny of the Mexican coastal leopard frogs of the Rana berlandieri group based on mtDNA sequences. Mol Phylogenet Evol 2004, 30(1):38–49.
6. Guzman-Speziale M, Valdes-Gonzalez C, Molina E, Gomez JM: Seismic activity along the Central America Volcanic Arc: Is it related to subduction of the Cocos plate? Tectonophysics 2005, 400(1–4):241–254.
7. Iturralde-Vinent MA, MacPhee RDE: Paleogeography of the Caribbean region: Implications for cenozoic biogeography. B Am Mus Nat Hist 1999, (238):1–95.
8. Bermingham E, Martin AP: Comparative mtDNA phylogeography of neotropical freshwater fishes: testing shared history to infer the evolutionary landscape of lower Central America. Mol Ecol 1998, 7(4):499–517.
9. Martin AP, Bermingham E: Systematics and evolution of lower Central American cichlids inferred from analysis of cytochrome b gene sequences. Mol Phylogenet Evol 1998, 9(2):192–203.
10. Mateos M: Comparative phylogeography of livebearing fishes in the genera Poeciliopsis and Poecilia (Poeciliidae : Cyprinodontiformes) in central Mexico. J Biogeogr 2005, 32(5):775–780.
11. Mateos M, Sanjur OI, Vrijenhoeck RC: Historical Biogeography of the Livebearing Fish genus Poeciliopsis (Poecilidae: Cyprinodontiformes). Evolution 2002, 56:972–984.
12. Perdices A, Bermingham EAM, Doadrio I: Evolutionary history of the genus Rhamdia (Teleostei: Pimelodidae) in Central America.Mol Phylogenet Evol 2002, 25:172–189.
13. Perdices A, Doadrio I, Bermingham E: Evolutionary history of the synbranchid eels (Teleostei: Synbranchidae) in Central America and the Caribbean

islands inferred from their molecular phylogeny. Mol Phylogenet Evol 2005, 37(2):460–473.

14. Bussing WA: Patterns of distribution of the Central American ichthyofauna. In The Great American Biotic Interchange. Edited by: Stehli FG, Webb SD. New York: Plenum Press, New York; 1985:453–473.

15. Reeves RG, Bermingham E: Colonization, population expansion, and lineage turnover: phylogeography of Mesoamerican characiform fish. Biol J Linn Soc 2006, 88(2):235–255.

16. Avise JC: Molecular Markers, Natural History and Evolution.New York, NY 1994.

17. Joseph L, Moritz C, Hugall A: Molecular Support for Vicariance as a Source of Diversity in Rain-Forest.Proc R Soc Lond [Biol] 1995, 260(1358):177–182.

18. Patton JL, Dasilva MNF, Malcolm JR: Gene Genealogy and Differentiation among Arboreal Spiny Rats (Rodentia, Echimyidae) of the Amazon Basin – a Test of the Riverine Barrier Hypothesis. Evolution 1994, 48(4):1314–1323.

19. Templeton AR, Routman E, Phillips CA: Separating Population-Structure from Population History – a Cladistic-Analysis of the Geographical-Distribution of Mitochondrial-DNA Haplotypes in the Tiger Salamander, Ambystoma-Tigrinum.Genetics 1995, 140(2):767–782.

20. Coates A, Oblando JA: The geologic evolution of the Central America Isthmus. In Evolution and Environmental in Tropical America. Edited by: Jackson JBC, Budd AF, Coates AG. Chicago: Chicago University Press, Chicago; 1996:21–56.

21. Myers GS: Derivation of Freshwater Fish Fauna of Central America. Copeia 1966, (4):766.

22. Concheiro Perez GA, Rican O, Orti G, Bermingham E, Doadrio I, Zardoya R: Phylogeny and biogeography of 91 species of heroine cichlids (Teleostei: Cichlidae) based on sequences of the cytochrome b gene. Mol Phylogenet Evol 2007, 43(1):91–110.

23. Rosen DE: Vicariance Model of Caribbean Biogeography. Syst Zool 1975, 24(4):431–464.

24. Rosen DE: Vicariant patterns and historical explanation in biogeography. Syst Zool 1978, 27:159–188.

25. Haq BU, Hardenbol J, Vail PR: CHRONOLOGY OF FLUCTUATING SEA LEVELS SINCE THE TRIASSIC. Science 1987, 235(4793):1156–1167.

26. Zeh JA, Zeh DW, Bonilla MM: Phylogeography of the harlequin beetle-riding pseudoscorpion and the rise of the Isthmus of Panama. Mol Ecol 2003, 12(10):2759–2769.

27. Crawford AJ, Smith EN: Cenozoic biogeography and evolution in direct-developing frogs of Central America (Leptodactylidae: Eleutherodactylus) as inferred from a phylogenetic analysis of nuclear and mitochondrial genes. Mol Phylogenet Evol 2005, 35(3):536–555.

28. Marko PB: Fossil calibration of molecular clocks and the divergence times of geminate species pairs separated by the Isthmus of Panama. Mol Biol Evol 2002, 19(11):2005–2021.

29. Baird SF, Girard CF: Descriptions of new species of fishes collected in Texas, New Mexico and Sonora, by Mr. John H. Clark, on the U. S. and Mexican Boundary Survey, and in Texas by Capt. Stewart Van Vliet, U.S.A. Proc Nat Acad Sci 1854:24–29.

30. Calcagnotto D, Schaefer SA, DeSalle R: Relationships among characiform fishes inferred from analysis of nuclear and mitochondrial gene sequences. Mol Phylogenet Evol 2005, 36(1):135–153.

31. Ortí G, Meyer A: The radiation of characiform fishes and the limits of resolution of mitochondrial ribosomal DNA sequences. Syst Biol 1997, 46(1):75–100.

32. Otero O, Gayet M: Palaeoichthyofaunas from the Lower Oligocene and Miocene of the Arabian Plate: palaeoecological and palaeobiogeographical implications. Palaeogeography Palaeoclimatology Palaeoecology 2001, 165(1–2):141–169.

33. Gayet M, Marshall LG, Sempere T, Meunier FJ, Cappetta H, Rage JC: Middle Maastrichtian vertebrates (fishes, amphibians, dinosaurs and other reptiles, mammals) from Pajcha Pata (Bolivia). Biostratigraphic, palaeoecologic and palaeobiogeographic implications. Palaeogeography Palaeoclimatology Palaeoecology 2001, 169(1–2):39–68.

34. Eschmeyer WN: Catalog of fishes. Updated database version of June 2007. FishBase 2007.

35. Lima FCT, Malabarba LR, Buckup PA, Pezzi Da Silva JF, Vari RP, Harold A, Benine R, Oyakawa OT, Pavanelli CS, Menezes NA, et al.: Genera Incertae Sedis in Characidae. In Checklist of the Freshwater Fishes of South and Central America. Edited by: Reis RE, Kullander SO, Ferraris CJ Jr. Porto Alegre Brasil: EDIPUCRS; 2003:106–168.

36. Strecker U, Bernatchez L, Wilkens H: Genetic divergence between cave and surface populations of Astyanax in Mexico (Characidae, Teleostei). Mol Ecol 2003, 12(3):699–710.

37. Strecker U, Faundez VH, Wilkens H: Phylogeography of surface and cave Astyanax (Teleostei) from Central and North America based on cytochrome b sequence data. Mol Phylogenet Evol 2004, 33(2):469–481.

38. Dowling TE, Martasian DP, Jeffery WR: Evidence for Multiple Genetic Forms with Similar Eyeless Phenotypes in the Blind Cavefish, Astyanax mexicanus. Mol Biol Evol 2002, 19(4):446–455.

39. Jeffery WR: Cave fish as a model system in evolutionary developmental biology. Developmental Biology 2001, 231:1–12.

40. Lozano-Vilano ML, Contreras-Balderas S: Astyanax armandoi, n. sp. from Chiapas, Mexico (Pisces, Ostariophysi: Characidae) with a Comparison to the nominal species A. aeneus and A. mexicanus. Universidad y Ciencia 1990, 7:95–107.

41. Paulo-Maya J: Análisis morfométrico del género Astyanax (Pisces: Characidae) en México. México, D. F. : Instituto Politécnico Nacional; 1994.

42. Bernatchez L, Chouinard A, Lu GQ: Integrating molecular genetics and ecology in studies of adaptive radiation: whitefish, Coregonus sp., as a case study. Biol J Linn Soc 1999, 68(1–2):173–194.

43. Brunner PC, Douglas MR, Osinov A, Wilson CC, Bernatchez L: Holarctic phylogeography of Arctic charr (Salvelinus alpinus L.) inferred from mitochondrial DNA sequences. Evolution 2001, 55(3):573–586.

44. Danley PD, Kocher TD: Speciation in rapidly diverging systems: lessons from Lake Malawi.Mol Ecol 2001, 10(5):1075–1086.

45. Valdez-Moreno ME: A checklist of the freshwater ichthyofauna from El Peten and Alta Verapaz, Guatemala, with notes for its conservation and management. Zootaxa 2005, (1072):43–60.

46. Rosen DE: A New Tetragonopterine Characid Fish From Guatemala.Am Mus Novit 1970, (2435):1–17.

47. Rosen DE: Origin of the Characid Fish Genus Bramocharax and Description of a Second, More Primitive, Species in Guatemala.Am Mus Novit 1972, (2500):1–21.

48. Schmitter-Soto JJ, Valdez-Moreno ME, Rodiles-Hernandez R, González-Díaz AA: Astyanax armandoi, a Junior Synonym of Astyanax aeneus (Teleostei: Characidae). Copeia 2008, (2):409–413.

49. Contreras-Balderas S, Lozano-Vilano ML: Problemas nomenclaturales de las formas mexicanas del género Astyanax (Pisces: Characidae). Zoología Informa 1988, 38:1–13.

50. Protas ME, Hersey C, Kochanek D, Zhou Y, Wilkens H, Jeffery WR, Zon LI, Borowsky R, Tabin CJ: Genetic analysis of cavefish reveals molecular convergence in the evolution of albinism. Nat Genet 2006, 38(1):107–111.

51. Wilkens H, Strecker U: Convergent evolution of the cavefish Astyanax (Characidae, Teleostei): genetic evidence from reduced eye-size and pigmentation. Biol J Linn Soc 2003, 80(4):545–554.

52. Miller R: Freshwater Fishes of México. Volume 1. Chicago: The University of Chicago Press; 2005.

53. Ferrari L, López-Martínez M, Aguirre-Díaz G, Carrasco-Núñez G: Space Time patterns of Cenozoic arc volcanism in central México: from the Sierra Madre Occidental to the Mexican Volcanic Belt. Geology 1999, 27:303–306.

54. Doadrio I, Perdices A: Phylogenetic relationships among the Ibero-African cobitids (Cobitis, cobitidae) based on cytochrome b sequence data. Mol Phylogenet Evol 2005, 37(2):484–493.

55. Murphy WJ, Collier GE: Phylogenetic relationships within the aplocheiloid fish genus Rivulus (Cyprinodontiformes, Rivulidae): Implications for Caribbean and Central American biogeography. Mol Biol Evol 1996, 13(5):642–649.

56. Bartoli G, Sarnthein M, Weinelt M, Erlenkeuser H, Garbe-Schonberg D, Lea DW: Final closure of Panama and the onset of northern hemisphere glaciation. Earth Planet Sci Lett 2005, 237(1–2):33–44.

57. Hrbek T, Seckinger J, Meyer A: A phylogenetic and biogeographic perspective on the evolution of poeciliid fishes. Mol Phylogenet Evol 2007, 43(3):986–998.

58. Collins L, Coates A, Berggren W, Aubry M, Zhang J: The late Miocene Panama isthmian strait. Geology 1996, 24(8):687–690.

59. Ferrari L, Conticelli S, Potrone CM, Manetti P: Late Miocene volcanism and intra-arc tectonics during the early development of the Trans-Mexican Volcanic Belt.Tectonophysics 2000, 318:161–185.

60. Ferrari L, Tagami T, Eguchi M, Orozco-Esquivel MT, Petrone CM, Jacobo-Albarran J, Lopez-Martinez M: Geology, geochronology and tectonic setting of late Cenozoic volcanism along the southwestern Gulf of Mexico: The Eastern Alkaline Province revisited. J Volcanol Geotherm Res 2005, 146(4):284–306.

61. Mulcahy DG, Mendelson JR: Phylogeography and Speciation of the Morphologically Variable, Widespread Species Bufo valliceps, Based on Molecular Evidence from mtDNA. Mol Phylogenet Evol 2000, 17(2):173–189.

62. Mulcahy DG, Morrill BH, Mendelson JR: Historical biogeography of lowland species of toads (Bufo) across the Trans-Mexican Neovolcanic Belt and the Isthmus of Tehuantepec. J Biogeogr 2006, 33(11):1889–1904.

63. Smith SA, Bermingham E: The biogeography of lower Mesoamerican freshwater fishes. J Biogeogr 2005, 32(10):1835–1854.

64. Guzman-Speziale M: Active seismic deformation in the grabens of northern Central America and its relationship to the relative motion of the North America-Caribbean plate boundary. Tectonophysics 2001, 337(1–2):39–51.

65. Seutin G, White BN, Boag PT: Preservation of Avian Blood and Tissue Samples for DNA Analyses.Can J Zool 1991, 69(1):82–90.

66. Sambrook J, Fritsch E, Maniatis T: Molecular cloning: A laboratory manual. New York: Cold Spring Laboratory; 1989.

67. Quenouille B, Bermingham E, Planes S: Molecular systematics of the damselfishes (Teleostei : Pomacentridae): Bayesian phylogenetic analyses of mitochondrial and nuclear DNA sequences. Mol Phylogenet Evol 2004, 31(1):66–88.

68. Huelsenbeck JP, Ronquist F: MrBayes: Bayesian inference of phylogeny. Bioinformatics 2001, 17:754–755.

69. Posada D, Crandall KA: MODELTEST: testing the model of DNA substitution. Bioinformatics 1998, 14(9):817–818.

70. Rambaut A, Drummond A: Tracer [computer program]. [http://tree.bio.ed.ac.uk/software/tracer;;].4th edition. 2007.

71. Swofford DL: PAUP*: Phylogenetic Analysis Using Parsimony (*and Other Methods). Sunderland, Massachussetts: Sinauer Associates; 1998.

72. Goloboff PA: NONA. Noname (a bastard son of Pee-Wee). In 2.0 (32 bit version) edn. New York Program and documentation. Computer program distributed by J.M. Carpenter, Department of Entomology, American Museum of Natural History New York, 1993; 1993.

73. Nixon KC: Winclada (BETA). 0.9.9 edition. New York: Published by the author, ITHACA, NY; 1999.

74. Goldman N: Simple Diagnostic Statistical Tests of Models for DNA Substitution.Journal of Molecular Evolution 1993, 37(6):650–661.

75. Sanderson MJ: A nonparametric approach to estimating divergence times in the absence of rate constancy. Mol Biol Evol 1997, 14(12):1218–1231.

76. Huelsenbeck JP, Rannala B: Phylogenetic methods come of age: Testing hypotheses in an evolutionary context. Science 1997, 276(5310):227–232.

77. r8s bootkit 2 [http://www.bergianska.se/index_forskning_soft.html]

78. Mesquite: a modular system for evolutionary analysis V. 2.01.

79. Díaz de Gamero ML: The changing course of the Orinoco River during the Neogene: A review. Palaeogeography Palaeoclimatology Palaeoecology 1996, 123(1–4):385–402.

80. Lundberg JG: The temporal context for the diversification of Neotropical Fishes. In Phylogeny and Classification of Neotropical Fishes. Edited by: Malabarba LR, Reis RE, Vari RP, Lucena ZMS, Lucena CAS. Porto Alegre, Brasil EDIPUCRS; 1998:49–68.

81. Lundberg JG, Machadoallison A, Kay RF: Miocene Characid Fishes from Colombia – Evolutionary Stasis and Extirpation. Science 1986, 234(4773):208–209.

82. Ferrari L, Lopez-Martinez M, Aguirre-Diaz G, Carrasco-Nunez G: Space-time patterns of Cenozoic arc volcanism in central Mexico: From the Sierra Madre Occidental to the Mexican Volcanic Belt. Geology 1999, 27(4):303–306.

83. Ward RD, Zemlak TS, Innes BH, Last PR, Hebert PDN: DNA barcoding Australia's fish species. Philos Trans R Soc Lond, B 2005, 360(1462):1847–1857.

84. Zardoya R, Doadrio I: Phylogenetic relationships of Iberian cyprinids: systematic and biogeographical implications. Proc R Soc Lond [Biol] 1998, 265(1403):1365–1372.

85. Palumbi S, Martin AP, Romano S, McMillan WO, Stice L, Grabowski G: The Simple Fool's Guide to PCR. 2.0th edition. Special publication, Honolulu: University Hawaii Press; 1991.

86. Murphy WJ, Thomerson JE, Collier GE: Phylogeny of the neotropical killifish family Rivulidae (Cyprinodontiformes, Aplocheiloidei) inferred from mitochondrial DNA sequences. Mol Phylogenet Evol 1999, 13(2):289–301.

Temperature-Dependent Sex Determination in Fish Revisited: Prevalence, a Single Sex Ratio Response Pattern, and Possible Effects of Climate Change

Natalia Ospina-Álvarez and Francesc Piferrer

ABSTRACT

Background

In gonochoristic vertebrates, sex determination mechanisms can be classified as genotypic (GSD) or temperature-dependent (TSD). Some cases of TSD in fish have been questioned, but the prevalent view is that TSD is very common in this group of animals, with three different response patterns to temperature.

Methodology/Principal Findings

We analyzed field and laboratory data for the 59 fish species where TSD has been explicitly or implicitly claimed so far. For each species, we compiled data on the presence or absence of sex chromosomes and determined if the sex ratio response was obtained within temperatures that the species experiences in the wild. If so, we studied whether this response was statistically significant. We found evidence that many cases of observed sex ratio shifts in response to temperature reveal thermal alterations of an otherwise predominately GSD mechanism rather than the presence of TSD. We also show that in those fish species that actually have TSD, sex ratio response to increasing temperatures invariably results in highly male-biased sex ratios, and that even small changes of just 1–2°C can significantly alter the sex ratio from 1:1 (males:females) up to 3:1 in both freshwater and marine species.

Conclusions/Significance

We demonstrate that TSD in fish is far less widespread than currently believed, suggesting that TSD is clearly the exception in fish sex determination. Further, species with TSD exhibit only one general sex ratio response pattern to temperature. However, the viability of some fish populations with TSD can be compromised through alterations in their sex ratios as a response to temperature fluctuations of the magnitude predicted by climate change.

Introduction

Sex determination mechanisms produce the sex ratio, a key demographic parameter crucial for population viability. In gonochoristic vertebrates, sex determining mechanisms can broadly be classified as genotypic (GSD) or temperature-dependent (TSD) [1], [2]. In species with TSD, there are no consistent genetic differences between sexes. The earliest ontogenetic difference between sexes is an environmental one because the ambient temperature during sensitive periods of early development irreversibly determines phenotypic sex and, therefore, the sex ratio [1], [2]. Thus, species with TSD have been proposed to be reliable indicators of the biological impact of global warming, since temperature-induced sex ratio shifts constitute a direct fitness response to thermal fluctuation [3].

So far, predicted effects of climate change on fish populations include distribution shifts [4], alterations in developmental time and larval dispersal [5], decrements in aerobic performance [6], and mismatches in species interactions [7]. Climate change effects on the sex ratio have already been inferred for some sea turtles with TSD [8], [9], but are lacking for fish. Thus, knowledge of the extent to which temperature affects sex ratios is relevant in order to gauge potential threats of rising temperatures on fish populations. Further, knowing the prevalence of TSD is essential for the correct theoretical and empirical study of the evolution of sex determining mechanisms [2], because otherwise inferences on the distribution and prevalence of a particular type of mechanism may be biased [10].

In fish, the first evidence of TSD was obtained in field and laboratory studies carried out in the Atlantic silverside, Menidia menidia (F. Atherinopsidae) [11]. Since then, TSD has been claimed in 59 different species (33 of them of the genus Apistogramma, F. Cichlidae, and all included in the same study) belonging to 13 families representative of many types of fishes. Fish with TSD have readily been grouped according to three patterns of sex ratio response to environmental temperature [12]–[16]: 1, more males at high temperature; 2, more males at low temperature; and 3 more males at extreme (high and low) temperatures (Fig. 1). However, a critical examination of sex ratio produced in response to temperature in fish has never been carried out. Based on all the available data on TSD in fish, it has been reported that 53–55 (including the 33 species of the genus Apistogramma), 2–4 and 2 of these species follow patterns 1, 2 and 3, respectively (Table 1). Note that what here are referred to as patterns 1 and 2 of fish essentially corresponds to what in reptiles are referred to as patterns Ib and Ia, respectively. However, pattern 3 of fish is not equivalent to pattern II of reptiles (female-biased sex ratios at low and high temperatures and male-biased sex ratios at intermediate temperatures) but it could be considered an inverse of it.

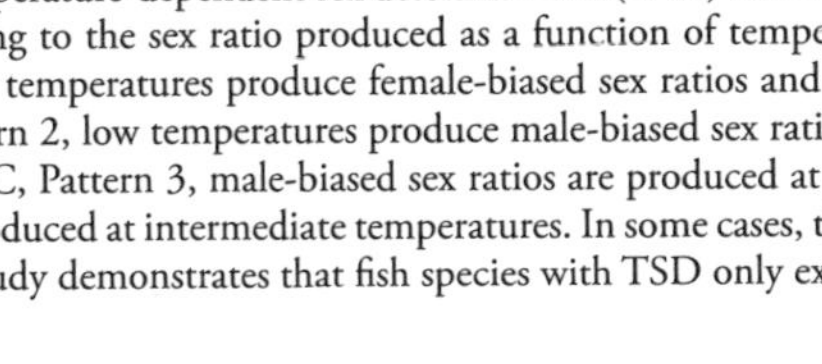

Figure 1. Patterns of temperature-dependent sex determination (TSD) in fish that had been recognized to date. They are defined according to the sex ratio produced as a function of temperature during the thermosensitive period. A, Pattern 1, low temperatures produce female-biased sex ratios and high temperatures produce male-biased sex ratios. B, Pattern 2, low temperatures produce male-biased sex ratios and high temperatures produce female-biased sex ratios. C, Pattern 3, male-biased sex ratios are produced at low and high temperatures, while balanced sex ratios are produced at intermediate temperatures. In some cases, the response may be partial (dashed line in A). The present study demonstrates that fish species with TSD only exhibit pattern 1.

GSD and TSD can be regarded as two discrete processes that give rise to a continuous pattern of sex determination mechanisms [2], or as two ends of a continuum [17]. In any case, the presence of TSD in a given species is not incompatible with the existence of genotype x environment interactions, which are common in fish, including Menidia [15], [18], [19]. However, too often assignment of TSD in many fish species has proceeded regardless of evidence such as the presence of sex chromosomes, which is strongly indicative of GSD [1], [2], [15]. Further, the Atlantic silversides (Menidia menidia and M. peninsulae) are the only fish species in which the existence of TSD has been demonstrated in the wild; in all other species, data were obtained from laboratory experiments [16]. Thus, evidence to support the presence of TSD has been obtained in many cases using temperatures in the laboratory that the species will rarely experience in nature. It has been pointed out that observed sex ratio shifts under these circumstances might be the consequence of thermal effects on GSD (GSD+TE) rather than proof of the presence of TSD [2], [16]. Thus, there is concern regarding the actual prevalence of TSD in fish. In particular, to discern true cases of TSD from GSD+TE [16]. Nevertheless, the existence of TSD in fish is now widely accepted, assumed to be widespread and expected to be found in more species as new studies become available [10], [12].

Table 1. Patterns of temperature-dependent sex determination in gonochoristic fish.

		Criteria used here			Confirmation by statistical analyses							
SPECIES	Pattern of TSD previously assigned*	Evidence for the presence of sex chromosomes [Reference]	Sex ratio shift within the RTD (see Suppl. Table 1)	Diagnosis	Lineal regression/F-test							New pattern of TSD proposed here
					n	Intercept	Slope	r^2	F	DFn/DFd	*P*	
Carassius auratus	(1)	Yes [49]	No	GSD+TE	-	-	-	-	-	-	-	0
Carassius carassius	1	Yes [15]	No	GSD+TE	-	-	-	-	-	-	-	0
Danio rerio	(1)	Yes [38]	No	GSD+TE	-	-	-	-	-	-	-	0
Gnathopogon caerulescens	1	Yes [50]	Yes	GSD+TE	-	-	-	-	-	-	-	0
Misgurnus anguillicaudatus	1	Yes [51]	Yes	GSD+TE	-	-	-	-	-	-	-	0
Ictalurus punctatus	2	Yes [52]	No	GSD+TE	-	-	-	-	-	-	-	0
Hoplosternum littorale	1	No	Yes	TSD	16	2.53	2.26	0.40	9.30	1/14	0.009	1
Oncorhynchus nerka	2	Yes [53]	No	GSD+TE	-	-	-	-	-	-	-	0
Menidia menidia	1	No	Yes	TSD	10	113.79	−995.21	−0.90	70.21	1/8	<0.0001	1
Menidia peninsulae	1	No	Yes	TSD	20	−27.81	2.74	0.58	24.61	1/18	0.0001	1
Odontesthes argentinensis	1	No	Yes	TSD	9	−55.62	3.91	0.67	14.42	1/7	0.0067	1
Odontesthes bonariensis	1	No	Yes	TSD	6	−182.40	9.39	0.98	242.90	1/4	<0.0001	1
Odontesthes hatcheri	1	No	No	GSD+TE	-	-	-	-	-	-	-	0
Oryzias latipes	(1)	Yes [54]	No	GSD+TE	-	-	-	-	-	-	-	0
Limia melanogaster	1	No	Yes	TSD	9	−22.15	2.60	0.69	15.29	1/7	0.0058	1
Poeciliopsis lucida	1	No	Yes	TSD	21	−139.50	7.21	0.76	60.91	1/19	<0.0001	1
Poecilia sphenops	(1)	Yes [55]	Yes	GSD+TE	-	-	-	-	-	-	-	0
Sebastes schlegeli	1	No	No	GSD+TE	-	-	-	-	-	-	-	0
Dicentrarchus labrax	1/2	No	No	GSD+TE	-	-	-	-	-	-	-	0
Apistogramma spp. (33 spp.)†	1×33	No	Yes	TSD	93	−75.93	4.78	0.75	283.60	1/91	<0.0001	1×33
Oreochromis aureus	1	Yes [56]	Yes	GSD+TE	-	-	-	-	-	-	-	0
Oreochromis niloticus	1/2	Yes [57]	Yes	GSD+TE	-	-	-	-	-	-	-	0
Oreochromis mosambicus	1	Yes [58]	Yes	GSD+TE	-	-	-	-	-	-	-	0
Paralichthys olivaceus	3	Yes [59]	Yes	GSD+TE	-	-	-	-	-	-	-	0
Paralichthys lethostigma	3	No	No	GSD+TE	-	-	-	-	-	-	-	0
Pseudopleuronectes yokohamae	1	Yes [60]	No	GSD+TE	-	-	-	-	-	-	-	0
Verasper moseri	1	No	No	GSD+TE	-	-	-	-	-	-	-	0

Abbreviations: TSD, temperature-dependent sex determination; GSD+TE, genotypic sex determination plus temperature effects; RTD, range of temperature during development under natural conditions; n, number of sex ratio datapoints (see Table S1 for references); r^2 is the correlation coefficient of the regression between temperature and sex ratio produced, whereas F, DFn, DFd and *P* indicate the value of the F-test, the degrees of freedom of the numerator and denominator and the significance, respectively, to determine whether the slope differs from zero, thus indicating that there was a significant effect of temperature on sex ratios. Notes: *Patterns of sex ratio response to temperature: 1, more males at high temperatures; 2, more males at low temperatures; 3, more males at low and high temperatures, as previously assigned based on refs. [12–16]; (1) indicates that pattern 1 was not explicitly assigned but that it could be deduced from the data (see Table S1); 0, TSD not supported by data, i.e., thermal effects on GSD (GSD+TE). †Average of the 33 species shown in Table S1.

The objective of this study was to assess the prevalence of TSD in fish by taking the species where this type of sex determining mechanism has been claimed and applying a series of proposed criteria to discern true cases of TSD from cases of GSD+TE. These included checking for the presence of sex chromosomes and determining whether the temperature used to elicit a change in sex ratios was ecologically relevant, i.e., a temperature that the species usually experiences in nature during the thermosensitive period. We found that TSD is far less widespread that currently thought. We also found that species who actually have TSD exhibit only one single response pattern, not three, producing highly male-biased sex ratios in response to even small increases in temperature. Thus, in one hand, by defining the species that actually have TSD, this study contributes to our understanding of the evolution of sex determining mechanisms. On the other hand, it reports previously unaccounted possible effects of global warming on fish sex ratios.

Materials and Methods

Species Selection

The 59 species analyzed in this study include all those gonochoristic fishes for which TSD has been explicitly or implicitly assumed as reported in published reviews on the subject [12]–[16], as well as in later publications in the primary literature. The species are representative of freshwater, estuarine and marine ecosystems. The only hermaphroditic species where TSD has been claimed, the self-fertilizing cyprinodont Kryptolebias (Rivulus) marmoratus, was not included in our study. In this species, there are no females; essentially all individuals develop as hermaphrodites. Exposure to low temperature during early development increases the proportion of gonochoristic males from ~3 to 72% [20]. Similarly, the Southern brook lamprey, Ichthyomyzon gagei, and the eels, including the American eel, Anguilla rostrata, were not included because the circumstantial evidence available so far points to growth-dependent sex differentiation [21] rather than to TSD in these species [22], [23].

Data Collection

For each species analyzed, field data, including the range of natural temperature in which the species can live (RNT), the range of temperature during development in the wild (RTD) as well as the lethal temperature (LT), when available, were obtained from ad hoc reviews, e.g., [24], Fishbase [25], or specific sources. Experimental (mostly laboratory) data were also compiled from the primary literature.

Diagnosis of Temperature-Dependent Sex Determination (TSD) as Opposed to Genotypic Sex Determination Plus Temperature Effects (GSD+TE)

To determine the actual prevalence of TSD in fish and to furnish robust patterns of sex ratio response to temperature, we have used a comparative analysis consisting of the application of two independent criteria to identify the presence of TSD (Fig. 2). The first is that of Valenzuela et al. [2], which: (i) stresses that the presence of chromosomal systems of sex determination such as XX/XY or WZ/ZZ, that imply consistent genetic differences between sexes, constitutes a very strong evidence of the presence of GSD, and thus it is extremely unlikely that species with these chromosomal systems have TSD. The evidence for sex chromosomes may have been obtained with direct (karyotyping, banding) or indirect methods (e.g., progeny analysis of sex-linked traits, mating experiments or crosses with sex-reversed fish); (ii) considers induced sex ratio shifts that occur only at extreme (but not defined), ecologically irrelevant temperatures, not proof of TSD. The second criteria, which complements the former, is that of Conover [16], which establishes that in order for a species to have TSD, sex ratio shifts in response to temperature fluctuations must occur within a certain range, defined as the range of natural temperature (RNT) in which the species lives. However, since the thermosensitive period in the vast majority of fish examined so far is usually located during early development, and particularly during the larval stages [12]–[16], a modification of the criterion in Conover [16] was used for final assignment of TSD to a given species. Therefore, only those species for which sex ratio shifts occurred not within the RNT but instead within the RTD -the range of temperatures during the period of development that usually includes the thermosensitive period- were considered candidates for having TSD. Particularly in seasonally breeding species of temperate latitudes, RTD is contained within RNT but the opposite is not true. Thus, response within the RNT is not enough evidence for TSD. Using the RTD instead of the RNT has the additional advantage of incorporating additional criteria of Valenzuela et al. [2] other than the absence of sex chromosomes, since it facilitates excluding cases of sex reversals induced at extreme temperatures, another possible source of confusion. When a species has a sex chromosomal system and/or sex ratio response to temperature occurring at extreme temperatures (sometimes close to the LT), and definitively outside the RTD (e.g., Fig. 3B), and hence ecologically irrelevant, then TSD is essentially very unlikely. These instances are more appropriately referred to as cases of naturally- or experimentally-induced alterations of genotypic sex determination or genotypic sex determination plus temperature effects (GSD+TE) [2], [16] rather than TSD. Thus, for any given species to have TSD, it should fulfill both of the following two conditions: 1) not having sex chromosomes, and 2) have sex ratio response to temperature within the RTD (Fig. 2). The possible error in proceeding in this manner is negligible and smaller than doing the opposite, i.e., classifying a species as having TSD that has sex chromosomes, which in most cases is strong evidence of GSD, and/or that exhibits sex ratio shifts at artificially high or low temperatures, which is ecologically irrelevant.

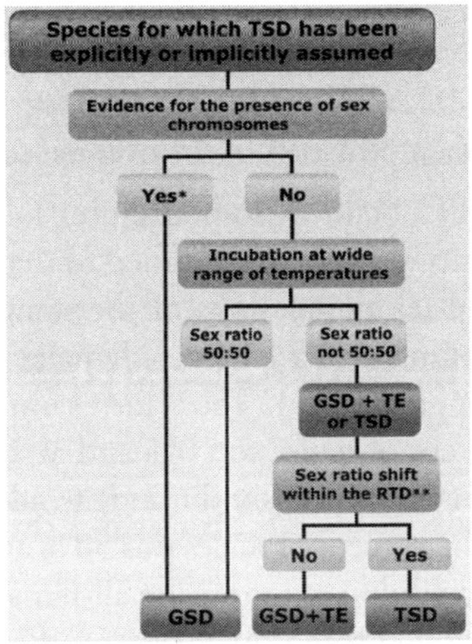

Figure 2. Set of criteria used to determine the presence of temperature-dependent sex determination (TSD) as opposed to genotypic sex determination (GSD), and to distinguish TSD from thermal effects on GSD (GSD+TE). This algorithm is based on the criteria of Valenzuela et al. (2003), and incorporates a modification of the criteria of Conover (2004). See text in the Materials and Methods section for a complete explanation. *Indicates that the evidence for a sex chromosomal system may come from direct (karyotyping, banding) or indirect methods (e.g., progeny analysis of sex-linked traits, mating experiments or crosses with sex-reversed fish). **Indicates that the sex ratio shift must occur within the range of developmental temperatures during development that includes the thermosensitive period (RTD) regardless of whether there is response within the range of natural temperatures where the species lives.

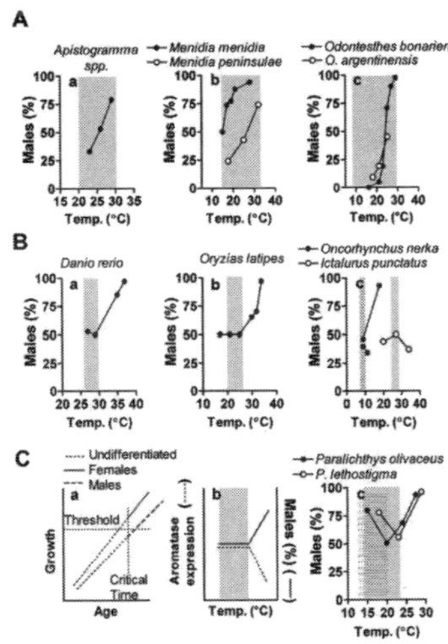

Figure 3. Patterns of sex ratio response to temperature in fish. A. Examples of authentic cases of TSD following pattern 1, more males with increasing temperatures. Sex ratio shifts occur within the range of temperature (shaded areas) normally experienced by fish in the wild. B, Examples of false cases of TSD. Sex ratio shifts only occur at extreme temperatures, and thus represent thermal effects on GSD (a, b). Formerly proposed pattern 2 (c), fewer males at high temperature, is not supported by re-analysis of data. C, Formerly proposed pattern 3, more males at extreme temperatures, can be explained from the combination of two effects unrelated to TSD: slow growing fish at low temperature differentiating as males (a), and the inhibition of aromatase at high temperature causing sex-reversal of genetic females (b). When combined, the two effects result in the observed pattern (c).

Statistical Analysis

Sex ratio deviations from 1:1 in Ictalurus punctatus were checked by applying the Chi-square 1test [26] to data provided in the original source [27].

Sex ratio data originally obtained from monosex (all-female) populations exposed to different temperatures were transformed to make them comparable with data obtained with mixed-sex populations of the same species by applying the following formula: Percent males in a 1:1 (male:female) population = 50+(percent males in the all-female population/2). Thus, for example, an all-female population that at 20°C the percent of males was 0% and at 28°C was 66% (indicating that two thirds of the females were masculinized) would be equivalent to an 1:1 population that at 20°C the percent of males was 50% and at 28°C was 50+(66/2) = 83%. Notice that the possibility of producing all-female stocks is indicative that the species in question has a chromosomal system of sex determination, usually of the XX/XY type, thus suggesting the presence of GSD rather than of TSD, as is demonstrated.

The presence of a significant sex ratio response to temperature within the RTD and the verification of the presence of TSD in species diagnosed as having such mechanism of sex determination after applying the criteria explained above was carried out as follows: First, we tested if there was a statistically significant relationship between sex ratio produced and temperature by using the Spearman rank correlation coefficient method. If so, then we compared the slope with the F-test [26] to check whether it was different from zero.

In a few instances, more than one intermediate temperature has been tested. For the regressions, all the available intermediate temperatures were used from the original sources. Likewise, each one of the 33 species of the genus Apistogramma studied by Römer and Beisenherz [28] was checked individually and the presence of TSD also confirmed statistically on a one-by-one basis, but for simplicity an average result representative of all of them is presented.

In all cases, sex ratio data expressed as percentages (i.e., 100·p, where p is the proportion of males) were arcsin transformed (arcsin of the square root of p) prior to statistical analysis [26]. Statistical analyses and graphs were carried out with the aid of StatGraphics v. 5.1 and Graphpad Prism Software v.4.0.

Results

Our results show that of the 53–55 species (depending on the authors) previously assigned to pattern 1, the 33 cichlid species of the genus Apistogramma indeed exhibit pattern 1 (Fig. 3A a; Table 1) fulfilling the criteria for the assignment of

TSD. However, only seven other species of the remaining 20–22 adhere to pattern 1 and have TSD (Fig. 3A b,c). In all but one of the species with TSD the best fit to the experimental data on sex ratio response to temperature was obtained with a linear regression (Y = a+bX). In Menidia menidia, however, the best fit was obtained with a reciprocal-X model (Y = a+b/X) (Fig. 4). Included among the species that did not pass the criteria to be diagnosed as true cases of TSD are some established research models such as the zebrafish (Danio rerio) and the medaka (Oryzias latipes) (Fig. 3B a,b).

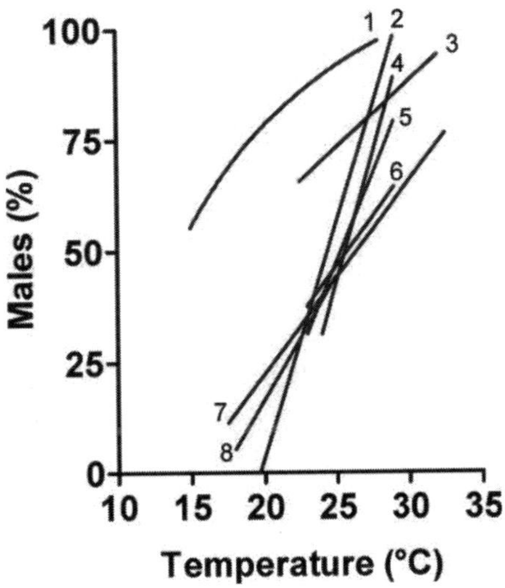

Figure 4. Patterns of sex ratio response to temperature in species of fish with TSD. In all cases, higher temperatures imply a higher number of males produced. Key: 1, Mendia menidia; 2, Odontesthes bonariensis; 3, Hoplosternum littorale; 4, Poeciliopsis lucida; 5, average of the 33 Apistogramma species; 6, Limia melanogaster; 7, Menidia peninsulae; 8, Odontesthes argentinensis.

Regarding pattern 2, analysis of the original data [27] of channel catfish (Ictalurus punctatus) in fact showed no differences with respect to the 1:1 sex ratio (Chi-square 1test = 1.42, P = 0.233) (Fig. 3B c). Likewise, additional experiments in sockeye salmon (Oncorhynchus nerka) reported in Azuma et al. [29] (Fig. 3B c) evidenced the presence of pattern 1 instead of pattern 2, as it had been previously suggested [30]. However, both the channel catfish and the sockeye salmon have sex chromosomes and tested temperatures fall outside the natural range (Table 1). Therefore, these are cases of GSD+TE, not of TSD.

Regarding pattern 3, the two flatfishes previously assigned to this pattern (Fig. 3C), the olive flounder (Paralichthys olivaceus) [31] and the Southern flounder (P. lethostigma) [32], each failed one of the TSD-determining criteria (Table 1).

Based on the relationship between temperature and sex ratio produced as shown in Table 1, we calculated that fish species with TSD exhibit an average (mean±S.E.M.) pivotal temperature (PT, temperature that produces balanced sex ratios) of 23.3±1.5°C (Table 2). Then, in the scenario of global warming, we took two temperature increases: 1.5 and 4°C, representative of a very likely increase in temperature of water bodies in the upcoming decades and of the maximum predicted increase by the end of this century [33], respectively. With an increase of just 1.5°C, the average number of males in the species with TSD would increase to 61.7±2.1%, and with an increase of 4°C, the average number of males would increase to 78.0±4.1%, i.e., the sex ratios (male:female) would shift from 1:1 to ~2:1 and to ~3:1, respectively (Table 2).

Table 2. Pivotal temperature in fish species with TSD and predicted sex ratio shifts with temperature increases.

Species	Pivotal temp. (°C)	Percent of sexes (♂:♀) at pivotal temp.+1.5°C	Percent of sexes (♂:♀) at pivotal temp.+4°C
*Apistogramma spp**	25.3	62 : 38	81 : 19
Hoplosternum littorale	18.8	56 : 44	65 : 35
Limia melanogaster	25.8	57 : 43	68 : 32
Menidia menidia	14.5	61 : 39	75 : 25
Menidia peninsulae	26.6	57 : 43	69 : 31
Odontesthes argentinensis	25.7	60 : 40	76 : 24
Odontesthes bonariensis	24.2	73 : 27	98 : 2
Poeciliopsis lucida	25.6	68 : 32	92 : 8
Pivotal temp. (mean±S.E.M.)	23.3±1.5	-	-
Percent males (mean±S.E.M.)	-	61.7±2.1	78.0±4.1

*Average of the 33 species shown in Table S1.

Discussion

Prevalence of TSD in Fish and Response Patterns

In reptiles, where TSD was first discovered in vertebrates, this mechanism of sex determination is now well established (see the book by Valenzuela and Lance [34], for reviews). In contrast, in fish, the absolute number of studies is more limited and, significantly, only few of them, concerning the Atlantic silversides, have been carried by samplings in the wild [16], while most have been carried out under controlled laboratory conditions. This may probably reflect the difficulty of sampling fish at different developmental stages in the wild and, especially, correlating environmental variables during critical thermosensitive periods with resulting sex ratios when adults. However, despite these limitations, this situation did not prevent that TSD was until now considered a widespread mechanism of sex determination in fish. Further, based on sex ratio response to temperature, fish species where TSD had been claimed had been grouped into three response patterns.

The analysis of sex ratio response to temperature, considering the scope of such response as well as the presence or not of sex chromosomes, carried out in the present study indicated that many species where TSD had been claimed before are in fact GSD species affected by temperature, i.e., cases of GSD+TE. In GSD+TE species, temperature rather than being the external environmental factor controlling sex determination is capable of affecting the process of gonadal sex differentiation under some circumstances. This distinction is not trivial nor semantic since, according to the canonical definition [1], in TSD species the first ontogenetic difference between sexes is an environmental one (temperature), whereas in GSD+TE species sex determination remains under genotypic control.

Our results support the presence of pattern 1 of sex ratio response to temperature (more males with increasing temperature) but the number of species with TSD is much lower than previously considered and concern mainly species of the families Cichlidae followed by species of the family Atherinopsidae. In addition, we have demonstrated that pattern 2 of sex ratio response to temperature does not exist in fish.

Regarding pattern 3, we propose that this pattern is the result of two independent effects unrelated to TSD (Fig. 3C). First, since exposure to low temperatures decreases growth rates in poikylothermic animals, the increase in males at low temperatures is likely the result of male development according to the threshold model for growth-dependent sex differentiation [21]. Briefly, applied here this model states that when a critical time is reached during development, a sexually undifferentiated gonad will develop as an ovary or as a testis depending on whether it has attained a certain size above or below a threshold, respectively (Fig. 3C a). In fact, a reduction in the number of females was observed among the lower growing fish in the olive flounder, one of the species previously assigned to pattern 3 [35]. Although initial exposure to low temperatures in some cases favors female sex differentiation (as in pattern 1), it is now known that if such exposure is prolonged, thus delaying growth, then male sex differentiation occurs [36]. The preponderance of males at low temperatures also coincides with the left half of pattern 2. Therefore, this pattern sometimes has been also erroneously assigned to species such as the sea bass (Dicentrarchus labrax) (Table 1), where growth-dependent sex differentiation occurs [36]. The other effect, the increase in males at high temperatures in species previously assigned to pattern 3, is likely the result of sex-reversal of females as a consequence of the inhibition of aromatase (Fig. 3C b), the enzyme that produces estrogens essential for female sex differentiation in fish [37]. When combined, the two effects produce pattern 3 (Fig. 3C c). In addition, the observed sex ratio response to temperature, especially in the Southern flounder, partly occurs outside the RTD, thus not being representative of true TSD. The inhibition of aromatase at high temperatures –and the consequent increase in the number of males- has also been reported in some species without TSD [15], [38], also explaining why they were assigned to pattern 1, and, interestingly, also seen in many species of reptiles [34], [39]. Thus, we find that only pattern 1 of sex ratio response to temperature is present in fish with TSD (Figure 4), since analysis of the available data does not support the existence of patterns 2 and 3, as accepted until now. This contrasts with the accepted existence of three response patterns in reptiles [34], although perhaps they should be revisited, as done in this study with fish. Further, it has been recognized that the prevalent pattern in reptiles with TSD is pattern Ib [40], which is the equivalent of pattern 1, found to be the only one actually present in fish.

The results of the present study have implications for our understanding of the evolution of vertebrate sex determining mechanisms. They still agree with the view that TSD has evolved independently many times [1], [2], [40], but we find TSD to be present in only four orders, which include only three of the seven used by Mank et al. [10] to discuss the evolution of sex determining mechanisms specifically in fish. Thus, there is no close relationship among the families where TSD is present (Fig. 5), and many species within the same families are well known for having GSD, suggesting that TSD is clearly the exception in fish sex determination. The phylogenetic distribution suggests that, when it occurs, TSD in fish is a derived rather than an ancestral mechanism. However, there are at least 27,977 known species of 1teleosts [41] and although admittedly the available data on sex determination are a good representation of the biodiversity, it has to be borne in mind that the number of species examined is still a minority so far. Thus, the picture shown here may change one day as new species are examined in regards to their sex determination mechanisms.

What is the reliability of the original data used to assign TSD in the different species that survived our analysis? In the species of the F. Atherinopsidae (silversides) the evidence seems robust [11], [13], [16], [18], [19], but it should be remembered that the species of the genus Odontesthes data has been obtained from laboratory experiments. In the genus Apistogramma (South American Cichlids), TSD was demonstrated in many species and thus also seems well established, although the evidence gathered so far originates from a single study [28]. The same situation applies for the atipa, Hoplosternum littorale, an Amazonian freshwater fish, where several batches of eggs were used and tested temperatures corresponded to the natural fluctuation; however, data originates from a single study [42]. In contrast, data concerning Poeciliopsis lucida, a freshwater fish from Mexico, not only comes from a single study [43] but also the two strains used were highly inbred, one responding to temperature and the other not. The former passed the criteria for being classified as TSD but whether similar results would be obtained with other strains remains to be determined. Further, this is a viviparous species, and viviparity seems incompatible with the requirements to develop TSD [2]. Thus, further research would be necessary to establish whether P. lucida has populations with GSD and others with TSD or whether it is a GSD+TE species.

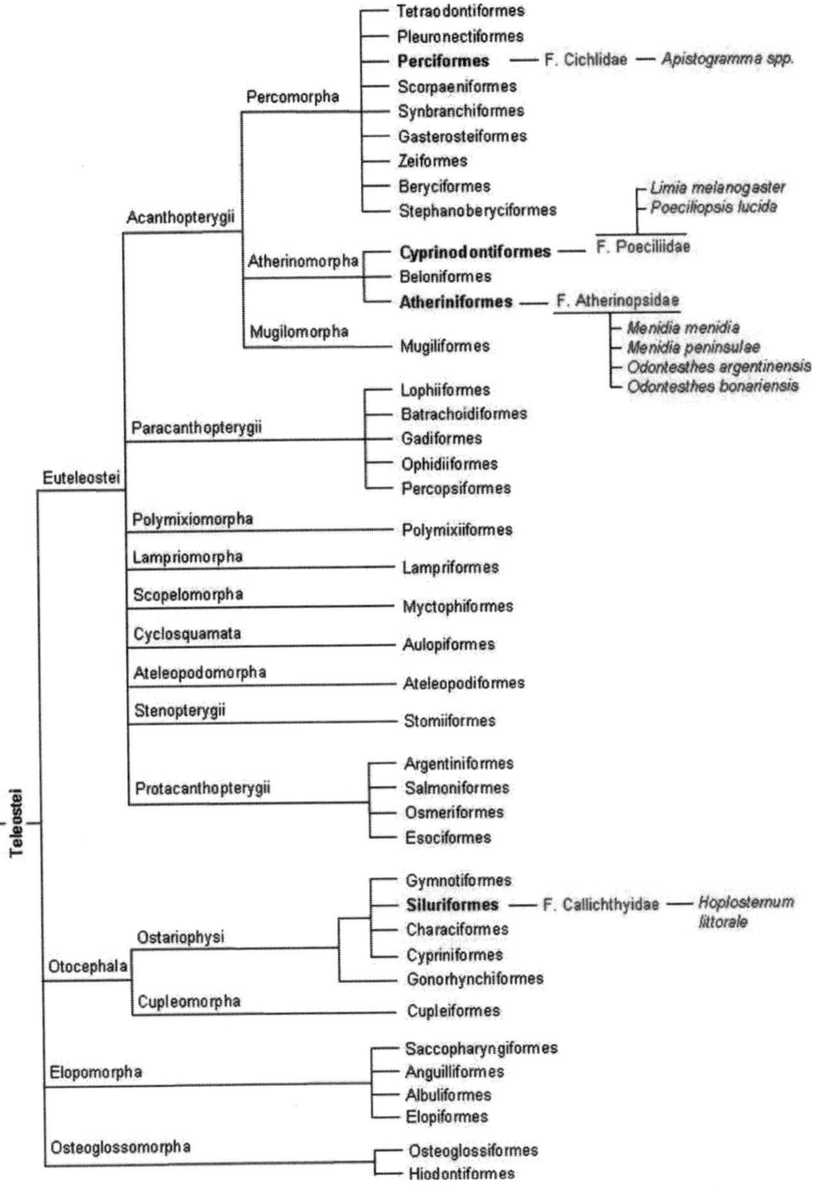

Figure 5. Distribution of temperature-dependent sex determination (TSD) in fish. Orders, families and species with TSD are marked in color. Teleost phylogeny based on Nelson [41].

The criteria used here allow the identification of the presence of TSD in a given species. However, this does not exclude the possibility that these species may also have populations with GSD. Therefore, populations with GSD and TSD may co-exist in a single species [16]. Here it is interesting to notice that even in these cases, the pattern of sex ratio response to temperature is invariably pattern 1. On the other hand, it should be noted that the identification of sex chromosomes, particularly if they are homomorphic, can depend on the sensitivity of the method used to search for them. Thus, the number of species with TSD may be further reduced in the future as new technical developments, such as new fluorescent molecular probes, increase our ability to detect sex chromosomes.

The tilapias (genus Oreochromis) deserve special attention, not only because their importance for aquaculture but also because some of them constitute established research models where many studies on the effects of temperature in fish sex differentiation have been carried out [12]–[15]. Tilapias did not pass our criteria to be considered TSD species because there are genetic differences between sexes that can be discerned with direct and indirect methods. In fact, currently the genetic sex determinism of tilapias is becoming well understood [44]. Further, recent studies have shown that some tilapia populations adapted to extreme conditions can tolerate temperatures close to 40°C and have rightly pointed out that high temperature influences the normal course of sex differentiation with the resulting masculinization of genetic females [45]. Thus, in accordance with the definitions used here and elsewhere [2], tilapias, then, are a prime example of GSD+TE species, but not of TSD species. To avoid confusion, then, if for a given species there is no compelling evidence of the presence of TSD is better to use the term "temperature effects on sex ratios" or "temperature effects on sex differentiation," but not "temperature-dependent sex determination."

TSD in Fish and Climate Change

How species with TSD will respond to current rapid climate change is a timely question [3], [9], [40]. Some data is available for sea turtles with TSD [8], [9] but is non-existent for fish. Based on the information gathered in the present study some predictions can be made, although it should be taken into account that they are based on a simple linear correlation between temperature and resulting sex ratio. However, in the absence of field data, they are the best educated guess one can make based on the data available so far.

The species identified as having TSD in this study constitute a heterogeneous group since they include both freshwater and marine species living also both in low and high latitudes. Some of them are typically eurithermal while others are stenothermal and, further, they exhibit different reproductive strategies.

Similarly, global warming is not a heterogeneous process, since it affects different parts of the Earth differently. Globally, however, mean temperatures of water bodies are projected to increase by up to ~4°C by the end of this century according to plausible global change scenarios [33]. Even modest changes of 1–2°C may significantly skew the sex ratio, as already shown in field studies with turtles [3] and sea turtles [9]. In fish, observations made with M. menidia eggs collected from the wild have shown that differences of 2°C during the thermosensitive period can result in sex ratio shifts from 50% to 69% males [19].

Thus, the number of females in species with TSD, some of which are of economic or recreational importance, could decrease. One such species is the Argentinean silverside (O. bonariensis), where recent studies suggest thermal effects on gonadal development already occurring in natural populations [46]. The species with the least pronounced slopes in the relationship between temperature and the sex ratio produced would be less affected or not affected at all. In O. bonariensis, an increase of just 1.5°C could shift the percent males from an average of 50% to ~73%, that is, from 1:1 to ~3:1. Since the reproductive potential of many fish communities is determined by the number of females available for egg production [47], highly male-biased sex ratios would likely affect population structure and the viability of sensitive stocks.

Potential temperature effects on sex ratios could be difficult to quantify if they are mitigated by other global warming-induced effects, including species distribution shifts [4]. In addition, skewed sex ratios may favor frequency-dependent selection of the less abundant sex, the evolution of TSD towards its disappearance or adjustments in the pivotal temperature [48]. In contrast to past, naturally occurring fluctuations of global temperature, the current climate change event with anthropogenic influences is characterized by its fast pace [33]. Thus, it has been suggested that sensitive species, including species with TSD, could not adapt fast enough to the rapid change in temperatures brought by the new thermal situation [3].

It should be noted that the impact of temperature on sex ratios could also affect species with identifiable sex chromosomes (by causing sex reversal) provided that those effects occur at temperatures within the natural range, or the new shifted range. However, at this point there is insufficient information to determine if, by virtue of their possible higher sensitivity to temperature, species with TSD are better indicators of the impacts of climate change on sex ratios than GSD+TE species.

Conclusions

In this study, we performed an analysis of field and laboratory data related to fish species for which TSD was assumed. By applying a series of criteria accepted to ascertain the actual presence of TSD, we can reasonably affirm that, excluding the species of the genus Apistogramma, in approximately 75% (19 out of 26) of the species considered to have TSD so far, observed sex ratio shifts at extreme temperatures are most likely the consequence of thermal effects on GSD rather than proof of the existence of TSD. Thus, there may be species in which TSD has not yet been discovered but, contrary to the prevailing view, TSD in fish is not as widespread as currently thought, and, importantly, only one general pattern of sex ratio response to temperature exists. However, species which do possess TSD, or species with GSD+TE, may compromise their viability by diminishing the number of females in response to even small increases in water temperatures.

Acknowledgements

We would like to thank the following persons: M. Blázquez, J. Cerdà, N. Mrosovsky, S. Sarre, M. Schartl and J. Viñas for helpful comments; F. Mayou for providing advice on statistical analyses; and J.I. Fernandino for providing information on Odontesthes sp.

Author Contributions

Conceived and designed the experiments: FP. Performed the experiments: NO. Analyzed the data: NO FP. Wrote the paper: FP.

References

1. Bull JJ (1983) Evolution of sex determining mechanisms. Menlo Park: Benjamin/Cummings.
2. Valenzuela N, Adams DC, Janzen FJ (2003) Pattern does not equal process: Exactly when is sex environmentally determined? Am Nat 161: 676–683.
3. Janzen FJ (1994) Climate change and temperature-dependent sex determination in reptiles. PNAS 91: 7487–7490.
4. Perry AL, Low PJ, Ellis JR, Reynolds JD (2005) Climate change and distribution shifts in marine fishes. Science 308: 1912–1915.
5. O'Connor MI, Bruno JF, Gaines SD, Halpern BS, Lester SE, et al. (2007) Temperature control of larval dispersal and the implications for marine ecology, evolution, and conservation. PNAS 104: 1266–1271.

6. Pörtner HO, Knust R (2007) Climate change affects marine fishes through the oxygen limitation of thermal tolerance. Science 315: 95–97.
7. Biro PA, Post JR, Booth DJ (2007) Mechanisms for climate-induced mortality of fish populations in whole-lake experiments. PNAS 104: 9715–9719.
8. Kamel SJ, Mrosovsky N (2006) Deforestation: risk of sex ratio distortion in Hawksbill sea turtles. Ecol Appl 16: 923–931.
9. Hawkes LA, Broderick AC, Godfrey MH, Godley BJ (2007) Investigating the potential impacts of climate change on a marine turtle population. Global Change Biol 13: 923–932.
10. Mank JE, Promislow DEL, Avise JC (2006) Evolution of alternative sex-determining mechanisms in teleost fishes. Biol J Linn Soc 87: 83–93.
11. Conover DO, Kynard BE (1981) Environmental sex determination - Interaction of temperature and genotype in a fish. Science 213: 577–579.
12. Baroiller JF, Guigen Y, Fostier A (1999) Endocrine and environmental aspects of sex differentiation in fish. Cell Mol Life Sci 55: 910–931.
13. Strüssmann CA, Patiño R (1999) Sex Determination, Environmental. In: Knobil E, Neill JD, editors. Encyclopedia of Reproduction. New York: Academic Press. pp. 402–409.
14. Baroiller JF, D'Cotta H (2001) Environment and sex determination in farmed fish. Comp Biochem Physiol C 130: 399–409.
15. Devlin RH, Nagahama Y (2002) Sex determination and sex differentiation in fish: an overview of genetic, physiological, and environmental influences. Aquaculture 208: 191–364.
16. Conover DO (2004) Temperature-dependent sex determination in fishes. In: Valenzuela N, Lance V, editors. Temperature-dependent sex determination in vertebrates. Washington: Smithsonian Books. pp. 11–20.
17. Sarre SD, Georges A, Quinn A (2004) The ends of a continuum: genetic and temperature-dependent sex determination in reptiles. Bioessays 26: 639–645.
18. Conover DO, Heins SW (1987) Adaptive variation in environmental and genetic sex determination in a fish. Nature 326: 496–498.
19. Conover DO, Heins SW (1987) The environmental and genetic components of sex ratio in Menidia menidia (Pisces, Atherinidae). Copeia 1987: 732–743.
20. Harrington RW (1967) Environmentally controlled induction of primary male gonochorists from eggs of the self-fertilizing hermaphroditic fish, Rivulus marmoratus Poey. Biol Bull 132: 174–199.

21. Kraak SBM, de Looze EMA (1992) A new hypothesis on the evolution of sex determination in vertebrates; big females ZW, big males XY. Neth J Zool 43: 260–273.
22. Beamish FWH (1993) Environmental sex determination in southern brook lamprey, Ichthyomyzon gagei. Can J Fish Aquat Sci 50: 1299–1307.
23. Krueger WH, Oliveira K (1999) Evidence for environmental sex determination in the American eel, Anguilla rostrata. Env Biol Fish 55: 381–389.
24. Jobling M (1981) Temperature tolerance and the final preferendum - rapid methods for the assessment of optimum growth temperatures. J Fish Biol 19: 439–455.
25. Froese R, Pauly D (2008) FishBase. Available: http://www.fishbase.org.
26. Zar JH (1984) Biostatistical analysis. New Jersey: Prentice-Hall.
27. Patino R, Davis KB, Schoore JE, Uguz C, Strüssmann CA, et al. (1996) Sex differentiation of channel catfish gonads: Normal development and effects of temperature. J Exp Zool 276: 209–218.
28. Römer U, Beisenherz W (1996) Environmental determination of sex in Apistogramma (Cichlidae) and two other freshwater fishes (Teleostei). J Fish Biol 48: 714–725.
29. Azuma T, Takeda K, Doi T, Muto K, Akutsu M, et al. (2004) The influence of temperature on sex determination in sockeye salmon Oncorhynchus nerka. Aquaculture 234: 461–473.
30. Craig JK, Foote CJ, Wood CC (1996) Evidence for temperature-dependent sex determination in sockeye salmon (Oncorhynchus nerka). Can J Fish Aquat Sci 53: 141–147.
31. Yamamoto E (1999) Studies on sex-manipulation and production of cloned populations in hirame, Paralichthys olivaceus (Temminck et Schlegel). Aquaculture 173: 235–246.
32. Luckenbach JA, Godwin J, Daniels HV, Borski RJ (2003) Gonadal differentiation and effects of temperature on sex determination in southern flounder (Paralichthys lethostigma). Aquaculture 216: 315–327.
33. IPCC (2007) Climate Change 2007: The Physical Science Basis.; In: Solomon S, Qin D, editors. Cambridge: Cambridge Univ. Press.
34. Valenzuela N, Lance V (2004) Temperature-dependent sex determination in vertebrates. Washington: Smithsonian Books.
35. Tabata K (1995) Reduction of female proportion in lower growing fish separated from normal and feminized seedlings of hirane Paralichthys olivaceus. Fish Sci (Tokyo) 61: 199–201.

36. Piferrer F, Blázquez M, Navarro I., Gonzalez A (2005) Genetic, endocrine, and environmental components of sex determination and differentiation in the European sea bass (Dicentrarchus labrax L.). Gen Comp Endocrinol 142: 102–110.

37. Piferrer F, Zanuy S, Carrillo M, Solar II, Devlin RH, et al. (1994) Brief treatment with an aromatase inhibitor during sex differentiation causes chromosomally female salmon to develop as normal, functional males. J Exp Zool 270: 255–262.

38. Uchida D, Yamashita M, Kitano T, Iguchi T (2004) An aromatase inhibitor or high water temperature induce oocyte apoptosis and depletion of P450 aromatase activity in the gonads of genetic female zebrafish during sex-reversal. Comp Biochem Physiol A 137: 11–20.

39. Crews D (1994) Temperature, steroids and sex determination. J Endocrinol 142: 1–8.

40. Janzen FJ, Krenz JG (2004) Phylogenetics: which was first TSD or GSD? In: Valenzuela N, Lance V, editors. Temperature-dependent sex determination in vertebrates. Washington: Smithsonian Books. pp. 121–130.

41. Nelson JS (2006) Fishes of the World, 4 ed. New Jersey: John Wiley and Sons.

42. Hostache G, Pascal M, Tessier C (1995) Influence de la température d'incubation sur le rapport mâle, femelle chez l'atipa, Hoplosternum littorale Hancock (1828). Can J Zool 73: 1239–1246.

43. Sullivan JA, Schultz RJ (1986) Genetic and environmental basis of variable sex-ratios in laboratory strains of Poeciliopsis lucida. Evolution 40: 152–158.

44. Cnaani A, Lee BY, Zilberman N, Ozouf-Costaz C, Hulata G, et al. (2008) Genetics of sex determination in tilapiine species. Sex Dev 2: 43–54.

45. Bezault E, Clota F, Derivaz M, Chevassus B, Baroiller JF (2007) Sex determination and temperature induced sex differentiation in three natural populations of Nile tilapia (Oreochromis niloticus) adapted to extreme temperature conditions. Aquaculture 272S1: S3–S16.

46. Cornejo AM (2003) Esterilidad en el pejerrey Odontesthes bonariensis en ambientes naturales. Biol Acuát 20: 19–26.

47. Parker K (1980) A direct method for estimating northern anchovy, Engraulis mordax, spawning biomass. Fish Bull 78: 541–544.

48. Conover DO, Voorhees DAV, Ehtisham A (1992) Sex ratio selection and changes in environmental sex determination in laboratory populations of Menidia menidia. Evolution 46: 1722–1730.

49. Yamamoto T, Kajishima T (1969) Sex-hormone induction of reversal of sex differentiation in the goldfish and evidence for its male heterogamety. J Exp Zool 168: 215–222.

50. Fujioka Y (1998) Survival, growth and sex ratios in gynogenetic diploid honmoroko. J Fish Biol 59: 851–861.

51. Nomura T, Arai K, Hayashi T, Suzuki R (1998) Effect of temperature on sex ratios of normal and gynogenetic diploid loach. Fish Sci 64: 753–758.

52. Davis KB, Simco BA, Goudie CA, Parker NC, Cauldwell W, et al. (1990) Hormonal sex manipulation and evidence for female homogamety in channel catfish. Gen Comp Endocrinol 78: 218–223.

53. Ueda T, Ojima Y (1984) Sex chromosomes in the Kokanee salmon Oncorhynchus nerka. Bull Jap Soc Sci Fish 50: 1495–1498.

54. Nanda I, Kondo M, Hornung U, Asakawa S, Winkler C, et al. (2002) A duplicated copy of DMRT1 in the sex-determining region of the Y chromosome of the medaka, Oryzias latipes. PNAS 99: 11778–11783.

55. Nanda I, Schartl M, Epplen JT, Feichtinger W, Schmid P (1993) Primitive sex chromosomes in poeciliid fishes harbor simple repetitive sequences. J Exp Zool 265: 301–308.

56. Mair GC, Scott AG, Penman DJ, Skibinski DOF, Beardmore JA (1991) Sex determination in the genus Orecochromis: 2. Sex reversal, hybridisation, gynogenesis and triploidy in O. aureus Steindachner. Theor Appl Genet 82: 153–160.

57. Harvey SC, Boonphakdee C, Campos-Ramos R, Ezaz MT, Griffin DK, et al. (2003) Analysis of repetitive DNA sequences in the sex chromosomes of Oreochromis niloticus. Cytogenet Genome Res 101: 314–319.

58. Pandian TJ, Varadaraj K (1990) Development of monosex female Oreochromis mossambicus broodstock by integrating gynogenetic technique with endocrine sex reversal. J Exp Zool 255: 88–96.

59. Tabata K (1991) Induction of gynogenetic diploid males and presumption of sex determination mechanisms in the hirame Paralichthys olivaceus. Bull Jpn Soc Sci Fish 57: 845–850.

60. Aida S, Arai K (1998) Sex ratio in the progeny of gynogenetic diploid marbled sole Limanda yokohamae. Fish Sci 64: 989–990.

Red Fluorescence in Reef Fish: A Novel Signalling Mechanism?

Nico K. Michiels, Nils Anthes, Nathan S. Hart, Jürgen Herler, Alfred J. Meixner, Frank Schleifenbaum, Gregor Schulte, Ulrike E. Siebeck, Dennis Sprenger and Matthias F. Wucherer

ABSTRACT

Background

At depths below 10 m, reefs are dominated by blue-green light because seawater selectively absorbs the longer, 'red' wavelengths beyond 600 nm from the downwelling sunlight. Consequently, the visual pigments of many reef fish are matched to shorter wavelengths, which are transmitted better by water. Combining the typically poor long-wavelength sensitivity of fish eyes with the presumed lack of ambient red light, red light is currently considered irrelevant for reef fish. However, previous studies ignore the fact that several marine organisms, including deep sea fish, produce their own red luminescence and are capable of seeing it.

Results

We here report that at least 32 reef fishes from 16 genera and 5 families show pronounced red fluorescence under natural, daytime conditions at depths where downwelling red light is virtually absent. Fluorescence was confirmed by extensive spectrometry in the laboratory. In most cases peak emission was around 600 nm and fluorescence was associated with guanine crystals, which thus far were known for their light reflecting properties only. Our data indicate that red fluorescence may function in a context of intraspecific communication. Fluorescence patterns were typically associated with the eyes or the head, varying substantially even between species of the same genus. Moreover red fluorescence was particularly strong in fins that are involved in intraspecific signalling. Finally, microspectrometry in one fluorescent goby, Eviota pellucida, showed a long-wave sensitivity that overlapped with its own red fluorescence, indicating that this species is capable of seeing its own fluorescence.

Conclusion

We show that red fluorescence is widespread among marine fishes. Many features indicate that it is used as a private communication mechanism in small, benthic, pair- or group-living fishes. Many of these species show quite cryptic coloration in other parts of the visible spectrum. High inter-specific variation in red fluorescence and its association with structures used in intra-specific signalling further corroborate this view. Our findings challenge the notion that red light is of no importance to marine fish, calling for a reassessment of its role in fish visual ecology in subsurface marine environments.

Background

At depths below 10 m, reefs are dominated by blue-green light because seawater selectively absorbs the longer, 'red' wavelengths (600 nm and more) from downwelling sunlight (Fig. 1)[1,2]. Consequently, many reef fish have visual pigments matched to shorter wavelengths, which are transmitted better by water [3-5]. In addition, ecological studies of fish vision must correct for the spectrum available at the depth where they live [1,6,7] and therefore routinely correct spectral sensitivity measurements from the laboratory for the available (mostly downwelling) light on the reef. This reduces the relevance of red light to reef fish even more. However, this procedure ignores the fact that several marine organisms, including deep sea fish, produce their own red bioluminescence and are capable of seeing it [8,9]. The purpose of this study was (1) to see "with our own eyes" whether there is indeed a lack of red light at depth in the euphotic zone during daytime and (2) to identify the observed sources of natural red fluorescence in fish in particular. This work combines results from several studies carried out on coral reefs in the Red Sea and the Great Barrier Reef and has been supplemented by observations and measurements on fish in the laboratory.

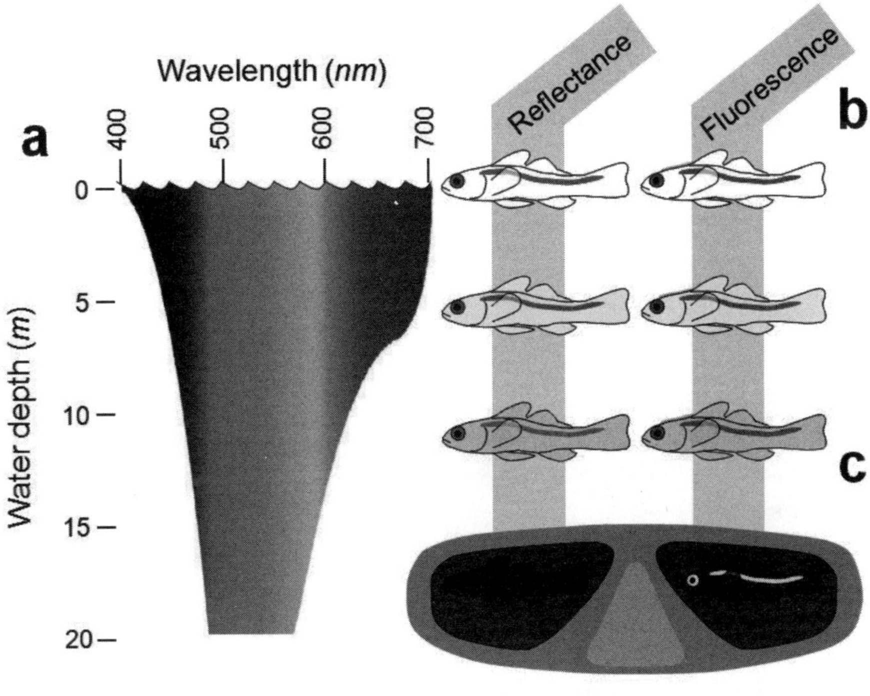

Figure 1. General introduction to light attenuation and observation of natural fluorescence in near-shore marine environments. a. The visual spectrum ranges from 400 to 700 nm at the water surface, but downwelling sunlight loses the red component (600–700 nm) rapidly within 10–15 m (modified from Pinet PR (2000) Invitation to Oceanography. Jones and Bartlett). UV and violet wavelengths are attenuated less rapidly. The attenuation with depth of spectral composition (and light intensity, not shown) varies strongly with the concentration of organic matter in the water column. b. Most red pigmentation is based on reflectance of the red component of ambient light and therefore only appears "red" when close to the surface during daytime or under broad spectral light (e.g. dive torch). Fish with this pigmentation appear dull grey in deeper water. Red fluorescent patterns, however, continue to appear reddish and bright, even in deeper water, where excitation of fluorescent pigments by shorter wavelengths induces redness. Note that red fluorescence is rarely perceived as pure red, but is mostly an enhancer of mixed colors such as pink, lilac or red brown. Even so, it remains clearly visible in deeper water as a contrast enhancer. Closer to the surface, fluorescent patterns are masked by reflective coloration (e.g. yellow and red in Eviota pellucida, Fig. 3). c. Since excitation frequencies (blue-green) are brighter than emission frequencies (red in our example) red fluorescence is best seen when viewed through a filter that blocks the excitation frequencies and only allows the emission frequencies to pass. When looking through a red filter in e.g. 20 m depth, all remaining red light must be "locally produced" through fluorescence or bioluminescence. Given that fluorescence exploits light energy from ambient light, it is more efficient than bioluminescence and therefore likely to be the mechanism of choice for diurnal fish.

Results

Seeing Red Fluorescence on Reefs

We separated excitation from emission wavelengths in the field under natural, day-time solar illumination by SCUBA-diving below the penetration depth of the red component of sunlight (15–30 m) using masks and cameras equipped with a red filter blocking wavelengths below 600 nm (Fig. 1). This revealed widespread, natural red fluorescence produced by many microorganisms, plants and invertebrates (Fig. 2). The latter included mostly corals [10-13], but also species for which red fluorescence has never been described before, such as a polychaete species, several sponges (not shown) and feather stars (details in Fig. 2).

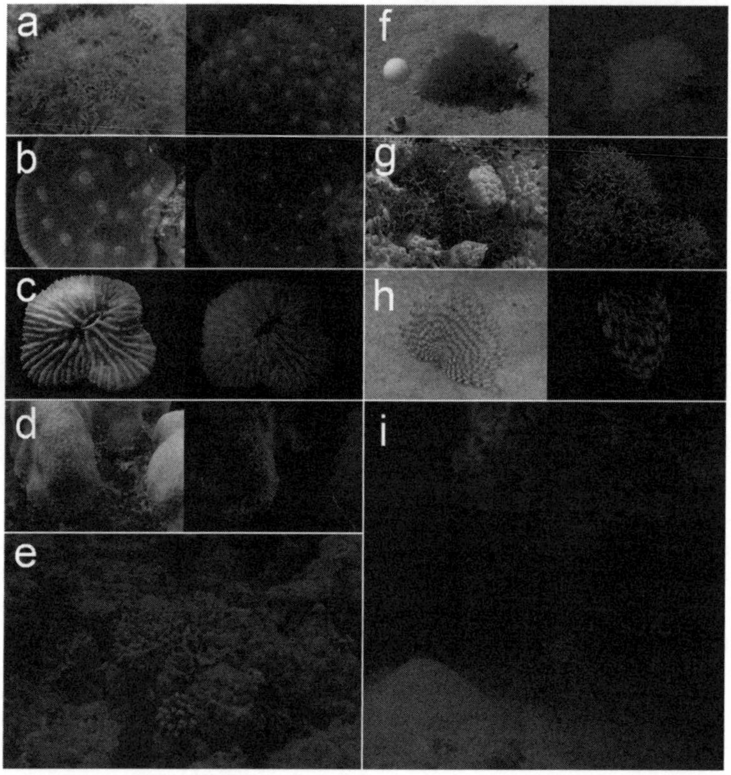

Figure 2. Examples of common red fluorescent invertebrates on coral reefs. a-d. Stony corals (a. Goniopora, b. Mycedium, c. Fungia, d. Porites). e. Reef scenery, as seen through a Lee Medium Red filter. f. Unidentified alga (white pingpong ball as reflectance reference). g. Calcareous alga Amphiroa. h. Polychaete worm Sabellastarte indica. i. Typical environment of S. indica under reef ledge. Pictures a-d and f-h show object under natural illumination (left) and as seen through a red filter (right). Pictures e and i show fluorescence in the field as seen by a digital camera. Most pictures taken in the field (Dahab, Egypt) under natural illumination between 14 and 17 m depth. Only c was photographed in the laboratory. Other reef invertebrates seen to fluoresce were sponges (e.g. Aaptos, Acanthella, Theonella) and feather stars (e.g. Colobometra, Oligometra).

Red Fluorescent Fishes

Of central importance here is our discovery of red fluorescence in reef fishes. Using the principle described above to distinguish "regular" red coloration from red fluorescence (Fig. 3) we identified at least 32 fish species belonging to 16 genera in 5 families that fluoresced visibly in red (Fig. 4, 5, Table 1). Fluorescent patterns usually included the eye ring and parts of the head or thorax and varied substantially between congeners (e.g. in the genera Eviota or Enneapterygius). Fins rarely fluoresced, except for the anal fin (some Gobiidae), the first dorsal fin (Tripterygiidae) or the tailfin (Syngnathidae). A 'whole body glow', including all fins, was present in the small wrasses Pseudocheilinus evanidus and Paracheilinus octotaenia. Visual and photographic evidence from the field was confirmed by extensive fluorescence microscopy and spectrometry of representative cases (Table 1). Fluorescence showed peak emission around 600 nm in most species (Fig. 6a, Table 1). Enneapterygius pusillus showed a second small peak at around 680 nm. P. evanidus differed from all others by having a double peak at 650 and 700 nm. We tested various light sources, including UV, but were not able to detect fluorescent emission at other (shorter) wavelengths in the fish described here. Anecdotal field observations suggest that yellow fluorescence may be present in other fishes (pers. obs.).

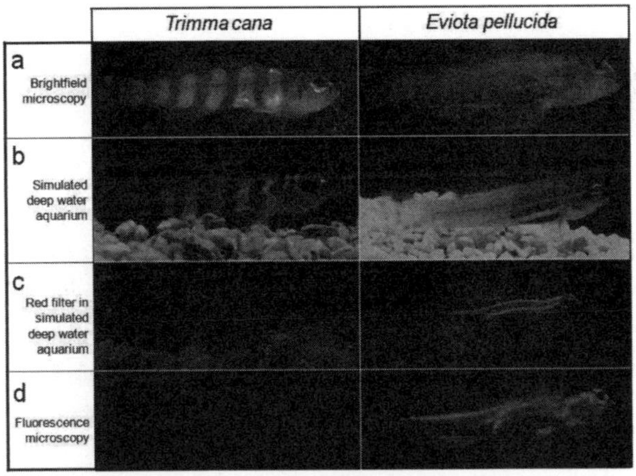

Figure 3. How to distinguish a red fish from a red fluorescent fish. Comparison between a non-fluorescent goby, Trimma cana (left), and a similar sized, red fluorescent goby, Eviota pellucida (right) under four viewing conditions. a. Artificial white light from a strong Schott KL 2500 LCD halogen cold light source under a Leica stereomicroscope (MZ 16F). b. In a halogen-illuminated aquarium with downwelling light filtered through Lee 729 Scuba-Blue filter (transmission range 400–550 nm, λ_{max} = 500 nm), thus simulating light at depth. c. Illumination as in b, but viewed through a red filter, revealing red fluorescence. d. Illumination as in a, seen under a Leica fluorescence stereomicroscope (MZ 16F) using green light for excitation, while viewing through red filter. The differences between the viewing conditions illustrate that red fluorescence can only be reliably seen when excitation and emission frequencies are separated, as at depth in the sea or under blue light, and by using a red filter for viewing.

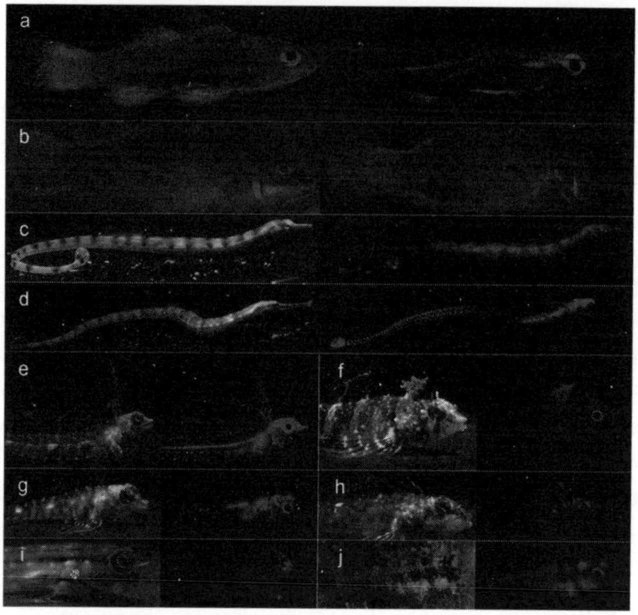

Figure 4. Red fluorescent representatives of five different reef fish families. a. Eviota pellucida (Gobiidae). b. Pseudocheilinus evanidus (Labridae). c. Corythoichthys flavofasciatus and d. C. schultzi (Syngnathidae), e. Enneapterygius pusillus, f. E. destai, g. E. abeli and h. Helcogramma steinitzi (Tripterygiidae). i. Ecsenius dentex and j. Crossosalarias macrospilus (Blenniidae). All pictures are from the laboratory, except for j (field). Left: broad spectrum illumination, right: red fluorescence under blue (laboratory) or natural (field) illumination.

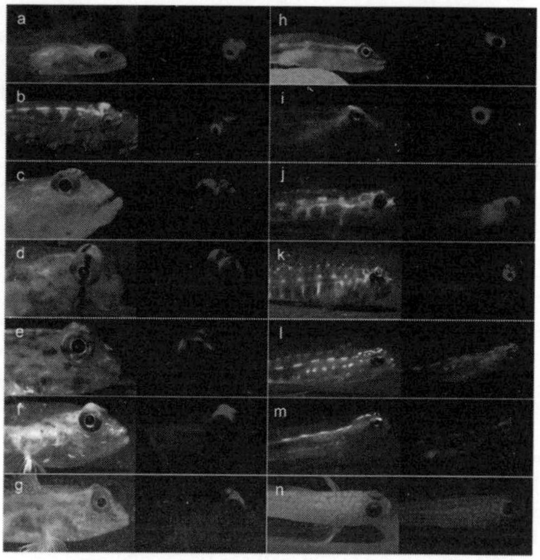

Figure 5. Diversity in red fluorescence in 14 goby species. a. Bryaninops natans. b. B. yongei. c. Ctenogobiops tangaroai. d. Gnatholepis anjerensis. e. Istigobius decoratus. f. Fusigobius duospilus. g. F. longispinus. h. Pleurosicya micheli. i. P. prognatha. j. Eviota guttata. k. E. prasina. l. E. zebrina. m. E. sebreei. n. Trimma avidori. All fish shown under broad spectrum illumination (left) in the laboratory (halogen) or field (a, b and h) and under blue illumination (right) with red filter.

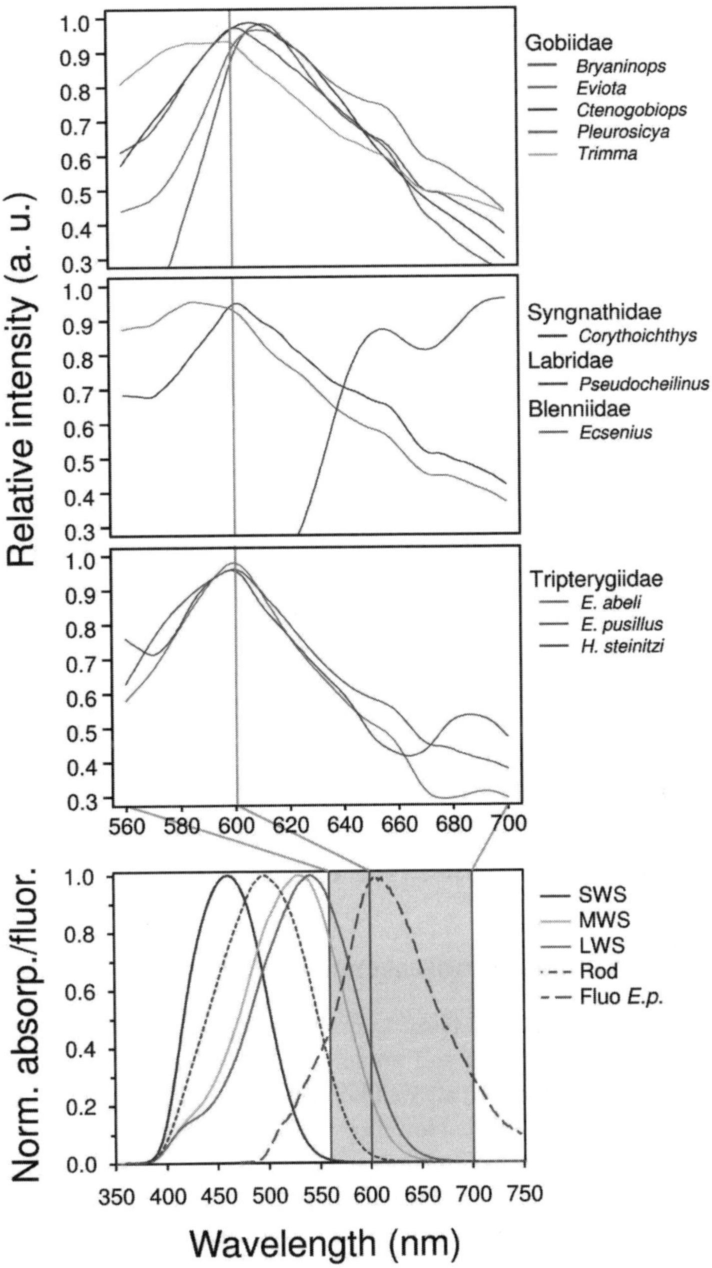

Figure 6. Spectrometric measurements confirm the emission of red light, and the ability to see it. a. Fluorescence emission spectra of five genera of Gobiidae (top), one genus of Syngnathidae, Labridae and Blenniidae each (middle) and three species of Tripterygiidae (bottom). b. Absorptance spectra of photoreceptor visual pigments found in Eviota pellucida with wavelengths of maximum absorptance (λ_{max}) at 497 (rods), 458 (SWS single cones), 528 (MWS, twin cones) and 540 nm (LWS, twin cones). The fluorescence emission spectrum of E. pellucida is included for comparison (dashed line).

Table 1. Reef fish species that were found to show red fluorescence

Family	Species	Site	Size (cm)	Red fluorescent body parts	Strength	Guanine crystals	Peak emission (nm)
Syngnathidae							
	Corythoichthys flavofasciatus	2	12	1, 2, 4, 8, 7	+++	yes	599–604
	C. schultzi	2	15	1, 2, 4, 7, 9	+++	yes	600–605
Labridae							
	Paracheilinus octotaenia	1, 2	9	4	++	-	653–658, 695–699
	Pseudocheilinus evanidus	1, 2, 4	8	4	++	no	-
Blenniidae							
	Crossosalarias macrospilus	3	8	1, 2, 4, 10	+++	-	-
	Ecsenius dentex	2	6	1, 2	+	yes	584–589
Tripterygiidae							
	Enneapterygius abeli	2	2.5	1, 2, 4	++	-	597–602
	E. destai	2	2	1, 5	++	-	-
	E. mirabilis	3	3.5	1, 5, 8, 10	+++	-	-
	E. pusillus	2	2	1, 2, 4, 5, 8, 10	+++	yes	596–601, 684–689
	Helcogramma steinitzi	2	5	1, 2, 5	+++	-	597–602
	Ucla xenogrammus	3	5.5	1, 10	+++	-	-
Gobiidae							
	Bryaninops natans	1, 2, 5	2.5	1	+++	yes	605–610
	B. ridens	2, 5	2	1	+	-	-
	B. yongei	2, 5	2.5	1	++	-	-
	Ctenogobiops maculosus	1, 2	7	1	+++	-	-
	C. tangaroai	4	7	1	+++	yes	605–610
	Eviota distigma	2	2	1, 11	+	-	too weak
	E. guttata	2	2.5	1, 2, 4, 6, 9	+++	-	605–610
	E. nigriventris	4	2.5	1, 2	+	-	-
	E. pellucida	3, 4	2.5	1, 2, 3, 6	+++	yes	604–609
	E. prasina	2	3	1, 12	+	-	596–601
	E. queenslandica	3	3	1	+		-
	E. sebreei	2	2.5	1	+	yes	585–590
	E. zebrina	3, 5	3	1, 2, 4, 9	+++	yes	600–604
	Fusigobius duospilus	2	6	1	+++	-	-
	F. longispinus	2	7	1	+	yes	600–605
	Gnatholepis anjerensis	2	8	1	+	yes	-
	Istigobius decoratus	2	12	1	+	yes	591–596
	Pleurosicya micheli	2, 5	3	1	++	yes	601–606
	P. prognatha	2, 5	2	1	+++	yes	608–613
	Trimma avidori	2	3	1, 2, 9	+	-	590–595

Study localities: 1 = El Quseir (Sharm Fugani, Egypt, field); 2 = Dahab Marine Research Centre (South Sinai, Egypt, field and laboratory), 3 = Lizard Island Research Station (Queensland, Australia, field and laboratory), 4 = Tübingen (laboratory), 5 = Collection JH, Vienna. Fluorescent body parts are encoded as follows: 1 = Eye rings, 2 = Head (dorsal), 3 = Lateral lines, 4 = Trunk, 5 = First dorsal fin, 6 = Anal fin, 7 = Tailfin, 8 = Snout, 9 = Dots on body, 10 = Spine, 11 = Gut. Strength of fluorescence is a consensus measure among observers based on visibility in the field and/or the laboratory: Strong (+++) means strikingly bright in the field as well as in the laboratory, easy to photograph and strong spectrometric signal; weak (+) means only visible under laboratory conditions, requiring long integration times for fluorescence measurement and long exposure times for photography (optimized for each specimen). Peak emission is shown as 5 nm interval.

Mechanism of Red Fluorescence in Fishes

Dissection revealed that red fluorescence was associated with guanine crystals in pipefish, triplefins, blennies and gobies (Fig. 7, Table 1). Guanine crystals are produced by iridophores and are well known as the source of silvery reflection and iridescence in bony fish[14]. However, they have never been described to show strong red fluorescence. In E. pellucida and Ctenogobiops tangaroai, only about half of the crystals isolated from the eye rings showed bright red fluorescence (Fig. 7b), suggesting that the fluorescing substance is produced or sequestered independently from the crystals. Preparations of crystals maintained strong fluorescence after prolonged storage in a dried or liquid form (70% EtOH or 4% formaldehyde), allowing us to confirm fluorescence in preserved gobies collected up to 5 years before (Collection JH, Vienna, Table 1). This is in striking contrast to reflective red pigmentation, which bleaches out within hours after fixation. We did not find fluorescent guanine crystals in P. evanidus. Here, microscopic investigation suggests the presence of a fluorescent pigment associated with the bony tissue of scales and fin rays (Fig. 7f). This pigment has a chemical stability in preserved specimens that matches that of guanine-linked red fluorescence.

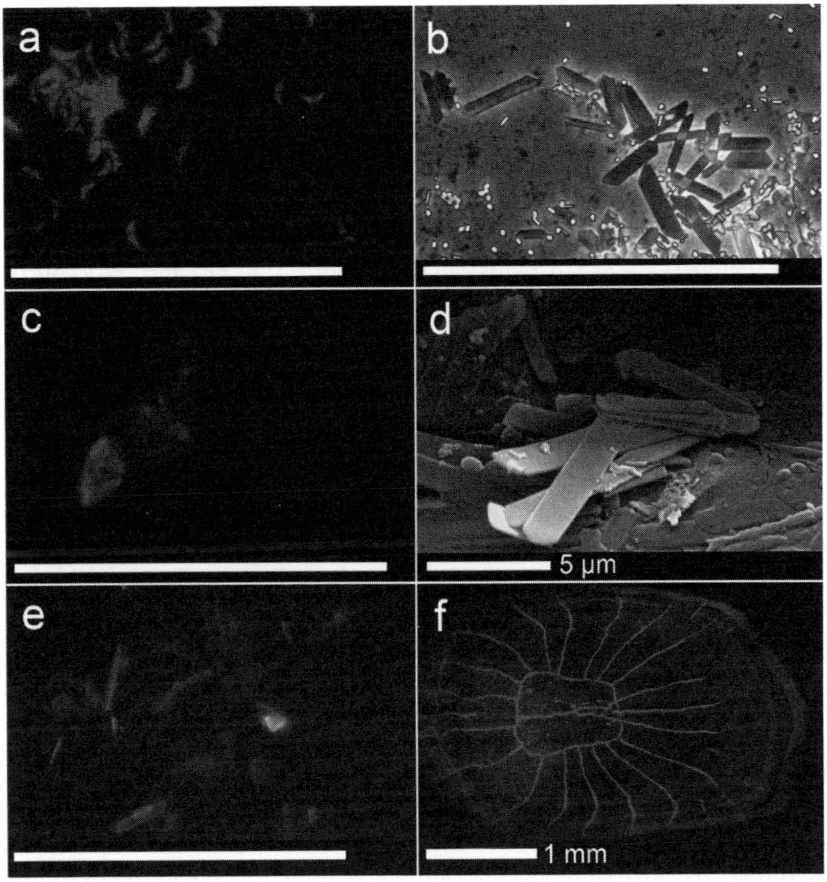

Figure 7. Sources of red fluorescence in fishes. a. Guanine crystals from the eye ring of Eviota pellucida (fluorescence microscopy). b. Same from Ctenogobiops tangaroai (fluorescence overlaid with phase contrast). c. As in b, diamond form (fluorescence microscopy). d. Guanine crystals falling apart in characteristic platelets (from eye ring of E. pellucida, scanning electron micrograph). e. Single red fluorescent guanine crystal among normal crystals in eye ring of the non-fluorescent goby Trimma cana (fluorescence microscopy, see also Fig. 3). f. Scale of Pseudocheilinus evanidus, in which the red fluorescent pigment is associated with bony scales and fin rays (fluorescence microscopy). Scale bar = 50 µm unless indicated otherwise.

As a control, we isolated guanine crystals from non-fluorescing fish. In T. cana (Fig. 3), Psetta maxima (turbot) and Engraulis encrasicolus (anchovy), we found less than 1% of the crystals to fluoresce in red (Fig. 7e), suggesting that a very low level of red fluorescence may be widespread and is not limited to reef fish.

Can Red Fluorescent Fish See Red Fluorescence?

With few exceptions [15-17] most marine fishes lack a sensitivity reaching into the red part of the spectrum[5]. But as far as we know, none of the fish genera listed in Table 1 has ever been tested for its spectral sensitivity. Here, we measured the retinal spectral sensitivity of wild-caught E. pellucida. Their retina contains rods and both single and twin cones as previously found in one other goby[18]. Mean wavelengths of maximum absorbance (λ_{max}) of the visual pigments in the rods and short-wavelength-sensitive (SWS) single cones were 497 and 458 nm, respectively. λ_{max} values in the twin cones were highly variable, ranging from 518–546 nm, although spectra were clustered into two groups with mean λ_{max} values at 528 (medium-wavelength-sensitive, MWS) and 540 nm (long-wavelength-sensitive, LWS; Fig. 6b). The majority of twin cones had the 528 nm pigment in both members (528/528 twins) although both 528/540 and 540/540 twins were also observed. These results suggest that there may be more than two distinct visual pigments in the twin cones, and/or co-expression of opsin genes within the same outer segment, as in other fish[19]. Regardless, Fig. 6b shows that there is considerable overlap between the red fluorescence emission spectrum and the absorbance spectra of the visual pigments in the twin cones. It is highly likely, therefore, that this species can see its own fluorescence. A similar result is expected for Corythoichthys pipefish, given that long wavelength sensitivity is known from other Syngnathids [17].

Discussion

The Function of Red Fluorescence

Why do some reef fish fluoresce strongly in red, whereas most do not? Although non-fluorescent (reflective) red pigmentation is widespread in reef fish[20], it appears grey or black at depth (Fig. 3), allowing fish to blend in with their background[21]. Red fluorescence, however, does the opposite: by emitting a color that is lacking from the environment, a fish contrasts more against its background. As a result, red fluorescence may function as a communication or attraction signal, as proposed for red-bioluminescent deep sea fishes[16] and siphonophores[22], ultraviolet-reflecting reef fishes[23] and green fluorescent mantis shrimp[24] and parrots[25].

We see four reasons why red fluorescence may be part of a private communication system in fish. Firstly, peaking mostly around 600 nm, red fluorescence is at the borderline of what is visible to many marine fishes, and due to rapid attenuation of red light by water, even those that can see red will be able to see it over short distances only. This is suggestive of an adaptive shift away from the "public area" of the visual spectrum into a bordering "private area." Secondly, most species found to fluoresce are small, benthic, pair- or group-living fishes, often with conspicuous intra-specific behaviors, but quite cryptic coloration in other parts of the visible spectrum. Thirdly, there is strong inter-specific variation within and between closely related genera suggestive of species-recognition (Fig. 4, 5). Finally, fluorescence is present in structures that are used in intra-specific signalling. This is true for the first dorsal fin in triplefins (Fig. 8a), the tailfin in pipefish (Fig. 8b) and possibly the anal fin in E. pellucida (Fig. 3) and E. guttata. Fluorescent eye rings may function as an indicator of presence (Fig. 8c) or reveal the direction of gaze (Fig. 4e–g, Fig. 8a, d). Since iridophores are involved in rapid color change in other fishes [26,27], fluorescent fish may also be able to change the strength of fluorescence.

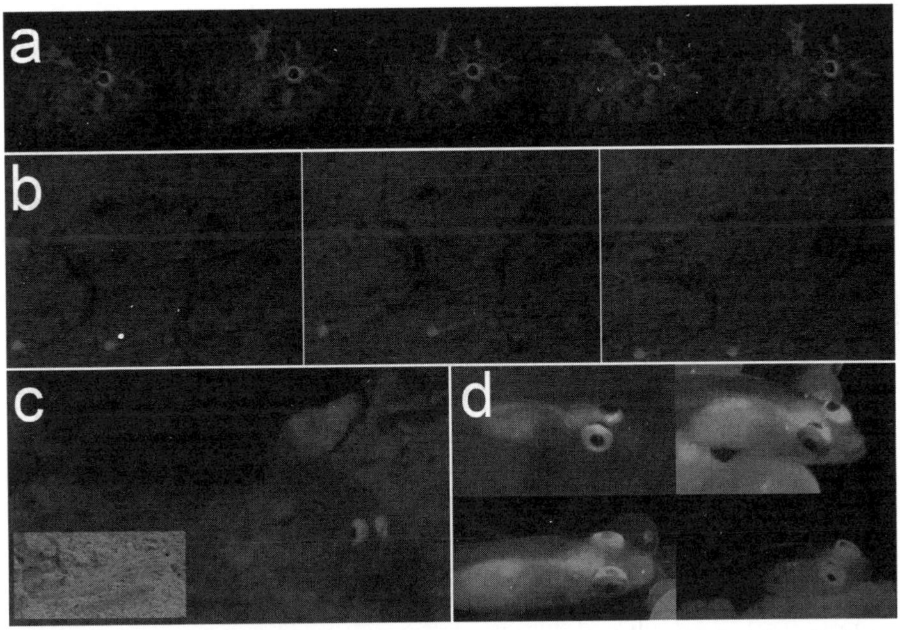

Figure 8. Red fluorescence in a signalling context. a-b. Frameshots of two videos suggestive of communication involving fluorescent fins. a. Enneapterygius destai waving its first dorsal fin when 'excited.' b. Corythoichthys schultzi pair interacting, displaying conspicuous red fluorescence on the tail plate (natural reef illumination, 20 m depth). c. Ctenogobiops maculosus is cryptic when sitting at its burrow entrance under natural illumination (insert), but shows conspicuous eyes when emphasizing red wavelengths. d. Bryaninops natans with a pupil-like black spot on the upper part of the fluorescent eye ring suggesting that fluorescence and (deceptive) gaze signalling may be linked in this species.

One reason why red fluorescence in fishes may have escaped attention is because fluorescence is usually observed during night dives using strong UV light sources. Although this reveals the spectacular bluish-green fluorescence typical of many corals, it is not a good strategy to visualize red fluorescence because excitation and emission wavelengths are far apart, making red fluorescence weak relative to the shorter wavelength emissions. Moreover, most fish are hiding at night, explaining why red fluorescence had not yet been described for fishes. Up till now we have not yet found a nocturnal fish that fluoresces in red, which fits well with widespread reflectant red coloration indicating that crypsis is more important in these species.

Final Remarks

For a correct interpretation of our results, it is important to stress that it is not required to look through a red filter to "see" red fluorescence during daytime, particularly not for the stronger cases. Using a filter merely facilitates its detection in the field. Without a filter, red fluorescence at depth is usually only one element in complex, multi-spectral, red-containing colors, such as orange-brown, red-brown, pink, lilac, violet or even bright white in some encrusting corals. Cases of pure fluorescence-based red are thus far limited to a few corals and sponges. Whether pure or mixed, red-containing colors cannot be generated by reflectance and therefore sources of red fluorescence can be identified without a filter at depth. This merely requires special attention by the diver: Although e.g. red-brown is prevalent on reefs, it is not perceived as unusual or unexpected by most. Consequently, while red masks are invaluable for initial detection, fluorescence photography is easiest using a regular mask and a camera with a red filter. An alternative is to refrain from a red filter altogether and to adjust the white balance of the camera manually to local light conditions using a white slate. This will emphasize reds, while suppressing, but not removing, shorter wavelengths. At depth, this will highlight any red fluorescence. Applying this to fish color vision, we suspect that, if red fluorescent fish can indeed see their fluorescence, they see it as enhanced contrast involving pink, red-brown or other red-containing, mixed colors in an otherwise blue-green environment. Consequently, behavioral experiments should not test the ability of fish to see weak red light, but their ability to distinguish between multi-spectral signals with and without a weak red component in a blue-green flooded environment.

Finally, we want to make a cautionary remark on diving safety. Diving with a red mask is similar to night diving, with dramatically reduced light intensities and viewing distances. Disorientation becomes a serious problem. Moreover, it takes several minutes to adapt to the darkness. Staying in a small, familiar area and moving slowly and carefully is crucial. Furthermore, it is essential to take a torch to read equipment. Dials, indicators and computer backlights are either reflectant or luminesce blue or green, making them effectively illegible at depth in the absence of a local white (red-containing) light source. To circumvent this problem, we also used the Oceanic DataMask which has a built-in dive computer that can be read irrespective of any filter attached to the front. Because of these unfamiliar restrictions, we recommend that only experienced divers use this procedure and that only one partner in a buddy team uses a red mask at any given time. We also recommend attaching filters in such a way that they can be instantly removed without having to change to a spare mask, which one should carry nevertheless.

Conclusion

We conclude that a considerable number of reef fishes have developed complex patterns of striking red fluorescence which may be used to enhance visual communication by exploiting a waveband invisible to most other fish. At least one species (E. pellucida) shows a retinal sensitivity to its own red fluorescence and many species show suggestive evidence that fluorescence is linked to signalling structures. Additional studies are required to confirm that this ability is also present in other fluorescent fish. Nevertheless, the prevalent assumption that red light is of low importance for reef fish[5] must be questioned. Obviously, a lack of downwelling red light is not a reason to stop seeing red, adding an exciting novel dimension to reef light ecology.

Methods

Material

Fish were collected in Dahab (Dahab Marine Research Centre, Egypt, with permit from NCS/EEAA) and at Lizard Island Research Station (Australia) (GBRMPA permit G05/13668.1 to UES) by anaesthetising individuals with clove oil (10 ml in 50 ml EtOH added to 200 ml sea water). Fish for laboratory work in Tübingen were obtained from the sustainable aquarium trade and kept in accordance with German animal care legislation. Species were identified using standard[20] and specialised [28-30] literature.

Digital Photography

Underwater pictures of fluorescence under natural illumination were taken with a Canon PowerShot G7 or G9, Canon Ixus 750 and Sony HD HDR-CX6EK using Lee Medium Red filters or Edmund Optics Y59-642 glass filters. Macro photography in the laboratory was done using Canon PowerShot G7 or G9 with blue LED light source (LUXEON K2 LXK2-PB14-P00, λ_{max} = 470 nm) and Edmund Optics Optical filter G43-943. All of these filters block most light below 600 nm, but allow some blue or green light to leak through (e.g. Fig. 4a–b). This blue-green hue was removed in most pictures by only showing the red channel (590–700 nm) of the RGB digital image. Other pictures were taken using a Leica MZ16F fluorescent stereo-microscope on anaesthetised, freshly killed or preserved animals. A Leica DM5000 fluorescence microscope was used to document guanine crystals.

Fluorescence Spectrometry

Fluorescent emission spectra were recorded using an AvaSpec-2048-USB2 spectrometer with a green laser (λ_{exc} = 532 nm) as excitation light source, yielding reproducible emission spectra of from live and anaesthetised fish. We supplemented these measurements with high-sensitivity spectrometry of fluorescent guanine crystals and scales of C. tangaroai, E. pellucida, E. sebreei, E. zebrina, F. longispinus, I. decoratus, P. micheli and P. evanidus. Here, emission spectra were induced using λ_{exc} = 473 nm laserlight (LDH – P400B Picoquant) on a custom-built Zeiss Axiovert 135 TV confocal laser scanning microscope equipped with an avalanche photodiode (APD, SPCM-AQR-14, Perkin Elmer) and a spectrometer with cooled CCD camera (Spec-10:100B, Princeton Instruments). Since these measurements coincided well with those obtained with the AvaSpec-2048-USB2, data from both methods were pooled (Fig. 6). Spectra were normalized by setting maximum emission equal to 1. Curves for genera (Fig. 6) are averages of the average spectra of each available species.

Scanning Electron Microscopy

For electron scanning micrographs, guanine crystals were isolated[14] from E. pellucida, coated with 20-nm Au/Pd and photographed using a Cambridge Stereoscan 250 Mk2 electron microscope (Fig. 7d).

Retina Microspectrometry

For microspectrometry of E. pellucida visual pigments, five gobies (13–18 mm standard length) were dark adapted for at least 1 hour and their retinas removed under infrared illumination with the aid of image converters. Each retina was dispersed mechanically and mounted in 340 mOsm kg-1 phosphate-buffered saline containing 10% dextran. Absorbance spectra (325–800 nm) of visual pigments housed in individual photoreceptor outer segments were measured using a wavelength-scanning microspectrophotometer, as described elsewhere[31] (Fig. 6b).

Author Contributions

NKM is the principal investigator and initiator and has been involved all aspects except for data collected in Australia. NA, JH, GS, DS and MFW collected field and laboratory data and contributed to data analysis and manuscript writing. GS was also responsible for spectrometry and digital recording. NSH and UES made the microspectrometric measurements of E. pellucida eyes and edited the manuscript. FS and AJM carried out the high-resolution spectrometry and provided technical background on fluorescence spectrometry. All authors have read and approved the final manuscript.

Acknowledgements

We thank Mathilde Bessert-Nettelbeck, Christoph Braun, Monica Gagliano, Rolanda Lange, Mark McCormick, Philip L. Munday, Christiane Schmidt, Jennifer Theobald, Manfred Walzl and Johanna Werminghausen for assistance in the field or the laboratory, Sergey Bogorodsky, Anne Hoggett, Martial Depczynksi, Wouter Holleman, John Randall and Marit Wagler for help with identification, Janine Wong and Andreas Oelkrug for fish keeping, Karl-Heinz Hellmer for electron microscopy, the Dahab Marine Research Centre with Heike Stürmer and Andy Tischer for logistical support, Redouan Bshary, Hans Fricke, Karen Hissmann, Hinrich Schulenburg, Hans-Joachim Wagner and Tony Wilson for comments and Moustafa Fouda (EEAA, Egypt) for the permission to work in Dahab.

References

1. Marshall NJ, Jennings K, McFarland WN, Loew ER, Losey GS: Visual biology of Hawaiian coral reef fishes. III. Environmental light and an integrated approach to the ecology of reef fish vision. Copeia 2003, 2003(3):467–480.
2. Loew ER, Zhang H: Propagation of visual signals in the aquatic environment: An interactive windows-based model. In Communication in Fishes. Volume 2. Edited by: Ladich F, Collin SP, Moller P, Kapoor BG. Enfield (NH): Science Publishers; 2006:281–302.
3. Losey GS, McFarland WN, Loew ER, Zamzow JP, Nelson PA, Marshall NJ: Visual biology of Hawaiian coral reef fishes. I. Ocular transmission and visual pigments. Copeia 2003, 2003(3):433–454.

4. Siebeck UE, Losey GS, Marshall J: UV Communication in Fish. In Communication in Fishes. Volume 2. Edited by: Ladich F, Collin SP, Moller P, Kapoor BG. Enfield (NH): Science Publishers; 2006:423–455.
5. Marshall J, Vorobyev M, Siebeck UE: What does a reef fish see when it sees a reef fish? Eating 'Nemo'©. In Communication in Fishes. Volume 2. Edited by: Ladich F, Collin SP, Moller P, Kapoor BG. Enfield (NH): Science Publishers; 2006:393–422.
6. Lythgoe JN, Muntz WRA, Partridge JC, Shand J, Williams DM: The ecology of the visual pigments of snappers (Lutjanidae) on the Great Barrier Reef. J Comp Physiol A 1994, 174(4):461–467.
7. Job SD, Shand J: Spectral sensitivity of larval and juvenile coral reef fishes: implications for feeding in a variable light environment. Mar Ecol Prog Ser 2001, 214:267–277.
8. Douglas RH, Mullineaux CW, Partridge JC: Long-wave sensitivity in deep-sea stomiid dragonfish with far-red bioluminescence: evidence for a dietary origin of the chlorophyll-derived retinal photosensitizer of Malacosteus niger. Philos T Roy Soc B 2000, 355(1401):1269–1272.
9. Bowmaker JK, Dartnall HJA, Herring PJ: Longwave-sensitive visual pigments in some deep-sea fishes: segregation of 'paired' rhodopsins and porphyropsins. J Comp Physiol A 1988, 163(5):685–698.
10. Mazel CH, Fuchs E: Contribution of fluorescence to the spectral signature and perceived color of corals. Limnol Oceanogr 2003, 48:390–401.
11. Limbaugh C, North WJ: Fluorescent, benthic, Pacific Coast coelenterates. Nature 1956, 178:497–498.
12. Field S, Bulina M, Kelmanson I, Bielawski J, Matz M: Adaptive evolution of multicolored fluorescent proteins in reef-building corals. J Mol Evol 2006, 62(3):332.
13. Fradkov AF, Chen Y, Ding L, Barsova EV, Matz MV, Lukyanov SA: Novel fluorescent protein from Discosoma coral and its mutants possesses a unique far-red fluorescence. FEBS Letters 2000, 479(3):127–130.
14. Levy-Lior A, Pokroy B, Levavi-Sivan B, Leiserowitz L, Weiner S, Addadi L: Biogenic guanine crystals from the skin of fish may be designed to enhance light reflectance. Cryst Growth Des 2008, 8(2):507–511.
15. Barry KL, Hawryshyn CW: Spectral sensitivity of the Hawaiian saddle wrasse, Thalassoma duperrey, and implications for visually mediated behavior on coral reefs. Environ Biol Fish 1999, 56(4):429–442.

16. Douglas RH, Partridge JC, Dulai K, Hunt D, Mullineaux CW, Tauber AY, Hynninen PH: Dragon fish see using chlorophyll. Nature 1998, 393(6684):423–424.

17. Mosk V, Thomas N, Hart NS, Partridge JC, Beazley LD, Shand J: Spectral sensitivities of the seahorses Hippocampus subelongatus and Hippocampus barbouri and the pipefish Stigmatopora argus. Visual Neuroscience 2007, 24(3):345–354.

18. Utne-Palm AC, Bowmaker JK: Spectral sensitivity of the two-spotted goby Gobiusculus flavescens (Fabricius): a physiological and behavioral study. J Exp Biol 2006, 209(11):2034–2041.

19. Shand J, Hart NS, Thomas N, Partridge JC: Developmental changes in the cone visual pigments of black bream Acanthopagrus butcheri. J Exp Biol 2002, 205(23):3661–3667.

20. Allen G, Steene G, Humann P, DeLoach N: Reef Fish Identificaton – Tropical Pacific. Jacksonville FL: New World Publications Inc; 2003.

21. Marshall NJ: Communication and camouflage with the same 'bright' colors in reef fishes.Philos T Roy Soc B 2000, 355(1401):1243–1248.

22. Haddock SHD, Dunn CW, Pugh PR, Schnitzler CE: Bioluminescent and red-fluorescent lures in a deep-sea siphonophore. Science 2005, 309(5732):263–263.

23. Siebeck UE: Communication in coral reef fish: the role of ultraviolet color patterns in damselfish territorial behavior. Anim Behav 2004, 68:273–282.

24. Mazel CH, Cronin TW, Caldwell RL, Marshall NJ: Fluorescent enhancement of signalling in a mantis shrimp. Science 2004, 303(5654):51.

25. Arnold KE, Owens IPF, Marshall NJ: Fluorescent signaling in parrots. Science 2002, 295(5552):92.

26. Odiorne JM: The Occurrence of guanophores in Fundulus.PNAS 1933, 19(7):750–754.

27. Mäthger LM, Land MF, Siebeck UE, Marshall NJ: Rapid color changes in multilayer reflecting stripes in the paradise whiptail, Pentapodus paradiseus.J Exp Biol 2003, 206(20):3607–3613.

28. Herler J, Hilgers H: A synopsis of coral and coral-rock associated gobies (Pisces: Gobiidae) from the Gulf of Aqaba, northern Red Sea.Aqua 2005, 10(3):103–132.

29. Holleman W: Fishes of the genus Helcogramma (Blennioidei: Tripterygiidae) in the Western Indian Ocean, including Sri Lanka, with descriptions of four new species. Smithiana Bulletin 2007, 7:51–81.

30. Holleman W: A review of the triplefin fish genus Enneapterygius (Blennioidei: Tripterygiidae) in the western Indian Ocean, with descriptions of four new species. Smithiana Bulletin 2005, 5:1–25. + 22 plates

31. Hart NS: Microspectrophotometry of visual pigments and oil droplets in a marine bird, the wedge-tailed shearwater Puffinus pacificus: topographic variations in photoreceptor spectral characteristics. J Exp Biol 2004, 207(7):1229–1240.

The Molecular Basis of Color Vision in Colorful Fish: Four Long Wave-Sensitive (LWS) Opsins in Guppies (*Poecilia reticulata*) are Defined by Amino Acid Substitutions at Key Functional Sites

Matthew N. Ward, Allison M. Churcher, Kevin J. Dick, Chris R. J. Laver, Greg L. Owens, Megan D. Polack, Pam R. Ward, Felix Breden and John S. Taylor

ABSTRACT

Background

Comparisons of functionally important changes at the molecular level in model systems have identified key adaptations driving isolation and speciation.

In cichlids, for example, long wavelength-sensitive (LWS) opsins appear to play a role in mate choice and male color variation within and among species. To test the hypothesis that the evolution of elaborate coloration in male guppies (Poecilia reticulata) is also associated with opsin gene diversity, we sequenced long wavelength-sensitive (LWS) opsin genes in six species of the family Poeciliidae.

Results

Sequences of four LWS opsin genes were amplified from the guppy genome and from mRNA isolated from adult guppy eyes. Variation in expression was quantified using qPCR. Three of the four genes encode opsins predicted to be most sensitive to different wavelengths of light because they vary at key amino acid positions. This family of LWS opsin genes was produced by a diversity of duplication events. One, an intronless gene, was produced prior to the divergence of families Fundulidae and Poeciliidae. Between-gene PCR and DNA sequencing show that two of the guppy LWS opsins are linked in an inverted orientation. This inverted tandem duplication event occurred near the base of the poeciliid tree in the common ancestor of Poecilia and Xiphophorus. The fourth sequence has been uncovered only in the genus Poecilia. In the guppies surveyed here, this sequence is a hybrid, with the 5' end most similar to one of the tandem duplicates and the 3' end identical to the other.

Conclusion

Enhanced wavelength discrimination, a possible consequence of opsin gene duplication and divergence, might have been an evolutionary prerequisite for color-based sexual selection and have led to the extraordinary coloration now observed in male guppies and in many other poeciliids.

Background

Understanding the molecular basis of characters shaped by selection is a major goal of evolutionary genetics. Of particular interest are genes that encode conspicuous secondary sexual traits in males and the genes that influence female preference for such traits [1]. Among fish; sticklebacks (genus Gasterosteus), cichlids, and poeciliids, including the guppy (Poecilia reticulata) and swordtails (genus Xiphophorus), are the most important models for the study of sexual selection driven by female choice. In each of these taxa, female mate choice is influenced by male coloration and in each group, male coloration and female preference have a genetic basis [2-6].

Mapping studies designed to uncover genes responsible for species- and population-level color variation in cichlids and sticklebacks are underway [7] but to date none have been identified. While it is also the case that no female preference loci have been uncovered in fish, many cichlid species and some populations possess unique opsin genes that provide strong candidates. Indeed, the only DNA sequences that have been found to differ among the 200 to 500 endemic Lake Victoria haplochromine species are long wave-sensitive (LWS) opsins [8,9]. In the cichlid genus Pundamilia, LWS opsin sequence and expression appears to be tuned to specific male color morphs [10]. Thus, it appears that variation in LWS opsin genes influences female mate choice and speciation in this family [11].

Opsin genes encode membrane-bound receptors that are expressed primarily in rod and cone cells of the retina. Each opsin protein is associated with a chromophore and when exposed to light, this complex changes shape leading to rod or cone cell hyperpolarization [12]. The detection of light at the receptor level requires input from just one type of opsin-chromophore receptor. However, discriminating among colors (wavelengths) involves the interpretation of signals from multiple adjacent retinal cone cells expressing different opsins. These different opsins often have names that reflect the wavelength of light to which they are most sensitive. For instance, short wave-sensitive (SWS), middle wave-sensitive (MWS), and long wave-sensitive opsins (LWS) are most sensitive to blue, green and red light, respectively. Gene duplication and divergence has generated this opsin diversity. For example, the human MWS opsin is a duplicate (or paralog) of the LWS opsin locus and now differs at three of the five amino acid positions known to influence wavelength sensitivity [13-16]. Zebrafish also have a pair of LWS opsin genes with different five key-site haplotypes [17].

The purpose of this study was to characterize LWS opsin gene sequence variation in guppies and in closely related species. We focused on this gene because microspectrophotometry (MSP) data indicated that guppies express more than one type of LWS opsin [18,19] and because orange is an important component of female mate choice for these fish [4,20-22]. While two recent studies have reported LWS opsin gene variation in guppies [23,24], one focused only on short amplicons from a single fish and both presented incomplete data on the key-site amino acids known to influence spectral sensitivity. Genomic sequences, transcript expression levels, and data from other poeciliids have also not been reported to date.

We show that guppies (Poecilia reticulata) and three species in the guppy sister group (Micropoecilia) have four LWS genes. Sequence variation at the five key sites indicates that three of these LWS opsins are most sensitive to different wavelengths of light providing Poecilia with a larger repertoire of LWS pigments than any other fish taxon. One of the guppy LWS opsins appears to be a single-exon gene, likely arising from a retrotransposition event. This gene was sequenced in all poeciliids surveyed except Tomeurus gracilis and has also been reported in the killifish, Lucania goodei (family Fundulidae). Two LWS opsins are linked, oriented in a tail-to-tail fashion, and separated by approximately 3.3 Kbp. The fourth is found only in the genus Poecilia. This is a hybrid or mosaic sequence in the guppies surveyed here from Cumaná Venezuela. All four LWS opsins in the guppy were amplified from RNA isolated from adult eyes, but qPCR experiments show much variation among these duplicates in the level of expression.

Methods

Genomic PCR and Sequencing

All species surveyed are in the family Poeciliidae. Long wave-sensitive (LWS) opsin genes were amplified from DNA isolated from Tomeurus gracilis (one individual), Xiphophorus pygmaeus (one individual), and from four species in the genus Poecilia: P. reticulata (14 individuals), P. picta (four individuals) P. parae (three individuals) and P. bifurca (three individuals). The genus Tomeurus is the sister group to a clade that includes Poecilia and Xiphophorus, and most other poeciliids [25,26]. Poecilia picta, P. parae, and P. bifurca, occur in the sister taxon to the guppy [27]. They were in a separate genus previous to Rosen and Bailey's [28] revision of the poeciliids and we refer to them collectively as Micropoecilia. Poecilia reticulata (the guppy) was sampled from a population collected in Cumaná, Venezuela and bred in our laboratory aquarium. The Cumaná guppy has also been referred to as Endler's guppy, but is closely related to other guppy populations [29]. PCR reactions were run using genomic DNA isolated from fish euthanized with buffered MS222 (Sigma® A5040) or from specimens preserved in 95% ethanol.

Initially, PCR and sequence data were obtained using primers ForBeg, Fw1a and Rev5, which are complementary to conserved regions of fish LWS opsin genes in exon I (ForBeg), exon II (Fw1a), and exon V (Rev5). After uncovering multiple LWS opsin sequences in Poecilia, we attempted to PCR-amplify DNA between guppy opsin genes. The between-gene PCR experiment was initiated because LWS opsins occur in tandem in human, zebrafish, and medaka. It employed the reverse complement of a forward primer close to the 5' end of the gene (Fw1a Comp) and the reverse complement of a reverse primer close to the 3' end of the gene (Rev8 Comp). Sequence data from amplicons derived from primers ForBeg, Fw1a and Rev5, and the success of between-gene PCR allowed us to develop gene-specific primers, including reverse primers complementary to 3' UTR sequences. Primers complementary only to guppy LWS 'variant 6' were designed from sequence data recently published by Weadick and Chang [24]. PCR amplicons were cut and purified from agarose gels using a QIAquick® Gel Extraction Kit and were cloned using the pGEM® – T Easy Vector System II kit (Promega™). Sequencing of insert-positive clones utilized labeled M13 forward and reverse primers and a LI-COR sequencer at the Centre for Biomedical Research at the University of Victoria.

Southern Blot Hybridization

DNA from one lab-reared Cumaná guppy was extracted using a Qiagen® DNeasy Tissue Kit and digested with the four-cutter restriction enzyme BfaI (New England Biolabs®). 423 bp DIG labeled probes were prepared from guppy genomic DNA using the PCR DIG Probe Synthesis Kit (Roche®) and primers Fw100 and Rev4. The amplicons from this PCR reactions were purified using the QIAquick® Gel Extraction Kit (Qiagen®). BfaI does not cut any of the LWS opsins in the region complementary to the probes. Southern blot hybridization was carried out using a modified protocol from the Roche® DIG application manual for filter hybridization. Digested DNA was blotted onto a Bio-Rad™ Zeta-Probe® Blotting membrane using the Bio-Rad™ Model 785 Vacuum Blotter. This was followed by UV exposure (120 mJ) and probe hybridization at 40°C overnight. The blot was washed in 2× SSC at room temperature and then in 1× SSC at 65°C and visualized using the DIG Luminescent Detection Kit for Nucleic Acids (Roche®).

LWS Opsin Gene Expression Using RT-PCR

Prior to quantitative PCR (qPCR) experiments (see below) we tested the hypothesis that all four LWS opsin loci were expressed using reverse transcriptase (RT)-PCR. Three guppies were euthanized in buffered MS222. A single eye from each individual was placed in 1.0 mL PureZOL™ (Bio-Rad®) with 3 mm tungsten carbide beads and homogenized for five minutes in a Retsch MM301 Mixer Mill. Total RNA was extracted using the Aurum™ Total RNA Fatty and Fibrous Tissue kit from BioRad®. The iScript™ kit (Bio-Rad®) was used to generate single-stranded cDNA. LWS opsin transcripts were PCR amplified and cloned using the pGEM® – T Easy Vector System II kit (Promega™) and then sequenced with labeled M13 primers. Primers reported by Meyer and Lydeard [25] that amplify XSrc were used as a positive control.

Quantitative PCR

Data from cichlids and zebrafish indicate that cone opsin expression is highest at the end of the day [30,31]. Guppies were maintained in a 14:10 hr light and dark cycle and qPCR experiments were performed on cDNA samples obtained from three fish in the last hour of the subjective day. Total RNA was extracted from the eyes of these fish (one adult male and two adult females) using the Retsch MM301 Mixer Mill and the Aurum™ Total RNA Fatty and Fibrous Tissue kit from BioRad®. Synthesis of cDNA for qPCR experiments utilized the SuperScript™ III First-Strand Synthesis SuperMix kit for qRT-PCR (Invitrogen™) and 1 μg of total RNA from the three samples. To determine the concentration of each transcript in the three cDNA samples, we used the Invitrogen™ SYBR® GreenER™ qPCR SuperMix Universal kit to prepare triplet qPCR reactions. qPCR was carried out in a Stratagene® Mx4000® Multiplex Quantitative PCR machine with the following locus-specific primer pairs: a/sExon2 and LWS1IntRev; A180SpecFwd and rev8; pExon2 and LWS2IntRev; and fw100 plus revA. A 1:10 ROX Reference Dye normalized the fluorescent reporter signal. qPCR conditions consisted of 1 cycle at 95°C (9 minutes); 50 cycles of 95°C (15 seconds), 60°C (30 seconds), 72°C (45 seconds); 1 cycle of 95°C (1 minute); and a 40-step melting curve analysis (initial temperature 55°C, increasing 1°C every 30 seconds). Each gene was also PCR amplified, cloned using the pGEM® – T Easy Vector System II kit (Promega™), sequenced to confirm identity, and then utilized for qPCR at concentrations of 1 ng, 1×10^{-3}ng and 1×10^{-5}ng per 16 μL reaction. Ct values from these plasmid templates were then used to generate a standard curve and to estimate qPCR efficiency ($qPCR_{eff} = [10(^{-1/slope}) - 1] \times 100$; Table 1). The plasmid template reactions were also run in triplicate. Dissociation curves (Fluorescence [-R'(T)] over T°C) and gel electrophoresis confirmed the presence of single amplicons in all qPCR reactions.

Table 1. Key-site haplotypes for LWS opsin duplicates in fish and humans.

Key Site Position	**Guppy LWS S180**	**Guppy LWS A180**	**Guppy LWS P180**	**Guppy LWS S180r**	Killifish LWS A	Killifish LWS B	Rice fish LWS A	Rice fish LWS B	Zebrafish LWS 1	Zebrafish LWS 2	Cave fish LWS g101	Cave fish LWS g103	Human LWS	Human MWS
180	S	A	P	S	S	S	S	S	A	A	A	A	S	A
197	H	H	H	H	H	H	H	H	H	H	H	H	H	H
277	Y	Y	F	Y	Y	Y	Y	Y	Y	F	F	F	Y	F
285	T	T	A	T	T	T	T	T	T	T	A	A	T	A
308	A	A	A	A	A	A	A	A	A	A	A	A	A	A
Expected λ_{max} (nm)	~560	~553	~531 +P	~560	~560	~560	~560	~560	~553	~546	~531	~531	~560	~531

The λ_{max} of each LWS opsin is estimated from the spectral shift predicted by each key-site substitution. The influence of the proline residue at position 180 is not known.

Phylogenetic Analyses

The Tetraodon nigroviridis LWS opsin amino acid sequence AAT38457.1 was employed as a query sequence in a BLASTp search [32] to identify homologs in the NCBI nr database. All hits with bitscores > 300 were aligned with one another and with the new data using the MPI version of ClustalW [33,34]. Subsequent sequence manipulations including multiple sequence alignments, toggle translations, hand editing, and delimitation of intron/exon boundaries utilized BioEdit v.7.0.5.3. [35]. A short nucleotide multiple sequence alignment (390 bp) that included coding sequences from exons IV and V was used to determine relationships among our new guppy LWS opsin genes and those from guppies of the Oropuche and Quare Rivers in Trinidad reported by Hoffmann et al. [23]: OR6-4 D09/DQ168660.1 and OR6-3 EO8/DQ168659.1 and QUEm5 LO6/DQ168661.1, and from the Paria River (Trinidad) guppy reported by Weadick and Chang [24]: DQ865167.1, DQ865168.1, DQ865169.1, DQ865170.1, DQ865171.1, and DQ865172.1. Maximum parsimony (MP) and Neighbor-joining (NJ) trees [36], which were based upon Tamura-Nei [37] distance estimates were reconstructed using MEGA v.4 [38]. Both analyses utilized all codon positions. Support for nodes was assessed using 1,000 bootstrap reiterations.

The MP and NJ analyses were repeated using an alignment of nucleotide sequences that varied in length from 619 to 1095 bp. LWS opsin sequences in this analysis included the following species and acquisition numbers: Zebrafish (Danio rerio), AB087803.1 and AB087804.1; Medaka (Oryzias latipes) AB223051.1 and AB223052.1; Bluefin killifish (Lucania goodei) AY296740.1 and AY296741.1; Blind cave fish (Astyanax mexicanus) M90075.1, U12024.1, and U12025.1; Sea chub (Girella punctata) AB158261.1; Nile tilapia (Oreochromis niloticus) AF247128.1; Fugu (Takifugu rubripes) AY598942.1; Spotted green pufferfish (Tetraodon nigroviridis) AY598943.1; Turbot (Scophthalmus maximus) AF385826.1; Winter flounder (Pseudopleuronectes americanus) AY631039.1; Goldfish (Carassius auratus) L11867.1; Coho salmon (Oncorhynchus kisutch) AY214145.1; Ayu smelt (Plecoglossus altivelis), AB098702.1 and AB107771.1; Atlantic halibut (Hippoglossus hippoglossus) AF316498.1; Carp (Cyprinus carpio) AB055656.1; human (Homo sapiens) NM_020061.3 and NM_000513.1; Arctic lamprey (Lethenteron japonicum) AB116381.1; and our new sequences from the Cumaná guppy (Poecilia reticulata), Picta or 'swamp guppy' (Poecilia picta), Parae (Poecilia parae), Bifurca (Poecilia bifurca), the Pygmy swordtail (Xiphophorus pygmaeus) and Tomeurus (Tomeurus gracilis).

Results

Hybrid or Mosaic Sequences

A large number of LWS opsin-like sequences (up to 17 per guppy) were uncovered after cloning and sequencing the products of PCR reactions utilizing primers Fw1a and Rev5. These sequences included suspected recombinants, that is, sequences that could have been generated by the ordered concatenation of fragments of other sequences produced in the same PCR reaction. Template switching during PCR and/or mismatch repair of cloned heteroduplex molecules has been shown to generate such artefacts [39-42]. To test the hypothesis that PCR and cloning could generate LWS opsin sequences not found in the guppy genome, we used primers Fw1a and Rev5 to re-amplify DNA from a two-sequence template (i.e., two insert-bearing plasmids). Five different sequences were uncovered from the two-template PCR reaction; one copy of each of the two templates and three recombinant sequences. These two-template experiments confirmed speculation by Hoffmann et al. [23] and Weadick and Chang [24] that LWS opsin genes in poeciliids are susceptible to PCR and/or cloning artefacts that generate artificial hybrid sequences.

To determine the minimum number of genuine LWS opsin sequences in our dataset we first considered variation at polymorphic positions. Two sequences (e.g., from two different loci or from two alleles at one locus) could serve as a PCR template for the generation of an enormous diversity of hybrid sequences via template switching or mismatch repair. However, among such a set of hybrid sequences there would be only two variants (substitutions or indels) at a given polymorphic site. LWS opsin sequences derived from individual fish using primers Fw1a and Rev5 included three different intron II haplotypes and a position in exon III that was polymorphic for three different nucleotides. Remarkably, this exon III variation translated into amino acid variation at position 180; the first of the five sites known to influence spectral sensitivity (see below). Gene duplication is the only explanation for the occurrence of three haplotypes in a single individual. We set out to strengthen this evidence for LWS opsin gene duplication by amplifying DNA between the genes (see next section).

Between-Gene PCR and Sequencing

PCR using primers designed to amplify between-gene DNA (Fw1a Comp and Rev8 Comp) produced a ~4 Kbp product in the guppy and in the three species of the guppy sister group Micropoecilia (P. picta, P. parae, and P. bifurca). These amplicons were cloned in three of these four species (cloning of this amplicon was unsuccessful in P. bifurca) and approximately 1500 bp were sequenced from each end of the clone insert. Each end of the insert contained the last intron and exon of an LWS opsin gene and approximately 790 bp beyond the stop codon. The explanation for this sequence pattern, confirmed by subsequent PCR experiments using only Rev8 Comp, was that this fragment was amplified with Rev8 Comp acting as a forward and a reverse primer and that it contained the ends of two LWS loci oriented in an inverted (tail-to-tail) fashion. The between-gene fragment did not amplify from X. pygmaeus or T. gracilis. In the guppy, additional primers were designed and the entire intergenic sequence was characterized. It was 3329 bp long, 66% A/T, and contained a short compound microsatellite; $(TGGA)_{10}(TA)_{9}$.

LWS Opsins in the Family Poeciliidae

Given the evidence for the artificial generation of opsin sequence variation during PCR or cloning, and the observation that artefacts produced by template switching and/or mismatch repair do not appear to be reproducible [41,43], only haplotypes recovered from multiple independent PCR and cloning experiments were assumed to represent genuine opsin sequences. Additional primers were designed from these reliable sequences and from the sequences obtained by the between-gene PCR experiments described above.

Initially, three different LWS opsin sequences were identified in guppy (P. reticulata). These three genes were delimited by variation at codon 180: TCT (serine), GCT (alanine), and CCT (proline), and by unique intron II and intron V mutations. Codon 180 is one of the five key positions that influence wavelength sensitivity [16] and the variation uncovered here is reflected in the names we have given to each of the loci; LWS S180 (S for serine), LWS A180 (A for alanine), and LWS P180 (P for proline).

Seven LWS S180 sequences were obtained from six different Cumaná guppies with five of these including the start codon, all exons and introns, and part of the 3' UTR. Thirteen LWS A180 sequences were obtained from seven guppies. Only one was full-length but six included sequence from exon II to the 3' UTR. The LWS A180 sequence appears to be a naturally occurring hybrid in the Cumaná guppy; the first five exons and four introns are most similar to LWS S180, whereas the last intron and exon are identical to the LWS P180 locus. As the two regions of this hybrid LWS A180 sequence will give conflicting phylogenetic signals, the LWS A180 sequences were truncated in the phylogenetic analyses reported below (i.e. only the first five exons were utilized). Seven LWS P180 sequences were obtained from seven guppies. The ForBeg primer, which includes the start codon, combined with any of the reverse primers, did not amplify the LWS P180 locus. Therefore, LWS P180 sequences spanned exon II to the 3' UTR. In addition to the proline residue at site 180, the LWS P180 locus has amino acids substitutions at two other key sites. In guppies, LWS P180 also possessed a variable-length tetranucleotide microsatellite in intron III. PCR experiments using Fw100 and a primer complementary only to Weadick and Chang's [24] variant 6 (RevA) uncovered a fourth LWS opsin gene. While we did not amplify or sequence the first exon or intron, we show that the rest of this gene is intronless, suggesting that variant 6 is a single-exon gene, arising from a retrotransposition. It has a serine at position 180 (codon: TCG) and is renamed LWS S180r (S for the serine at position 180 and r for retrotransposition). Finally, our southern blot shows four bands (Fig. 1), consistent with the PCR-based hypothesis that the Cumaná guppy has four LWS loci. These four LWS opsins encode the following five key-site haplotypes: SHYTA (LWS S180 and LWS S180r), AHYTA (LWS A180), and PHFAA (LWS P180) and are thus expected to be most sensitive to three different wavelengths of light [16] (Table 1).

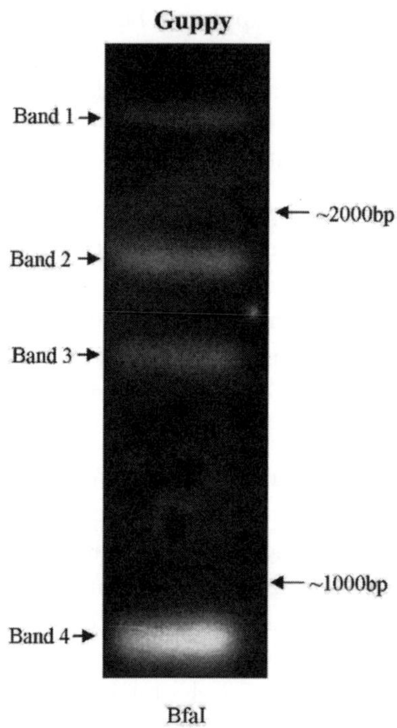

Figure 1. Southern blot hybridization for determination of LWS opsin copy-number in Poecilia reticulata. Four bands (labeled 1–4) correlate with four LWS loci from a single Cumaná guppy. BfaI (New England Biolabs®) was the restriction enzyme used. A generic DIG-labeled probe was designed to target all LWS loci (see methods). Two size markers (given in base-pairs) are shown to the right of the blot.

Long portions of the four LWS genes found in the guppy were also amplified and sequenced from P. picta and P. bifurca. Three of these opsins, LWS S180, LWS S180r, and LWS P180, were sequenced from P. parae. We did not obtain the 3' end of LWS A180 from any of these three species. Therefore the mutation producing the hybrid sequence consistently recovered from Cumaná guppy cannot yet be mapped onto the poeciliid phylogeny. Xiphophorus pygmaeus had three LWS opsin genes: LWS S180, LWS S180r, and LWS P180. Only one LWS opsin sequence (LWS S180) was recovered from Tomeurus gracilis.

All sequences have been deposited in GenBank under accession numbers EU329428 – EU329486.

Phylogenetic Analysis of LWS Opsin Gene Duplication in Poeciliidae

The guppy LWS opsin sequences obtained here and those reported by Hoffman et al. [23] were added to the 390 bp alignment reported by Weadick and Chang [24]. Phylogenetic analyses sorted these guppy LWS opsins into three well-supported clades; LWS S180r, LWS P180 and LWS S180 plus LWS A180. Weadick and Chang's [24] variant 6 clustered with the single exon gene LWS S180r, and variant 5 clustered with the LWS P180 gene. This last result was anticipated before phylogenetic reconstruction because both sequences encode a phenylalanine at position 277 and an alanine at position 285. The remaining guppy LWS opsin genes; variants 1–4 from Weadick and Chang [24], the three LWS opsins from Hoffman et al. [23] (LWS_OR6-4_D09, LWS_OR6-3_E08, LWS_QUEm5_L06) and our LWS S180 and LWS A180 genes, formed the third clade. Over this 390 bp alignment, these genes are almost identical (mean percent identity = 98.6%). We suspect that the Weadick and Chang [24] variants 1–4 and the three Hoffman et al. [23] LWS genes include alleles at the LWS A180 and LWS S180 loci. This guppy-only LWS opsin sequence comparison also revealed that Weadick and Chang's [24] variant 5 is a recombinant or mosaic sequence; the first 221 bp are identical to variant 4, and the last 170 bp are identical to our LWS P180 sequence. In our phylogenic analysis, this gene occurred in the LWS P180 clade because the region where it is identical to variant 4 has few phylogenetically informative characters.

Maximum parsimony analysis of the longer multiple sequence alignment with the new poeciliid sequences and LWS opsins from a diversity of ray-finned fish produced a single tree (Fig. 2). The NJ tree included all of the nodes from the MP tree that had bootstrap support >65% and many of the nodes with lower support. Unlike the MP tree in Fig. 2, the NJ analysis placed the Tomeurus LWS opsin as the sister sequence to a clade with the Xiphophorus and Poecilia LWS P180, LWS S180 and LWS A180 genes. This reconstruction makes more sense than the MP tree with respect to poeciliid taxonomy; morphological and molecular data indicate that Poecilia and Xiphophorus are more closely related to one another than either is to Tomeurus. However, we present the MP tree because the neighbor-joining tree also placed the bluefin killifish LWSA gene at the base of the LWS S180r clade and the Xiphophorus LWS S180 gene at the base of the LWS P180 clade. The number of gene duplication events and gene losses required to reconcile such a topology with the well-supported taxonomic relationships among these species makes these components of the NJ topology very unlikely.

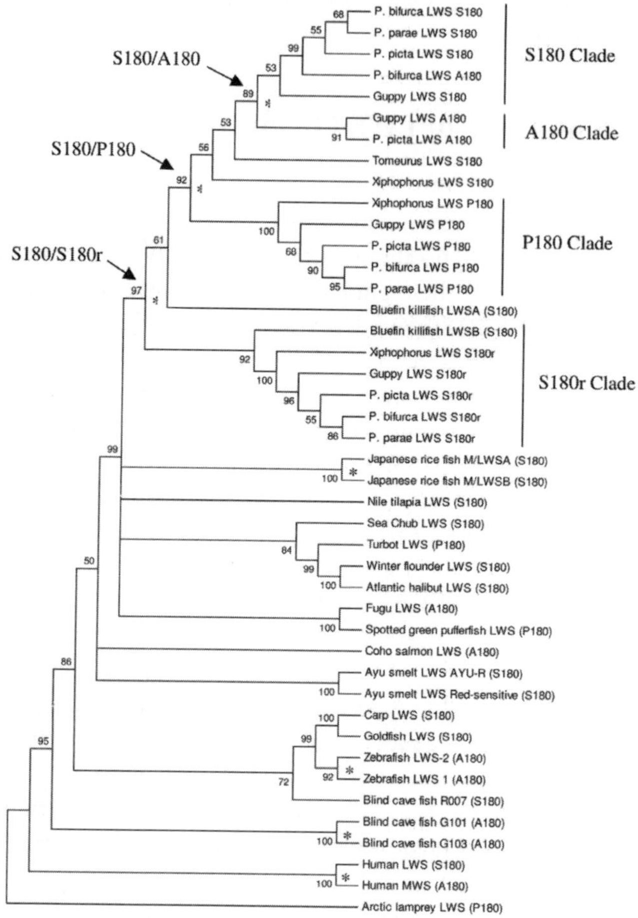

Figure 2. Phylogenetic analysis of LWS opsin genes. A Maximum Parsimony (MP) bootstrap consensus tree of long wavelength-sensitive (LWS) opsins from representative ray-finned fish lineages. The percentage of trees in which the associated taxa clustered in the bootstrap re-analyses (1000 replicates) is shown. Nodes in less than 50% of the bootstrap trees were collapsed. All codon positions were included and gaps treated as missing data. There were 1101 positions in the alignment and 520 were parsimony informative. The tree was rooted with the arctic lamprey LWS gene. Duplication events marked with an asterix. Phylogenetic analyses utilized MEGA4 [77].

The MP tree indicates that the guppy LWS opsin repertoire is a consequence of gene duplication events that occurred i) prior to the divergence of families Fundulidae and Poeciliidae, ii) in the common ancestor of Xiphophorus and Poecilia, and iii) within the genus Poecilia. Mechanisms of LWS opsin gene duplication include retrotransposition (producing LWS S180r) and inverted tandem duplication (producing the gene pair, LWS S180 and LWS P180). Formation of the hybrid LWS A180 locus may have involved quasipalindrome correction [44].

These three duplication events have provided species in this genus with a larger repertoire of LWS opsin pigments than any other fish taxon.

LWS Gene Evolution in Teleosts

Relationships among higher taxonomic groups were well resolved in the tree reconstructed from LWS opsin sequences. There is high (>75%) bootstrap support for monophyly of Cyprinidontiformes (the bluefin killifish and all poeciliids), Pleuronectiformes (sea chub, turbot and flounder), Percomorpha, and the family Cyprinidae (goldfish, carp and zebrafish). One of the blind cavefish (Astyanax mexicanus) LWS genes (R007) is the sister sequence to those from goldfish, carp and zebrafish, which is not surprising as all species are in the taxon Ostariophysi. However, there were also two cavefish LWS genes at the base of the actinopterygian clade (G101 and G103). These genes might be derived from a gene produced during the fish-specific whole genome duplication event [45]. Long Branch Attraction (LBA) occurs when rapidly evolving sequences are attracted to the base of a tree [46] and is an alternative explanation for the position of the cavefish duplicates in our analysis. However, LBA is an artefact that is usually correlated with poor taxonomic sampling, which is not the case here.

Among LWS sequences available for all ray-finned fishes, there is much variation at the five sites that influence spectral sensitivity most. Three different amino acids were observed at position 180 (A, S, or P), two at position 277 (Y or F), and two at position 285 (T or A). SHYTA is believed to be the ancestral five-site haplotype for vertebrates [16]. Serine to alanine substitutions at position 180 are common, but only guppies, turbot, and the spotted green pufferfish have a proline in this position. Lamprey, though not a ray-finned fish, also has a proline at position 180. Amino acid variation at positions 197 and 308 are also known to influence spectral sensitivity, however, all ray-finned fish surveyed to date possess only H197 and A308. LWS opsins have been duplicated at least six times in ray-finned fishes; twice within Poeciliidae, once prior to the divergence of Fundulidae and Poeciliidae, once in medaka (Oryzias latipes), once in zebrafish (Danio rerio), and again in the blind cave fish (Astyanax mexicanus). Only in poeciliids and zebrafish (and humans), has the duplication been followed by a substitution at one or more of the key sites (Table 1).

LWS Opsin Gene Expression

RT-PCR experiments show that all four transcripts were expressed at the same time in the eyes of adult guppies. We then used qPCR to compare transcript copy numbers. In three adult guppies, LWS A180 was expressed at a much higher level than LWS S180 and LWS S180r (approximately 26 and 127 times greater, respectively). LWS P180 was expressed at very low levels, with approximately 5338 times less transcript abundance than LWS A180 (Fig. 3). The amount of cDNA in the three samples was estimated by comparing critical threshold (Ct (dR)) values between samples and standard curves prepared from plasmids containing each transcript (Tables 2 and 3). Standard curve log fit values are shown in Table 3.

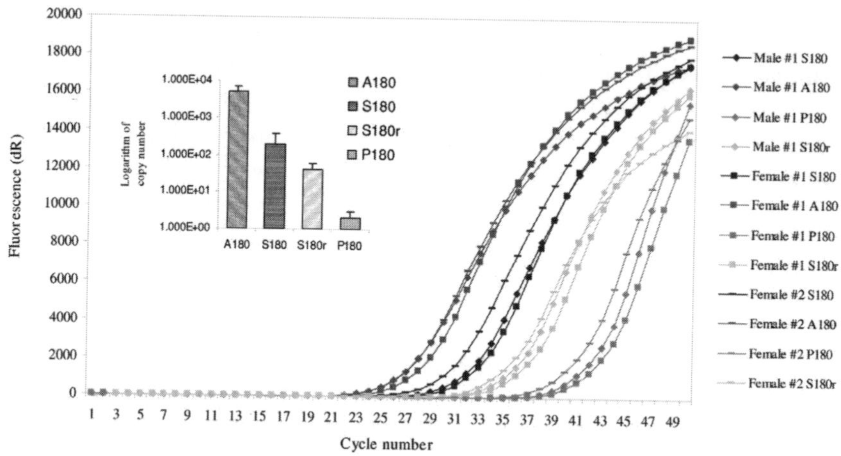

Figure 3. Quantitative PCR amplification plot and logarithm histogram of original copy number. Four LWS transcripts from the cDNA of three adult guppies were quantified using qPCR with the amplification plot shown. Ct (dR) values are given in Table 2. A histogram showing the averaged logarithm value of original transcript copy number is shown (with standard error bars included).

Table 2. Quantitative PCR data.

Individual	Transcript	Ct (dR)	Original Copy Number
Male	*A180*	24.35	6758
Male	*S180*	29.54	121
Male	*S180r*	33.12	39
Male	*P180*	39.01	1+
Female #1	*A180*	25.46	3126
Female #1	*S180*	30.10	78
Female #1	*S180r*	33.90	24
Female #1	*P180*	39.58	1+
Female #2	*A180*	24.49	6131
Female #2	*S180*	27.93	416
Female #2	*S180r*	32.36	63
Female #2	*P180*	37.51	1+

Ct (dR) values are shown for each LWS opsin transcript from three adult guppies. Original copy number was calculated using Ct (dR) values and the standard curve data shown in Table 3.

Table 3. Standard curve log fit values for quantitative PCR.

(Plasmid) Standard Curve	Linear Equation	Efficiency (%)	RSq
A180	Y = -3.315(logX) + 37.05	100.3	0.999
S180	Y = -2.992(logX) + 35.77	115.9	0.996
S180r	Y = -3.670(logX) + 38.96	87.3	0.985
P180	Y = -3.319(logX) + 36.68	100.1	0.999

The linear equation, efficiency and r-squared (RSq) values for the qPCR standard curve are shown. Plasmid copy numbers used as standards were determined by calculating the number of moles in each sample multiplied by Avogadro's number. A standard curve of Ct (dR) was used to calculate original copy number of mRNA transcripts in the diluted cDNA samples.

Discussion

Variation at Key Sites Among LWS Opsin Gene Duplicates in Poeciliidae

The guppy (Poecilia reticulata) and species in its sister group Micropoecilia possess four LWS opsin genes that we have named LWS S180, LWS S180r, LWS A180, and LWS P180. The first two genes encode proteins with the five key-site haplotype, SHYTA, and the second two genes encode proteins with key-site haplotypes, AHYTA and PHFAA, respectively. Three of these genes; LWS S180, LWS S180r, and LWS P180, were also amplified and sequenced from pygmy swordtail (Xiphophorus pygmaeus) genomic DNA. We found only the LWS S180 gene in Tomerus gracilis. Serine (S) and alanine (A) are common residues at position 180, but proline (P) is rare. This proline residue encoded by LWS P180 might disrupt the transmembrane domain [47,48] and compromise opsin protein function [48,49]. However, several observations suggest that it is functional. First, LWS P180 is at least 44 million years old, as it evolved before the divergence of Poecilia and Xiphophorus [26], and it has no other amino acid substitutions that are expected to disrupt function. Second, the LWS P180 locus has diverged from paralogous LWS opsins in ways that are expected to enhance color vision; positions 277 and 285 have experienced a tyrosine to phenylalanine and threonine to alanine substitution, respectively. These are the same key-site substitutions involved in the evolution of an MWS opsin from an LWS opsin in humans. Third, this gene is expressed, albeit at very low levels. Finally, LWS opsins from arctic lamprey, turbot, and the spotted green pufferfish also have a proline at position 180 and no other substitutions likely to disrupt protein function.

All four LWS opsins uncovered in this study are predicted to have unique roles in color vision. With three different five key-site haplotypes, they are predicted to be most sensitive to three different wavelengths of light. Also, despite encoding a gene with the same key-site haplotype as LWS S180, the LWS S180r opsin differs from all other LWS opsins at amino acid positions known to play a role in binding and activating transducin [24].

Southern blot experiments in our study revealed four bands (Fig. 1) consistent with the hypothesis that Cumaná guppies have four LWS opsin loci. Hoffman et al. [23] produced a southern blot with only three bands and suggested that guppies have a minimum of two LWS opsin genes. Variation in LWS opsin gene number among populations may be another trait guppies share with humans [50-52].

Two of the four LWS opsin genes described here (LWS S180 and LWS A180) were reported in Hoffman et al.'s [23] study of guppies from the Quare and Oropuche Rivers in Trinidad. Although sequence data reported by Weadick and Chang [24] did not include all five key sites, our phylogenetic analysis indicates that Weadick and Chang [24] sequenced portions of all four loci from their Paria River guppy. The phylogenetic relationships among guppy LWS opsin paralogs reported by Weadick and Chang [24] differ from those shown here. Both topologies were produced using maximum parsimony, but by surveying more individuals, more species and by obtaining longer sequences we have produced a larger set of parsimony-informative characters. This is most evident when considering the relationship between LWS P180 and Weadick and Chang's variant 5, and variant 4. The large number of differences between variant 5 and variant 4 are apparent in Weadick and Chang's [24] tree where maximum likelihood branch lengths have been superimposed on the MP topology. Nonetheless, in their analysis, these two sequences form a monophyletic group. However, the sister sequence relationship between variant 5 and variant 4 disappears with the addition of LWS P180 genes from other Poecilia species and Xiphophorus pygmaeus because many of the unique nucleotides (autopomorphies) in variant 5 become synapomorphies (shared derived traits) in this larger dataset. Also, two of the three characters that had united variant 5 and variant 4 in Weadick and Chang's [24] MP analysis (adenines at positions 126 and 213 in their alignment), appear to be homoplasious when compared to a much larger set of LWS P180 and LWS S180 sequences. The origin of these apparent homoplasies is intriguing and is discussed below.

Mechanisms of LWS Duplication in Poeciliids

The first duplication event that expanded the guppy LWS opsin repertoire produced two genes that have retained SHYTA five key-site haplotype: LWS S180 and LWS S180r. The later duplicate is missing introns II-V and is likely a product of retrotransposition. Partial cDNA sequences for each gene have been reported [24] but their intron-exon structure was unknown until now. It is not clear how LWS S180r has retained retinal expression, although it may occur in the vicinity of other LWS opsins (and their regulatory modules) in the guppy genome. Of interest, gene duplication by retrotransposition has also produced a pair of fish rhodopsin 1 (RH1) genes called errlo and single-exon rho [53]. errlo and rho occur on different chromosomes. Therefore, the observation that rho is expressed in the retina (rod cells) [53] demonstrates that upstream regulatory elements can be retained during retrotransposition.

In medaka (Oryzias latipes) and zebrafish (Danio rerio), duplicated LWS opsins are linked and oriented in a head-to-tail manner [54,55]. Phylogenetic analysis shows that independent mutations produced these gene pairs [54]. Our study also characterized an LWS opsin tandem duplication event. Duplication of the LWS S180 gene early during the evolution of poeciliids, produced an LWS opsin that retained the SHYTA haplotype (LWS S180) and an LWS opsin that evolved a PHFAA haplotype (LWS P180). These two genes are linked, but in an inverted (i.e., tail-to-tail) orientation. Several models have been proposed to explain the formation of inverted duplicates. Secondary rearrangement after duplication by unequal sister chromatid exchange is one. Another is intra-chromosomal replication slippage in trans [56]. This occurs when the DNA polymerase reverses direction using either the nascent strand (intra-molecular strand switch) or opposite strand (inter-molecular strand switch) as a template. By running backwards, a duplicate of the just-completed sequence is produced in an inverted orientation before the polymerase switches back to the correct template. The DNA downstream of LWS S180 has strings of adenines and thymines (data not shown) that might have facilitated strand switching by the polymerase during DNA replication [57]. The inverted arrangement of LWS opsins in the Poecilia genome might make them even more prone to additional duplication events [58]. Therefore, variation in LWS gene number (among species or populations) would not be surprising. Xiphophorus pygmaeus also has the LWS P180 gene. Post-duplication gene transposition or the expansion of intergenic DNA are possible explanations for our failure to amplify DNA between LWS P180 and LWS S180 in this species.

The third and most recent LWS gene duplication uncovered in our study lead to the production of LWS S180 and LWS A180. In the Cumaná guppy, the first five exons and four introns of LWS A180 are most similar to LWS S180, and the last intron, exon, and 3' UTR are identical to LWS P180. These data suggest that in this population LWS A180 is a hybrid gene; its formation might have been facilitated by the inverted tandem orientation of LWS S180 and LWS P180 [44]. As mentioned above, variant 5, reported by Weadick and Chang [24] is also a hybrid sequence, with approximately one half of the sequence identical to the LWS S180 gene and the other half identical to LWS P180. As is the case for LWS A180 reported here in the Cumaná guppy, the position of variant 5 in the tree based upon the 390 bp alignment depends upon which fragments are used in the phylogenetic analysis (i.e., it would occur in the LWS S180/A180 clade if only the first 220 bp were utilized). As there are many more phylogenetically informative characters in the second half of this sequence, variant 5 was placed in the LWS P180 clade when the entire sequence was used, despite being identical to LWS A180 sequences over the first 220 bp. The observation that variant 5 has not diverged from either of its progenitor sequences and that it was recovered only once by Weadick and Chang [24] leads us to the conclusion that while both halves of the sequence can be found in the guppy genome, their concatenation is an artefact produced by template switching or mismatch repair during cloning.

In the larger alignment (Fig. 2), LWS S180 and LWS A180 are not partitioned into monophyletic clades. For instance, the P. bifurca sequence that clusters with the LWS S180 genes has an alanine at position 180. One explanation for this observation is that the P. bifurca LWS A180 sequence is an allele of the LWS S180 locus. A similar situation occurs in non-African humans where a common allele of the LWS opsin locus (which typically has the SHYTA haplotype) has an alanine in position 180 and thus, an AHYTA five key-site haplotype [59,60].

The Evolutionary Consequences of LWS Opsin Duplication in Guppies

The evolutionary implications of opsin gene duplication and divergence depend largely upon the expression patterns of these genes. In several species, the possession of a large opsin family allows the retina to be spectrally tuned for different environments and/or life stages. For example, eels (Anguilla anguilla) have two rhodopsins, each tuned to slightly different wavelengths. They express a green-shifted locus as juveniles in fresh water and a paralogous blue-shifted locus when they return to the ocean and mature [61]. The lamprey (G. australis) also adjusts its spectral sensitivity by changing opsin gene expression as it moves between marine and riverine environments [62]. In cichlids, opsin gene expression varies during development [63]. Of particular note is the observation that variation in LWS opsin sequence and expression is associated with variation in water turbidity [9,64]. This has lead to the hypothesis that species- and population-level differences in opsin gene sequence and expression represent adaptations for foraging in either turbid or clear water and that these differences in spectral sensitivity may drive and/or maintain divergence in male coloration via sexual selection [11]. Guppies, however, do not move very far during their lifetime [65] and thus differential use of opsin gene duplicates in different habitats is an unlikely explanation for the evolution of LWS opsin gene diversity in this taxon.

The simultaneous expression of opsin paralogs with different sensitivities might expand the region of the spectrum where the guppy possesses high sensitivity and expand the range of detectable wavelengths. This enhancement and broadening of wavelength sensitivity can occur when individual cone cells express more than one opsin gene [66] or when adjacent cone cells express different opsins (e.g., humans and transgenic mice – see below). MSP data in guppies showing cells with a broad range of sensitivities in the long wave region of the spectrum [18,19] are consistent with the hypothesis that the sensitivity of some cones is a consequence of the co-expression of different LWS opsins. By providing guppies with a broad region of maximum wavelength sensitivity, LWS opsin gene duplication and divergence might make multi-colored male guppies appear brighter (more conspicuous) [67] to other guppies, but not to predators with wavelength sensitivity limited (by LWS opsin gene copy number) to a narrower region of the spectrum.

Expression of different LWS opsins in adjacent cones not only improves overall spectral sensitivity, but is the basis of wavelength discrimination. Observations from humans and mice are consistent with the hypothesis that opsin gene duplication and divergence can lead to better color discrimination even without any associated revisions to neuroanatomy. Among human women who are heterozygous at either the LWS or MWS locus, some appear to have a pattern of X-inactivation that leads to tetrachromacy. These women see an average of 10 colors in a spectrum, whereas trichromatic women typically see seven [[68], but see reference [69]]. In mice, the hypothesis that extra opsin genes can improve wavelength discrimination was supported by data from females expressing an SWS opsin and two LWS opsins (an endogenous LWS gene on one X chromosome and a human LWS opsin on the other). These knock-in mice performed better in wavelength discrimination tests than wild-type mice with a single LWS gene [70].

Sexual selection in guppies favors males with more red, orange, and yellow color patches [71-73] suggesting that females use color diversity (chroma) to evaluate males. This may be because males with more chroma are more conspicuous [67]. However, the 'extra' chroma is a consequence of the guppy visual system and this conspicuousness may not, therefore, apply to predators. Finally, if LWS opsin gene duplication improves motion detection, as proposed by White et al. [74] then female guppies might also be 'pre-adapted' to evaluate the well-characterized sigmoid display, a behavior that consists of the male arching its body into a S-shape and oscillating the long axis of the body both horizontally and vertically [75].

LWS Expression in the Guppy

Gene expression data will help us to test alternative hypotheses about the adaptive value of LWS opsin diversity. To expand the range of maximum sensitivity and enhance wavelength discrimination, it is necessary that different opsins be expressed at the same time. All four LWS opsin gene transcripts were amplified from cDNA derived from adult eyes in our lab and by Weadick and Chang [24]. However, our qPCR experiments on three adults (1 male, 2 females) showed that most of the LWS opsin mRNA in the Cumaná guppy retinas was derived from the LWS A180 gene. Human SWS (blue) cone cells make up only 15% of the retina cone cell repertoire, yet play an important role in wavelength discrimination. Therefore, qPCR data showing unequal expression among LWS opsin paralogs in three adults do not rule out a role for LWS opsin gene duplication and divergence in wavelength discrimination in guppies, but does indicate the need for further investigation. We are currently using qPCR to examine LWS expression in a larger sample of adults and in fish at different stages of development. Finally, duplicated opsins are sometimes expressed in different regions of the retina [55,76]. In-situ hybridization experiments are underway to test the hypothesis that different LWS opsin paralogs have unique expression domains within the guppy retina, as is the case in zebrafish [76].

Conclusion

Gene duplication and divergence has provided Poecilia and its close relatives with four distinct LWS opsins; a larger repertoire than any other fish. Phylogenetic analyses suggest that three of these LWS opsins (LWS S180r, LWS P180 and LWS S180) were present very early in poeciliid evolution and are predicted to occur in all of the approximately 239 species of the subfamily Poeciliinae. Adult guppies express all four LWS paralogs simultaneously, albeit at varying levels. As a consequence of these gene duplications, the potential for a broad region of high sensitivity and/or enhanced wavelength discrimination in the long-wave portion of the visible spectrum, may have facilitated a red-orange color bias for sexual selection within the guppy.

Author Contributions

JST supervised the study. MNW and JST designed and implemented wet-lab experimentation and data-analysis. AMC discovered the LWS P180 locus and carried out the template switching/mismatch repair experiment. KJD carried out the between-gene PCR and sequencing experiments. CRJL and GLO assisted with RT-PCR, qPCR and sequencing of the S180r locus. FB and JST obtained all samples. MDP and PRW assisted in RT-PCR. MNW and JST wrote the manuscript with editing assistance from FB.

Acknowledgements

The authors thank Roderick Haesevoets for help with DNA sequencing, Shelby Temple and two anonymous referees for comments on the manuscript and

Gustavo Ybazeta for help with the phylogenetic analysis. The Environmental Protection Agency of Guyana, Ministry of Agriculture of Guyana, and Animal Husbandry and Fisheries of Suriname provided research and collecting permits. Fish were collected in Venezuela in collaboration with D. Taphorn of UNELLEZ (permit no. 0497). This work was funded by grants from the Canadian Foundation for Innovation and the British Columbia Knowledge Development Fund (JST), Natural Sciences and Engineering Research Council of Canada (JST and FB) and the National Geographic Society (FB).

References

1. Andersson M: Sexual Selection. Princeton University Press, Princeton, New Jersey; 1994:599.
2. Breden F, Stoner G: Male predation risk determines female preference in the Trinidad guppy. Nature 1987, 329:831–33.
3. Bakker TCM: Positive genetic correlation between female preference and preferred male ornament in sticklebacks. Nature 1993, 363:255–57.
4. Houde A: Sex, Color, and Mate Choice in Guppies. Princeton University Press, Princeton, New Jersey; 1997:210.
5. Haesler MP, Seehausen O: Inheritance of female mating preference in a sympatric sibling species pair of Lake Victoria cichlids: implications for speciation. Proc R Soc B 2005, 272:237–45.
6. Salzburger W, Niederstätter H, Brandstätter A, Berger B, Parson W, Snoeks J, Sturmbauer C: Color-assortative mating among populations of Tropheus moorii, a cichlid fish from Lake Tanganyika, East Africa.Proc R Soc B 2006, 273:257–66.
7. Streelman JT, Albertson RC, Kocher TD: Genome mapping of the orange blotch pattern in cichlid fishes. Mol Ecol 2003, 12:2465–71.
8. Terai Y, Mayer WE, Klein J, Tichy H, Okada N: The effect of selection on a long wavelength-sensitive (LWS) opsin gene of Lake Victoria cichlid fishes. Proc Natl Acad Sci USA 2002, 99:15501–06.
9. Terai Y, Seehausen O, Sasaki T, Takakashi K, Mizoiri S, Sugawara T, Sato T, Watanabe M, Konijnendijk N, Mrosso HDJ, Tachida H, Imai H, Schichida Y, Okada N: Divergent selection on opsins drives incipient speciation in Lake Victoria cichlids. PLoS Biol 2006, 4:2244–51.
10. Carlton KL, Parry JWL, Bowmaker JK, Hunt DM, Seehausen O: Color vision and speciation in Lake Victoria cichlids of the genus Pundamilia. Mol Ecol 2006, 14:4341–53.

11. Maan ME, Hofker KD, van Alphen JJM, Seehausen O: Sensory drive in cichlid speciation. Am Nat 2006, 167:947–54.

12. Masland RH: The fundamental plan of the retina.Nat Neurosci 2001, 4:877–86.

13. Nathans J, Thomas D, Hogness DS: Molecular genetics of human color vision: The genes encoding blue, green, and red pigments. Science 1986, 232:193–232.

14. Dulai KS, von Dornum M, Mollon JD, Hunt DM: The evolution of trichromatic color vision by opsin gene duplication in New World and Old World primates. Genome Res 1999, 9(7):629–38.

15. Yokoyama S, Radlwimmer FB: The molecular genetics of red and green color vision in mammals. Genetics 1999, 153(2):919–32.

16. Yokayama S, Radlwimmer FB: The molecular genetics and evolution of red and green color vision in vertebrates. Genetics 2001, 158(4):1697–1710.

17. Chinen A, Hamaoka T, Yamade Y, Kawamura S: Gene duplication and spectral diversification of cone visual pigments of Zebrafish. Genetics 2003, 163(2):663–75.

18. Archer SN, Endler JA, Lythgoe JN, Partridge JC: Visual pigment polymorphism in the guppy, Poecilia reticulata. Vis Res 1987, 27:1243–52.

19. Archer SN, Lythgoe JN: The visual pigment basis for cone polymorphism in the guppy, Poecilia reticulata.Vis Res 1990, 30:225–33.

20. Bourne GR, Breden F, Allen TC: Females prefer carotenoid males as mates in the pentamorphic livebearing fish, Poecilia parae.Naturwissenschaften 2003, 90:402–05.

21. Endler JA: Natural selection on color patterns in Poecilia reticulata. Evol 1980, 34:76–91.

22. Körner KE, Schlupp I, Plath M, Loew ER: Spectral sensitivity of mollies: comparing surface- and cave-dwelling Atlantic mollies, Poecilia mexicana. J Fish Biol 2006, 69:54–65.

23. Hoffmann M, Tripathi N, Henz SR, Lindholm AK, Weigel D, Breden F, Dreyer C: Opsin gene duplication and diversification in the guppy, a model for sexual selection. Proc R Soc Lond B 2007, 274:33–42.

24. Weadick CJ, Chang BSW: Long-wavelength sensitive visual pigments of the guppy (Poecilia reticulata): six opsins expressed in a single individual. BMC Evol Biol 2007, 7(suppl 1):S1–S11.

25. Meyer A, Lydeard C: The evolution of copulatory organs, internal fertilization, placentae and viviparity in killifishes (Cyprinodontiformes) inferred

from a DNA phylogeny of the tyrosine kinase gene X-src. Proc Biol Sci 1993, 245:153–62.

26. Hrbek T, Seckinger J, Meyer A: A phylogenetic and biogeographic perspective on the evolution of poeciliid fishes. Mol Phylogen Evol 2007, 43:986–98.

27. Breden F, Ptacek MB, Rashed M, Taphorn D, Figueiredo CA: Molecular phylogeny of a live-bearing fish genus Poecilia (Poeciliidae: Cyprinodontiformes). Mol Phylogen Evol 1999, 12:95–104.

28. Rosen DE, Bailey RM: The poeciliid fishes (Cyprinodontiformes), their structure, zoogeography and systematics. Bull Am Mus Nat Hist 1963, 126:1–176.

29. Alexander HJ, Breden F: Sexual isolation and extreme morphological divergence in the Cumaná guppy: a possible case of incipient speciation. J Evol Biol 2004, 17:1238–54.

30. Halstenberg S, Lindgren KM, Samagh SPS, Nadal-Vicens M, Balt S, Fernald RD: Diurnal rhythm of cone opsin expression in the teleost fish Haplochromis burtoni. Vis Neurosci 2005, 22:135–41.

31. Li P, Temple S, Gao Y, Haimberger TJ, Hawryshyn CW, Li L: Circadian rhythms of behavioral cone sensitivity and long wavelength opsin mRNA expression: a correlation study in zebrafish. J Exp Biol 2005, 208:497–504.

32. Altschul SF, Gish W, Miller W, Myers EW, Lipman DJ: Basic local alignment search tool. J Mol Biol 1990, 215(3):403–10.

33. Thompson JD, Higgins DG, Gibson TJ: ClustalW: improving the sensitivity of progressive multiple sequence alignment through sequence weighting, position- specific gap penalties, and weight matrix choice.Nuc Acids Res 1994, 22:4673–4780.

34. Li K-B: ClustalW-MPI: ClustalW analysis using distributed and parallel computing. Bioinfor 2003, 19:1585–86.

35. Hall TA: BioEdit: a user-friendly biological sequence alignment editor and analysis program for Windows 95/98/NT.Nucl Acids Symp Ser 1999, 41:95–98.

36. Saitou N, Nei M: The neighbor-joining method: A new method for reconstructing phylogenetic trees. Mol Biol Evol 1987, 4(4):406–25.

37. Tamura K, Nei M: Estimation of the number of nucleotide substitutions in the control region of mitrochondrial DNA in humans and chimpanzees. Mol Biol Evol 1993, 10(3):512–26.

38. Kumar S, Tamura K, Nei M: MEGA4: Integrated software for Molecular Evolutionary Genetics Analysis and sequence alignment. Briefings in Bioinfor 2004, 5:150–63.

39. Saiki RK, Gelfand DH, Stoffel S, Scharf SJ, Higuchi R, Horn GT, Mullis KB, Erlich HA: Primer-directed enzymatic amplification of DNA with a thermostable DNA polymerase. Science 1988, 239:487–91.

40. Oldenberg SJ, Weiss RB, Hata A, White R: Template-switching during DNA synthesis by Thermus aquaticus DNA polymerase I.Nuc Acids Res 1995, 23:2049–57.

41. Zylstra P, Rothenfluh HS, Weiller GF, Blanden RV, Steele EJ: PCR amplification of murine immunoglobulin germline V genes: strategies for minimization of recombination artifacts. Immun Cell Biol 1998, 76:395–405.

42. Shammas FV, Heikkilä R, Osland A: Fluorescence-based method for measuring and determining the mechanisms of recombination in quantitative PCR. Clinica Chimica Acta 2001, 304:19–28.

43. Cronn R, Cedroni M, Haselkorn T, Grover C, Wendel JF: PCR-mediated recombination in amplification products derived from polyploid cotton. Theor Appl Genet 2002, 104:482–89.

44. van Noort V, Worning P, Ussery DW, Rosche WA, Sinden RR: Strand misalignments lead to quasipalindrome correction. Trends Genet 2003, 19:365–69.

45. Taylor JS, Braasch I, Frickey T, Meyer A, Peer Y: Genome duplication, a trait shared by 22,000 species of ray-fined fish. Genome Res 2003, 13:382–90.

46. Bergsten J: A review of long-branch attraction.Cladistics 2005, 21:163–93.

47. Pakula AA, Sauer RT: Genetic analysis of protein stability and function. Annu Rev Genet 1989, 23:289–310.

48. Weitz CJ, Miyake Y, Shinzato K, Montag E, Zrenner E, Went LN, Nathans J: Human tritanopia associated with two amino acid substitutions in the blue-sensitive opsin. Am J Hum Genet 1992, 50(3):498–507.

49. Gunther KL, Neitz J, Neitz M: A novel mutation in the short-wavelength-sensitive cone pigment gene associated with a tritan color vision defect. Vis Neurosci 2006, 23:403–09.

50. Jagla WM, Jägle H, Hayashi T, Sharpe LT, Deeb SS: The molecular basis of dichromatic vision in males with multiple red and green visual pigment genes. Hum Mol Genet 2002, 11(1):23–32.

51. Macke JP, Nathans J: Individual variation in the size of the human red and green visual pigment gene array. Invest Ophthalmol Vis Sci 1997, 38(5):1040–43.

52. Wolf S, Sharpe LT, Hans-Jûrgen AS, Knau H, Weitz S, Kioschis P, Poustka A, Zrenner E, Lichter P, Wissinger B: Direct visual resolution of gene copy number in the human photopigment gene array. Inv Ophtha & Vis Res 1999, 40(7):1585–89.

53. Bellingham J, Tarttelin EE, Foster RG, Wells DJ: Structure and evolution of the teleost extraretinal rod-like opsin (errlo) and ocular rod opsin (rho) genes: is teleost rho a retrogene? J Exp Zool 2003, 297(B):1–10.

54. Matsumoto Y, Fukamachi S, Mitani H, Kawamura S: Functional characterization of visual opsin repertoire in Medaka (Oryzias latipes). Gene 2006, 371:268–78.

55. Chinen A, Matsumoto Y, Kawamura S: Reconstruction of ancestral green visual pigments of zebrafish and molecular mechanism of their spectral differentiation. Mol Biol Evol 2005, 22:1001–10.

56. Chen J-M, Chuzhanova N, Stenson PD, Férec C, Cooper DN: Intrachromosomal serial replication slippage in trans gives rise to diverse genomic rearrangements involving inversion. Hum Mut 2005, 26:362–73.

57. Hyrien O, Debatisse M, Buttin G, de Saint Vincent BR: The multicopy appearance of a large inverted duplication and the sequence at the inversion joint suggests a new model for gene amplification. EMBO J 1988, 7:407–17.

58. Passananti C, Davies B, Ford M, Fried M: Structure of an inverted duplication formed as a first step in a gene amplification event: implications for a model of gene amplification. EMBO J 1987, 6:1697–1703.

59. Verrelli BC, Tishkoff SA: Signatures of selection and gene conversion associated with human color vision variation. Am J Hum Genet 2004, 75:363–75.

60. Winderickx J, Battisti L, Hibiya Y, Motulski AG, Deeb SS: Haplotype diversity in the human red and green opsin genes: evidence for frequent sequence exchange in exon 3. Hum Mol Genetics 1993, 2(9):1413–21.

61. Archer S, Hope A, Partridge JC: The molecular basis for the green-blue sensitivity shift in the rod visual pigments of the European eel. Proc R Soc Lond B 1995, 262:289–95.

62. Davies WL, Cowing JA, Carvalho LS, Potter IC, Trezise AEO, Hunt DM, Collin SP: Functional characterization, tuning, and regulation of visual pigment gene expression in an anadromous lamprey. FASEB J 2007, 21:2713–24.

63. Spady TC, Parry JWL, Robinson PR, Hunt DM, Bowmaker JK, Carleton KL: Evolution of the cichlid visual palette through ontogenetic subfunctionalization of the opsin gene arrays. Mol Biol Evol 2006, 23(8):1538–47.

64. Carelton KL, Kocher TD: Cone opsin genes of African cichlid fishes: Tuning spectral sensitivity by differential gene expression. Mol Biol Evol 2001, 18(8):1540–50.

65. Crispo E, Bentzen P, Reznick DN, Kinnison MT, Hendry AP: The relative influence of natural selection and geography on gene flow in guppies. Mol Ecol 2006, 15:49–62.

66. Lukáts Á, Dkhissi-Benyahya O, Szepessy Z, Röhlich P, Vigh B, Bennett NC, Cooper HM, Szél Á: Visual pigment coexpression in all cones of two rodents, the Siberian hamster, and the pouched mouse. Invest Ophthalmol Vis Sci 2002, 43(7):2468–73.

67. Endler JA: Variation in the appearance of guppy color patterns to guppies and their predators under different visual conditions. Vis Res 1991, 31(3):587–608.

68. Jameson KA, Highnote SM: Richer color experience in observers with multiple photopigment opsin genes. Psychon Bull & Rev 2001, 8(2):244–61.

69. Hood SM, Mollon JD, Purves L, Jordan G: Color discrimination in carriers of color deficiency. Vis Res 2006, 46:2894–2900.

70. Jacobs GH, Williams GA, Cahill H, Nathans J: Emergence of novel color vision in mice engineered to express a human cone photopigment. Science 2007, 315:1723–25.

71. Endler JA: Natural and sexual selection on color patterns in poeciliid fishes. Env Biol Fish 1983, 9:173–90.

72. Houde AE: Mate choice based upon naturally occurring color pattern variation in a guppy population. Evol 1987, 41:1–10.

73. Houde AE, Endler JA: Correlated evolution of female mating preferences and male color patterns in the guppy, Poecilia reticulata. Science 1990, 248:1405–08.

74. White EM, Church SC, Willoughby LJ, Hudson SJ, Partridge JC: Spectral irradiance and foraging efficiency in the guppy, Poecilia reticulata. An Beh 2005, 69(3):519–27.

75. Luyten PH, Liley NR: Geographic variation in the sexual behavior of the guppy, Poecilia reticulata (Peters). Behav 1985, 95:164–79.

76. Takechi M, Kawamura S: Temporal and spatial changes in the expression pattern of multiple red and green subtype opsin genes during zebrafish development. J Exp Biol 2005, 208:1337–45.

77. Nei M, Kumar S: Molecular evolution and phylogenetics. New York: Oxford University Press; 2000.

78. Felsenstein J: Confidence limits on phylogenies: An approach using the bootstrap. Evol 1985, 39:783–91.

79. Tamura K, Dudley J, Nei M, Kumar S: MEGA4: Molecular Evolution Genetics Analysis (MEGA) software version 4.0.Mol Biol Evol 2007, 24:1596–99.

Plasticity of Electric Organ Discharge Waveform in the South African Bulldog Fish, *Marcusenius pongolensis*: Tradeoff Between Male Attractiveness and Predator Avoidance?

Susanne Hanika and Bernd Kramer

ABSTRACT

Background

In adult male Marcusenius pongolensis the duration of their Electric Organ Discharge (EOD) pulses increases with body size over lifetime (267 to 818 μs, field-measured). Spawning males have been observed to exhibit an additional, temporary pulse duration increase which probably betters their mating success but increases predation risk by electroreceptive catfish. We here study the question of how the additional pulse duration increase is triggered and for how long it persists, in an attempt to understand the compromise between opposing selective forces.

Results

Here, we demonstrate short-term plasticity in male EOD waveform in 10 captive M. pongolensis. An increase in EOD duration was experimentally evoked in two different ways: by exchanging the familiar neighbours of experimental subjects for stranger males that were separated by plastic mesh partitions, or by separating familiar fish by plastic mesh partitions introduced into their common tank. Both treatments evoked an increase of male EOD duration. Values exceeded those found in the non-reproductive season in nature. In one male the increase of EOD duration was 5.7 fold, from 356 μs to 2029 μs. An increase in EOD duration was accompanied by a high level of aggression directed against the neighbours through the plastic mesh. With conditions remaining constant, EOD duration receded to 38 – 50% of the maximum EOD duration after 10 weeks, or, more rapidly, when sensory contact between the fish was severely restricted by the introduction of a solid plastic wall.

Conclusion

The short-term increase of EOD duration evoked by experimental manipulation of sensory contact with conspecifics through the plastic mesh, as reported here, resembled the changes in EOD waveform that accompanied reproduction in two captive males. Plasticity of the male EOD in pulse duration seems to be an adaptation for (1) securing a higher fitness by a sexually "attractive" long-duration EOD, while (2) limiting the risk of detection by electroreceptive predators, such as the sharptooth catfish, by receding to a shorter EOD as soon as reproduction is over.

Background

Mormyrid fish communicate by continuously emitting weak electric organ discharges (EODs; for reviews see [1-6]). Waveform of EODs and rhythm of discharge are both important factors in mormyrid communication. The rhythm of discharge is highly variable, accompanying the moment-to-moment fluctuations of social context (e.g., [7-10]). Individual waveforms of EOD are usually constant over long periods of time [11-13] and signal species identity; they may also vary between the sexes and signal individual identity [14]; reviews, [1,3,15].

Although genuinely sexually dimorphic waveforms of EOD are rare (sensu two alternative forms [14]), in some genera of mormyrid fish, such as Marcusenius, male EOD waveforms of certain species are more diverse than those of females [16]. This is also true in some members of the Marcusenius macrolepidotus, or Bulldog, species complex, such as Marcusenius pongolensis. However, differences between the sexes were less distinct than those observed in an allopatric population of Bulldog fish, M. altisambesi from the Upper Zambezi River (Namibia), the males of which generate EODs of up to 4700 µs, or 11× the female average, in the wet season (the two forms of Bulldog fish are now considered to have differentiated on the species level; [17,18]). The EOD duration values of M. pongolensis females were very similar amongst each other and did not change over lifetime. Juvenile males' EODs resembled those of females, but, on turning mature, increased statistically significantly with standard length over lifetime, up to 3 fold in the biggest males.

Experimental playback of long-duration EODs of sympatric males via an electric dipole fish model evoked a significant increase in male aggression towards the stimulus dipole in M. pongolensis males [19]. This result suggests that a long EOD pulse duration signals size, fighting ability, or reproductive status of males, and may thus confer fitness advantages.

EODs of long duration may not only attract females but also electroreceptive predators. The common sharptooth catfish Clarias gariepinus detects M. altisambesi male EODs with pulse duration greater than 2 ms at considerable distance, but not EODs of females and males with EOD duration as found in the non-reproductive season [20,21]. Accordingly, Bulldogs were found to represent the main prey of C. gariepinus and certain other catfish in the Okavango/Upper Zambezi River systems during certain periods of the year [22,23]. In contrast, male M. pongolensis from South Africa studied in the field were not found to form part of these catfishes' diet [24]. A possible reason for this difference is that in M. pongolensis EOD duration (hence spectral low-frequency content of EOD pulses) remains well below the critical detection threshold of C. gariepinus [21].

In two M. pongolensis males an EOD duration increase that was limited to the short period of actual spawning has been observed in the laboratory [10,18], demonstrating short-term plasticity of EOD waveform within the male sex. Here, we characterise short-term plasticity of male EOD waveform during territorial encounters more clearly, and identify the conditions and limits of an EOD duration increase.

Results

Behavior

Under Treatment 1, one or two 'familiar' male neighbours kept in compartments adjacent to an experimental subject were exchanged for unfamiliar male stranger fish. After an exchange, all three males in their serially adjoining compartments (Fig. 1, tanks A and B) usually behaved aggressively towards their respective neighbour(s) at the plastic mesh partitions even during the day. The most frequent aggressive act was Parallel Swimming that was often accompanied by vigorous attempts to butt or bite through the plastic mesh (described in [9]). Presumably due to the presence of partitions, no fish became dominant, and aggression remained on a high level during the whole observation period of 150 days.

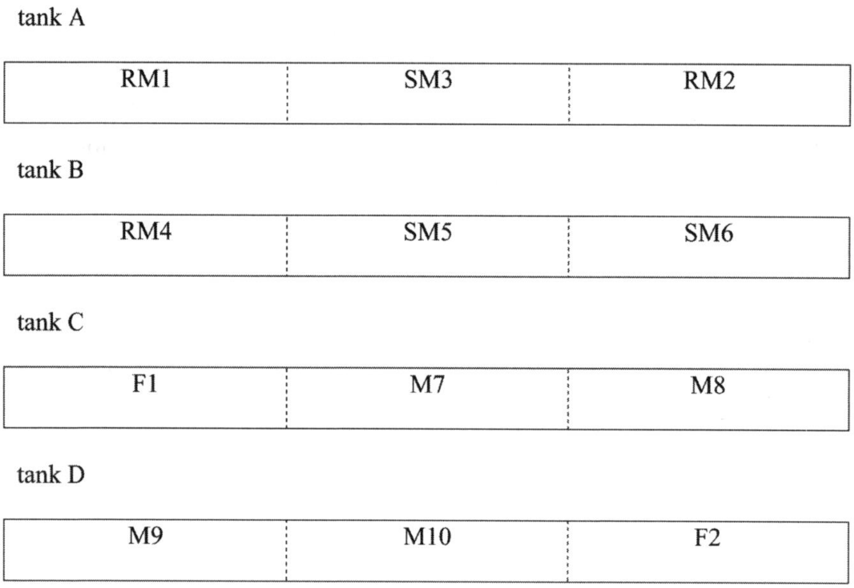

Figure 1. Tanks A and B (Treatment 1). Stage 1: three fish individually kept in three serially adjoining compartments per tank, separated by plastic mesh partitions. Stage 2: certain resident males were exchanged for stranger males. Tanks C and D (Treatment 2): communal tanks for 3 fish each, after 3 weeks plastic mesh partitions separated the fish. Individuals coded RM, "resident" males; SM, "stranger" males; M, males; F, females. F and M were kept in groups even before Treatment 2 started.

Under Treatment 2 (tanks C and D, Fig. 1), three fish that had shared a communal tank for three weeks were suddenly separated by two plastic mesh partitions that allowed electrical and limited physical contact. Both males and females showed strong aggression towards their neighbours immediately after introduction of the plastic mesh partitions even during the generally inactive diurnal period. Similar to tanks A and B, aggression did not decrease during the observation periods of 18 and 54 days; presumably because of the partitions there was no loser. After termination of the experiments it was not possible to keep these individuals together in one tank with partitions removed, because the stronger fish continued to attack the weaker as it takes a few days until EOD durations began to recede. Fish had to be kept in new tanks together with unfamiliar fish.

EOD-Waveform

Under Treatments 1 and 2, increased male aggression accompanied a strong increase of EOD pulse duration in all fish, combined with an increase of P/N amplitude ratio that was due to an amplitude decrease of N (Fig. 2A, B, referring to tank B).

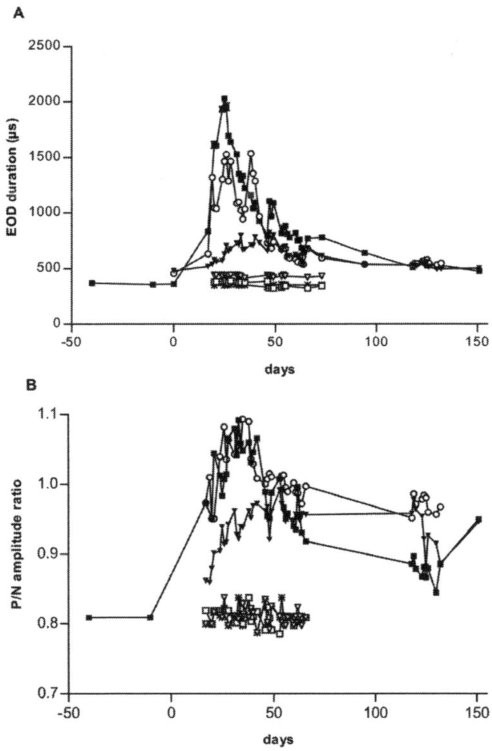

Figure 2. Change of EOD duration (A) and P/N amplitude ratio (B) in male RM4 (■) over 164 days, and males SM5 (o) and SM6 (▼) over 124 days (Treatment 1). Day-40: three then mutually stranger males (only RM4 shown) were placed separately in the three compartments of one common tank. Day 0: the two males that are not shown were replaced by SM5 (o) and SM6 (▼). Note EOD waveform changes in fish RM4, SM5 and SM6, in contrast to individuals M11( ), M12 (□) and F3(*) that received No Treatment.

In tank A, the other three males' EOD duration was 364, 452, and 461 μs before treatment. Evoked by Treatment 1, EOD duration increased 2.3-, 1.3-, and 1.4-fold in these fish (Table 1).

Table 1. Change of EOD duration in male M. pongolensis under Treatments 1 and 2

tank	fish	D1 (μs)	D2 (μs)	Increase in %	after days	D3 (μs)	D4 (μs)	increase in %
A	RM1	364 ± 6.8	833.0 ± 9.8	129	21	-	-	-
	RM2	452 ± 5.9	578.0 ± 11.8	27.9	21	-	-	-
	SM3	461 ± 7.3	637.0 ± 14.5	38.2	21	-	-	-
B	RM4	356 ± 2.1	2029 ± 55	470	25	-	-	-
	SM5	456 ± 7.8	1527 ± 32	234	26	-	-	-
	SM6	480 ± 5.6	793 ± 3.5	65	33	-	-	-
C	M7	456 ± 7	1056 ± 18	131.6	19	578.2 ± 9	1511 ± 34	161.4
	M8	360 ± 12	1259 ± 14	249.7	17	467.5 ± 6.2	2148 ± 61	359.5
	F1				data listed in Table 2			
D	M9	401 ± 4.5	2178 ± 57	443	23	1098 ± 14	2408 ± 69	119
	M10	519 ± 24.7	1574 ± 63	203.3	33	652.6 ± 5.6	1134 ± 72	73.8
	F2				data listed in Table 2			

D1, mean EOD duration ± SD, N = 5 before exchanging animals or introducing plastic mesh partitions; D2, maximum EOD duration ± SD, N = 5 evoked by treatment; D3 minimum EOD ± SD, N = 5 duration after stopping treatment (introduction of solid plastic walls); D4 maximum EOD duration ± SD, N = 5 after onset of a second treatment.

In tank B, EOD duration of tank mates RM4, SM5 and SM6 increased 5.7-, 3.3-, and 1.7-fold to maximum values of 2029 μs, 1527 μs and 793 μs, respectively (Fig 2a, Table 1). However, the amplitude ratio of positive over negative EOD phases (P/N ratio; with P = 1) increased with a delay relative to EOD duration. P/N ratios peaked only 8 days (fish RM4), 6 days (fish SM5), and 9 days (fish SM6) after EOD maximum duration (Fig. 2B). P/N ratios increased to a maximum of 1.091 (+26%; fish RM4), 0.992 (+15%; fish SM5) and 1.093 (+14%; fish SM6).

Before start of Treatment 2 in tanks C and D, EODs were monitored for 21 days. During this period the dominant males M7 and M10 displayed EODs of longer duration than the subdominant fish M8 and M9 (by +27% and +29%, respectively; Table 1). Dominant fish had territories of at least 1.5 fold size of the subdominant fish. Subdominant fish also tried to avoid dominant fish and escaped when attacked. EOD duration differed only slightly within individuals during an observation period of three weeks preceding the experimental treatment.

Upon introduction of plastic mesh partitions under Treatment 2 that allowed unrestricted sensory contact, EOD duration increased in all males. It decreased again when the partitions were exchanged for solid plastic walls that severely restricted sensory contact (Figs 3 and 4). However, EOD duration did not recede completely to the initial values preceding Treatment 2. When plastic mesh partitions were introduced for a second time, the EOD duration increase was even stronger than the one observed the first time. By contrast, EOD duration did not change noticeably in females (Figs 3 and 4, Table 2).

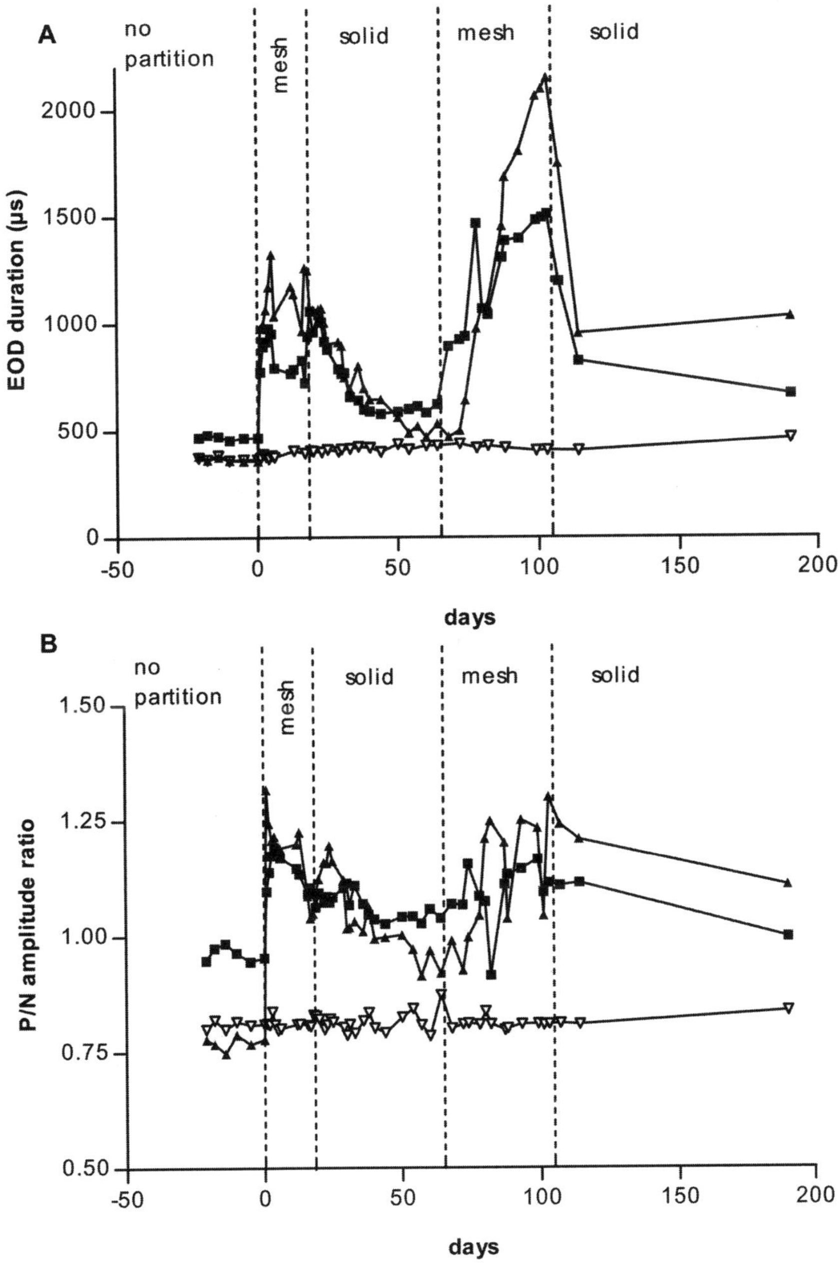

Figure 3. Time course of EOD duration (A) and P/N amplitude ratio (B) changes in males M7 (■) and M8 (▲), and female F1 (, all Treatment 2). Mesh partitions ("mesh") or solid plastic walls ("solid") separating the three fish as indicated. Note reproducible EOD duration increase upon introduction of plastic mesh partitions in communal tank ("no partition"), and decrease evoked by solid plastic walls severely reducing sensory contact.

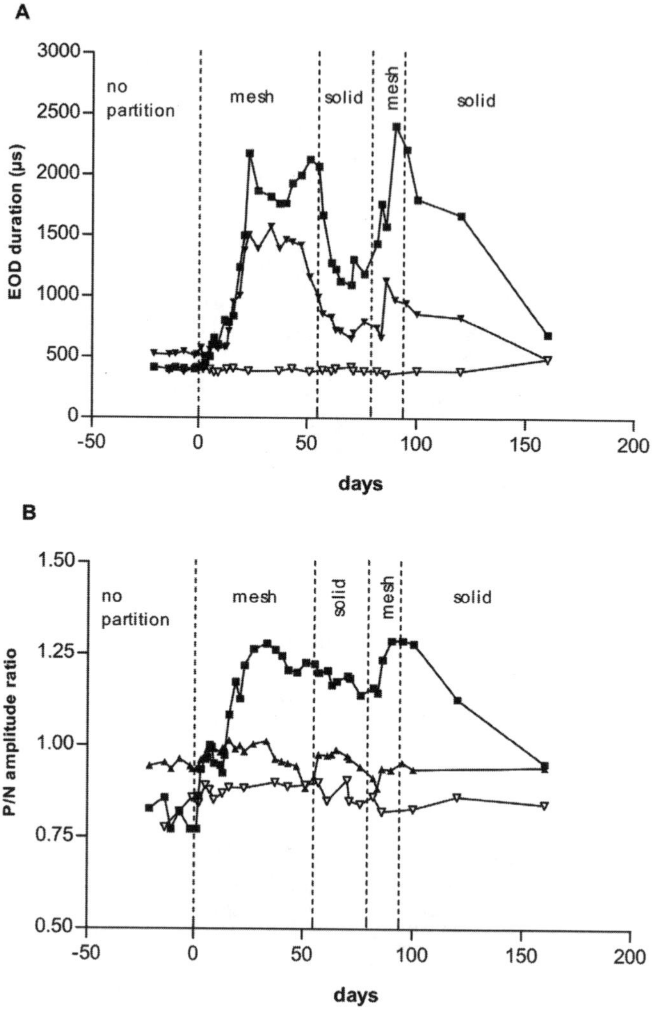

Figure 4. Time course of EOD duration (A) and P/N amplitude ratio (B) changes in males M9 (■) and M10 (▼), and female F2 (, all Treatment 2). As Fig. 3, but tank D.

Table 2. Variability of EOD duration in male control fish (No Treatment) and female M. pongolensis

	EOD duration first day (µs)	maximum duration (µs)	minimum duration (µs)	No. of observed days	ΔEOD duration min/max in %
F1	384.2 ± 18	405 ± 12.6	361 ± 7.2	120	12.2
F2	360.57 ± 12.3	435.8 ± 13.38	360.57 ± 12.3	114	20.8
M11	440 ± 3.5	441 ± 3.6	413 ± 0.4	35	6.8
M12	348 ± 1.5	358 ± 2.1	329 ± 2.5	35	8.8
M13	435.7 ± 1.9	439 ± 1.1	420 ± 1.7	21	4.5
M14	324 ± 1.7	343 ± 292	322 ± 0.78	21	6.5
F3	379 ± 2.0	387.8 ± 2.4	322.4 ± 0.8	35	20

Mean EOD duration ± SD, N = 5.

The EODs of control fish, another five males and one female M. pongolensis (M11–M15, and F3), which received No Treatment were recorded simultaneously with Treatments 1 and 2. Their EOD waveforms were stable over time (Fig. 2A, Table 2).

Discussion

In mormyrid weakly electric fish, individual EOD waveforms are documented as temporally stable in several studies [12,13]; see also Table 2 for fish M11–M15 that were not exposed to new neighbours). But there is also evidence for EOD waveform plasticity. In captive Brienomyrus brachyistius, alpha males in social groups (three males and three females) increased their total EOD duration by 20% [25]. Similar results were found in captive Gnathonemus petersii, the EOD duration of which was influenced by social interactions. Dominant fish increased their EOD duration from about 280 to 340 µs in male-male, female-female and mixed-sex pairs over an observation period of 4 days [26].

In M. pongolensis, a permanent characteristic of male EOD duration to increase with fish size, established in the wild, is superimposed by short-term plasticity evoked by social interaction and reproduction, observed in captivity. The permanent characteristic was seen in field recordings in the non-reproductive dry season, when EOD duration in male M. pongolensis was positively correlated with fish length to up to 818 µs in the largest males. This is in contrast to females where no such dependency was found [17,18]. A superimposed short-term waveform plasticity (that is the subject of the present study) was first documented during reproduction of two males in captivity. EOD duration increased only for a short period of actual spawning (by around 70% in one male [10]; in another male, by 75%: from 511 µs EOD duration as measured in the field to a maximum of 894 µs during spawning in the laboratory [18]). A longer-lived short-term plasticity was described in the present study in the context of social interaction. By experimentally manipulating territory and neighbour identity, a surprisingly high degree of short-term plasticity in EOD waveform in male M. pongolensis was evoked. An increase in EOD duration was observed in ten males; in one male by 470% or almost 2500 µs total duration. The EOD pulse duration increase was reversible and reproducible, and was correlated with aggressive interactions with both males and females, accompanied by escalating fights at the territory boundary. Before experimental treatment, EOD duration in males that became dominant was greater by about 30% than that of submissive ones, similar to the results of [26] and [25].

Fluctuation in EOD waveform has been studied in detail in gymnotiform weakly electric fish. In the family Hypopomidae EOD amplitude and duration seem to depend on social status and time of the day [27-29]. The masculinization of male EOD waveform (an increase in certain EOD parameters) is greatest in dominant males (Brachyhypopomus pinnicaudatus, unpublished data reviewed in [30]), and may develop within tens of minutes in response to aggressive contacts (review [31]). Similar to the present results, an increase in EOD duration was independent of social status when males were separated by plastic mesh [30]. As partitions prevented escalating fights, stronger males probably could not prove their dominance.

In gymnotiform fish three types of plasticity of EOD duration are discussed (reviews [30-32]). Apart from two rapid mechanisms there exists a mechanism of slow change in EOD waveform that is hormonally triggered. Also in certain mormyrids, EOD waveform duration increased upon androgen administration (e.g., [33,34]). In the mormyrid B. brachyistius, endogenous 11-ketotestosterone level increased with social status [25]. Among other effects, testosterone is known to raise aggression. An increase in testosterone titre evoked by our treatments may have caused the high level of aggression and the elongation of EOD duration.

Conclusion

The strong increase in male EOD duration evoked by social interactions is in line with the assumption that EOD duration in M. pongolensis is an important factor in social context. (1) In adult male fish studied in the wild, EOD duration is positively correlated with standard length [18]. Larger males probably enjoy a higher reproductive success, as larger males are more successful in competing for females, have better access to high-quality feeding and spawning sites. Also in many other species, females prefer larger males, as body size may indicate fitness [35,36]. (2) Playback EODs of long duration evoked stronger aggression from resident male M. pongolensis [19], probably because they signalled a greater threat.

Males may be particularly vulnerable to predation when emitting long-duration EODs because these are detected more easily at lower threshold by electroreceptive predatory catfish [20,21]. A possible way out of this male dilemma may be short-term plasticity of EOD waveform, that is, to signal long-duration EODs for a limited period of time only, when competing for mates.

To restrict EODs of long duration to the period of reproduction, if not only actual spawning, appears to be a superior adaptation to predator pressure. South African Bulldogs (M. pongolensis) seem to have found a way to signal resource holding potential in male-male interactions without alerting predatory catfish for longer than absolutely necessary.

Methods

Animals and Experimental Procedure

The specimens used were 14 male and three female Marcusenius pongolensis from the Incomati River System (South Africa, Mpumalanga Province, 25°30'35"S, 31°11'58"E, coll. F.H. van der Bank & J. Engelbrecht, 14 February 1997). These fish represent the nominal species Gnathonemus pongolensis Fowler, 1934 that was synonymised with M. macrolepidotus by Crass (1960), and resurrected as Marcusenius pongolensis (Fowler, 1934) by [18]. Marcusenius pongolensis is synonymous with "Marcusenius macrolepidotus (South African form)" in previous papers from our group [9,10,19]. All fish were beyond 40% of the maximum species size which, for males, is the approximate minimum size for sexual maturity [14]. Fish in the same tank were of similar size (from 15 cm-18 cm SL). Experimental aquaria (210 cm × 60 cm × 50 cm) were divided into three compartments of equal size using two plastic-mesh partitions that allowed electrical, visual, and limited mechanical interaction. Every compartment contained a porous pot as a hiding place for the fish. The light/dark cycle was 12:12 h, water conductivity 100 ± 3 μS/cm and temperature 23 ± 0.5°C.

We followed three experimental protocols (one a control, No Treatment, see below). (1) Treatment 1: in tanks A and B three males each were individually kept in three adjacent compartments, separated by plastic mesh partitions, for at least seven weeks. After this time, observation of EOD waveform started, and one or two males per tank were exchanged for unfamiliar ones ('stranger males', SM). The male of the middle compartment was exchanged for the stranger male SM3 in tank A, with resident males RM1 and RM2 remaining in place. In tank B, the two males occupying the middle and the right compartments were exchanged for males SM5 and SM6, whereas fish RM4 occupying an end compartment remained in place (Fig. 1). (2) Treatment 2: in tanks C and D, we monitored EOD waveforms in groups of three fish per tank (two males, one female), each group kept in a communal tank without any partitions, for three weeks prior to experiment onset. We then introduced two plastic mesh partitions in each tank such that each fish occupied an individual compartment (experimental subjects M7, M8 and F1 in tank C, and M9, M10 and F2 in tank D; Fig. 1). After 18 days (tank C) and 54 days (tank D), the plastic mesh partitions were replaced by solid, tightly fitting plastic walls for 25 days (tank D) and 49 days (tank C), to reduce all visual, mechanical, and electrical stimuli between the fish as much as possible. Subsequently, the walls were replaced again by plastic mesh partitions for another 18 days (tank D) and 37 days (tank C). During the whole observation period, waveforms were sampled daily or at least twice per week. (3) No Treatment: as a control we sampled EOD waveforms from five males and one female over a period of 53 days. Specimens M11, M12 and F3 were kept together in a communal tank, whereas specimens M13, M14 and M15 were kept completely isolated in separate tanks.

Waveform Measurement

EOD waveforms were recorded during the day when fish were resting in their shelter, at 23 ± 0.5°C water temperature and 100 ± 3 μS/cm conductivity, using low-impedance carbon electrodes. EODs were differentially amplified (× 10; 0.2 Hz – 100 000 Hz), digitised (TDS oscilloscope model 420, Tektronix Inc., Heerenveen, Holland), sampling rate 250 kHz, 11 bit vertical resolution), and stored on computer disk. As already described [9,17,18], EOD waveforms of all Marcusenius pongolensis specimens used in the present study were biphasic, with a head-positive phase (P) followed by a negative one (N, Fig. 5). EOD duration was estimated after normalising the P peak amplitude to 1 V, using the software package Famos (IMC, Berlin). We used a ± 2% threshold criterion relative to P peak amplitude = 100% for estimating onset and termination of an EOD waveform.

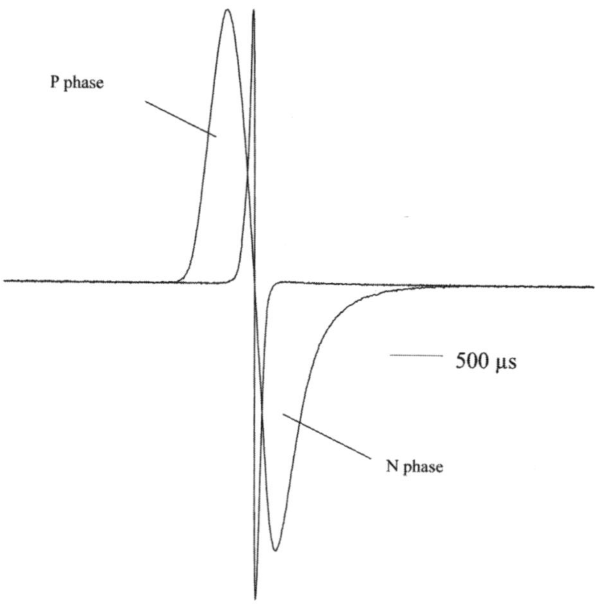

Figure 5. EOD in resident male RM4 before (short duration, 356 μs) and after exchanging its two familiar neighbours for two stranger males. EODs normalised to the same positive peak amplitude P. Voltage over time (baseline = 0 V), head-positivity is upwards. Time in μs as indicated by time bar. The EOD of longer duration (2029 μs) was sampled 25 days after introducing the stranger males. Note an increase of both P and N (head-negative) phase duration of the EOD.

Abbreviations

EOD: electric organ discharge; SL: standard length (from the tip of the mouth to the midbase of the forked tail fin); RM: resident male; SM: stranger male; M: male; F: female; P: head-positive phase of the EOD; N: negative phase.

Competing Interests

The authors declare that they have no competing interests.

Author Contributions

SH designed and carried out the behavioral studies and waveform measurements. BK provided the animals, carried out the waveform measurements in the field, participated in the experimental design and helped to draft the manuscript. Both authors read and approved the final manuscript.

Acknowledgements

We would like to thank Dr. F.H. van der Bank (Rand Afrikaans University, Auckland Park, South Africa) and Dr. J. Engelbrecht (Mpumalanga Parks Board, South Africa) for sampling and exporting our fish. Supported by the Deutsche Forschungsgemeinschaft (Grants Kr446/11-3 and -4) to B.K All fish were captured and exported by, or in close collaboration with, Mpumalanga Parks Board (who granted permits and gave permissions to export live fish). The work presented here complies with current regulations covering experimentation and care of animals in Germany.

References

1. Kramer B: Electrocommunication in teleost fishes: behavior and experiments. Berlin, Springer; 1990.
2. Kramer B: Electroreception and communication in fishes. Stuttgart, Gustav Fischer; 1996.
3. Moller P: Electric fishes. History and behavior. London, Chapman and Hall; 1995.
4. Turner RW, Maler L, Burrows M: Electroreception and electrocommunication. J Exp Biol 1999., 202:

5. Bullock TH, Hopkins CD, Popper AN and Fay RR (Eds): Electroreception. New York, Springer; 2005.
6. Ladich F, Collin SP, Moller P and Kapoor BG (Eds): Communication in fishes. Enfield, NH, USA, Science Publishers Inc.; 2006.
7. Scheffel A, Kramer B: Electric signals in the social behavior of sympatric elephantfish (Mormyridae, Teleostei) from the Upper Zambezi River. Naturwissenschaften 2000, 87:142–147.
8. Scheffel A, Kramer B: Intra- and interspecific electrocommunication among sympatric mormyrids in the Upper Zambezi River. In Communication in fishes. Edited by: Ladich F, Collin SP, Moller P and Kapoor BG. Enfield, NH, USA, Science Publishes Inc.; 2006:733–751.
9. Werneyer M, Kramer B: Intraspecific agonistic interactions in freely swimming mormyrid fish, Marcusenius macrolepidotus (South African form). J Ethol 2002, 20:107–121.
10. Werneyer M, Kramer B: Electric signalling and reproductive behavior in a mormyrid fish, the bulldog Marcusenius macrolepidotus (South African form). J Ethol 2005, 23:113–125.
11. Kramer B, Westby GWM: No sex difference in the waveform of the pulse type electric fish, Gnathonemus petersii (Mormyridae). Experientia 1985, 41:1530–1531.
12. Bratton BO, Kramer B: Intraspecific variability of the pulse-type discharges of the African electric fishes, Pollimyrus isidori and Petrocephalus bovei (Mormyridae, Teleostei) and their dependence on water conductivity. Exp Biol. 1988, 47(4):227–238.
13. Crawford JD: Sex recognition by electric cues in a sound-producing mormyrid fish Pollimyrus isidori. Brain Behav Evol 1991, 38:20–38.
14. Kramer B: A field study of African elephantfish (Mormyridae, Teleostei): electric organ discharges in Marcusenius macrolepidotus (Peters, 1852) and Petrocephalus catostoma (Günther, 1866) as related to sex. J Afr Zool 1997, 111:313–341.
15. Paintner S, Kramer B: Electrosensory basis for individual recognition in a weakly electric, mormyrid fish, Pollimyrus adspersus (Günther, 1866). Behav Ecol Sociobiol 2003, 55:197–208.
16. Scheffel A, Kramer B: Electrocommunication and social behavior in Marcusenius senegalensis (Mormyridae, Teleostei). Ethology 1997, 103:404–420.

17. Kramer B, Van der Bank FH, Skelton PH: Two new species of snoutfish (Mormyridae) from South Africa: evidence from electric organ discharges. In Paradi Conference. Grahamstown (RSA); 1998:50.

18. Kramer B, Skelton PH, Van der Bank FH, Wink M: Allopatric differentiation in the Marcusenius macrolepidotus specis complex in southern and eastern Africa: the resurrection of M. pongolensis and M. angolensis, and the description of two new species (Mormyridae, Teleostei). J Nat Hist 2007, 41: 647–708.

19. Hanika S, Kramer B: Intra-male variability of its communication signal in the weakly electric fish, Marcusenius macrolepidotus, and possible functions. Behav 2005, 142:145–166.

20. Hanika S, Kramer B: Electric organ discharges of mormyrid fish as a possible cue for predatory catfish. Naturwissenschaften 1999, 86:286–288.

21. Hanika S, Kramer B: Electrosensory prey detection in the African sharptooth catfish Clarias gariepinus (Clariidae), of a weakly electric mormyrid fish, the bulldog (Marcusenius macrolepidotus). Behav Ecol Sociobiol 2000, 48:218–228.

22. Merron GS: Pack-hunting in two species of catfish, Clarias gariepinus and C. ngamensis in the Okavango delta, Botswana. J Fish Biol 1993, 43:575–584.

23. Winemiller KO, Kelso-Winemiller LC: Comparative ecology of the African pike, Hepsetus odoe, and tigerfish, Hydrocynus forskahlii, in the Zambezi River foodplain. J Fish Biol 1994, 45:211–225.

24. Bruton MN: The food and feeding behavior of Clarias gariepinus in Lake Sibaya, South Africa, with emphasis on its role as a predator of cichlids. Trans Zool Soc Lond 1979, 35:47–114.

25. Carlson BA, Hopkins CD, Thomas P: Androgen correlates of socially induced changes in the electric organ discharge waveform of a mormyrid fish. Horm Behav 2000, 38:177–186.

26. Terleph TA, Moller P: Effects of social interaction on the electric organ discharge in a mormyrid fish, Gnathonemus petersii (Mormyridae, Teleostei). J Exp Biol 2003, 206:2355–2362.

27. Hagedorn M: The electric fish Hypopomus occidentalis can rapidly modulate the amplitude and duration of its electric organ discharges. Anim Behav 1995, 49:1409–1413.

28. Franchina CR, Stoddard PK: Plasticity of the electric organ discharge waveform of the electric fish Brachyhypopomus pinnicaudatus. I. Quantification of day-night changes. J Comp Physiol A 1998, 183:759–768.
29. Franchina CR, Salazar VK, Volmar CH, Stoddard PK: Plasticity of the electric organ discharge waveform of male Brachyhypopomus pinnicaudatus. II. Social effects. J Comp Physiol A 2001, 187:45–52.
30. Stoddard PK, Zakon HH, Markham MR, McAnelly L: Regulation and modulation of electric waveforms in gymnotiform electric fish. J Comp Physiol A Neuroethol Sens Neural Behav Physiol 2006, 192(6):613–624.
31. Bass AH, Zakon HH: Sonic and electric fish: At the crossroads of neuroethology and behavioral neuroendocrinology. Horm Behav 2005, 48:360–372.
32. Zakon HH, McAnelly L, Smith TG, Dunlap K, Lopreato G, Oestreich J, Few WP: Plasticity of the electric organ discharge: implications for the regulation of ionic currents. J Exp Biol 1999, 202:1409–1416.
33. Landsman RE, Harding CF, Moller P, Thomas P: The effects of androgens and estrogen on the external morphology and electric organ discharge waveform of Gnathonemus petersii (Mormyridae, Teleostei). Horm Behav 1990, 24:532–533.
34. Herfeld S, Moller P: Effects of 17a-methyltestosterone on sexually dimorphic characters in the weakly discharging electric fish Brienomyrus niger (Günther, 1866)(Mormyridae): electric organ discharges, ventral body wall indentation, and anal-fin ray expansion. Horm Behav 1998, 34:303–319.
35. Andersson M: Sexual Selection. Princeton, New Jersey, Princeton University Press; 1994.
36. Ryan MJ, Keddy-Hector A: Directional patterns of female mate choice and the role of sensory biases. American Naturalist 1992, 139:4–35.

A Fish Eye Out of Water: Ten Visual Opsins in the Four-Eyed Fish, *Anableps anableps*

Gregory L. Owens, Diana J. Windsor, Justin Mui and John S. Taylor

ABSTRACT

The "four-eyed" fish Anableps anableps has numerous morphological adaptations that enable above and below-water vision. Here, as the first step in our efforts to identify molecular adaptations for aerial and aquatic vision in this species, we describe the A. anableps visual opsin repertoire. We used PCR, cloning, and sequencing to survey cDNA using unique primers designed to amplify eight sequences from five visual opsin gene subfamilies, SWS1, SWS2, RH1, RH2, and LWS. We also used Southern blotting to count opsin loci in genomic DNA digested with EcoR1 and BamH1. Phylogenetic analyses confirmed the identity of all opsin sequences and allowed us to map gene duplication and divergence events onto a tree of teleost fish. Each of the gene-specific primer sets produced an amplicon from cDNA, indicating that

A. anableps possessed and expressed at least eight opsin genes. A second PCR-based survey of genomic and cDNA uncovered two additional LWS genes. Thus, A. anableps has at least ten visual opsins and all but one were expressed in the eyes of the single adult surveyed. Among these ten visual opsins, two have key site haplotypes not found in other fish. Of particular interest is the A. anableps-specific opsin in the LWS subfamily, S180γ, with a SHYAA five key site haplotype. Although A. anableps has a visual opsin gene repertoire similar to that found in other fishes in the suborder Cyprinodontoidei, the LWS opsin subfamily has two loci not found in close relatives, including one with a key site haplotype not found in any other fish species. A. anableps opsin sequence data will be used to design in situ probes allowing us to test the hypothesis that opsin gene expression differs in the distinct ventral and dorsal retinas found in this species.

Introduction

Anableps anableps is an active surface feeder found in the murky intertidal regions, oceanic shore waters, and freshwater streams of Central America and northern South America [1]. It can jump out of water to catch flying insects, but also feeds on floating material and diatoms in riverbank mud [2]. A. anableps eyes have morphological adaptations that allow for simultaneous vision above and below water. For example, its cornea is separated into two parts by a pigment stripe that prevents glare [3], with the above-water portion flatter than its ventral counterpart [4]. This difference appears to compensate for the fact that light entering the cornea from the aerial environment is refracted much more than light entering from the aquatic environment [5]. While most fish have spherical lenses, in A. anableps the lens is oval-shaped. This allows light from the aerial field to pass through a relatively flat portion of the lens, similar to the lens of a land animal, and light from the aquatic environment to pass through a portion of the lens with a curvature more typical for an aquatic animal [4]. Finally, the retina is divided into dorsal and ventral portions, which receive light from the aquatic and aerial environment respectively.

There are two other species in the genus Anableps, A. microlepis (the finescale four-eyed fish), which is found on the Atlantic coasts of Central and South America, and A. dowei from the Pacific coast of Central America. All three Anableps species possess the unusual eye morphology. The other taxa in the family Anablepidae, genus Jenynsia (sister taxon to Anableps with 12 species) and genus Oxyzygonectes (with one species O. dovii), have typical teleost eyes with a single cornea and pupil, a spherical lens, and a cup-shaped retina [6].

Beyond morphology, vision can also be examined at a molecular level. Light receptors expressed in rod and cone cells of the retina are called opsins. Each opsin protein is associated with a chromophore and when exposed to light this complex changes conformation leading to rod or cone cell hyperpolarization [7]. The detection of light requires input from just one type of opsin-chromophore receptor. However, no single opsin receptor is sensitive to all wavelengths of visible light. Furthermore, wavelength discrimination (color vision) involves the interpretation of signals from different subpopulations of cone cells expressing opsins with different spectral sensitivities [8].

Gene duplication and divergence events are the evolutionary source of opsins with different spectral sensitivities. In vertebrates there are SWS1, SWS2, RH2, and LWS cone opsins. SWS opsins are Short Wavelength Sensitive opsins that are most sensitive to UV and blue light. RH2 opsins (Rhodopsin-like) are most sensitive to wavelengths in the middle of the visible light spectrum (i.e. green light) and LWS (Long Wavelength Sensitive) opsins are most sensitive to orange and red light. Rod cells, which function primarily in dim light, express RH1 genes (Rhodopsin) that encode a green light absorbing pigment [9]. Opsin subfamilies have been expanded or lost in different vertebrate lineages. For example, while dogs have one LWS opsin, humans have two, and guppies (Poecilia reticulata) have four [10], [11]. Placental mammals have lost both the RH2 and SWS2 opsin subfamilies and the coelacanth (Latimeria chalumnae) has lost all but the RH1 and RH2 opsins [10], [12].

Individual opsins vary in their spectral sensitivity among and within subfamilies. This variation is a result of changes at key amino acid sites, which are sites that have a disproportional effect on spectral sensitivity and are often found at locations where the opsin contacts the chromophore [13]. Previous work has quantified the contribution of each site to the overall wavelength of maximal sensitivity (λmax) and it is therefore possible to identify opsins within subfamilies with different spectral sensitivities by comparing their amino acid sequences. In fish there are two types of opsin-associated chromophore, A1 and A2, and depending on which is used the spectral sensitivity can differ by up to 50 nm [14]. Some species tune their vision by switching from one chromophore to the other in response to environmental or developmental changes [15], [16]. While chromophore use is not considered here, previous study has shown homogenous use of A1 in the A. anableps retina [17].

Microspectrophotometry (MSP) is a technique that estimates wavelength sensitivity (λmax) at the cellular level. An MSP study detected only three different classes of cones cells in A. anableps [17]. However, phylogenetic data from close relatives, guppy and bluefin killifish (Lucania goodei), suggests that it has many more. MSP data might not reflect the four-eyed fish's true repertoire if only a subset of loci are expressed in adults or if multiple opsins are expressed in the same photoreceptor, as has been shown in mice, eels, and salamanders among others [18]–[20].

In addition to gene number and sequence, opsin expression also varies among species, populations and even within individuals at different periods of development. For example, in cichlids LWS opsin expression varies with water turbidity and it appears that population-level variation in wavelength sensitivity has played a role in variation in male coloration [21]. At the individual level, European eels (Anguilla anguilla) have two RH1 opsins, each tuned to slightly different wavelengths. They express a green-shifted locus as juveniles in fresh water and a paralogous blue-shifted locus when they return to the ocean [22]. The lamprey (Geotria australis) also adjusts its spectral sensitivity by switching from the expression of one opsin paralog to another as it moves between marine and freshwater environments and Zebrafish (Danio rerio) have two LWS opsins that are expressed at different times of development and in different regions of the retina [23], [24]. Given these observations, we hypothesized that the morphological adaptations leading to simultaneous aerial and aquatic vision in A. anableps would be accompanied by changes in opsin gene number and/or sequence and by changes in opsin expression patterns.

Here we report the results of a PCR-based survey of A. anableps opsins using primers complementary to regions of each locus that are conserved in closely related species. We also used Southern blotting probes to identify the number of opsin loci in the A. anableps genome. These techniques revealed that A. anableps has ten visual opsins, including representatives from each opsin subfamily.

Results

Visual Opsin Sequences

Transcripts of eight opsin genes (SWS1, SWS2A, SWS2B, RH2-1, RH2-2, RH1, LWS S180α, and LWS S180r) were amplified and sequenced using primers listed in Table 1 from cDNA derived from a single A. anableps eye. Southern blotting experiments utilized LWS, SWS1, SWS2, RH2 and RH1 opsin gene probes and two samples of A. anableps genomic DNA, one digested with EcoR1 and the other digested with BamH1. These experiments indicated there might be two SWS1 and RH1 loci and three LWS loci (Table 2). We used PCR to survey A. anableps genomic DNA to test the hypothesis that there were additional loci in these three subfamilies not detected in cDNA. Five clones with inserts derived from RH1-specific primers and five clones with inserts derived from SWS1-specific primers were sequenced and all had the same sequence as the original cDNA amplicon.

Table 1. Primers used for cDNA and genomic PCR and Southern blot probe synthesis.

Opsin category	Primer Name	Sequence
SWS1	SWS1Fw1	5′- AACTACATCYTGGTMAACATCTCC-3′
	SWS1Fw2	5′- TGGGCSTTCYACCTGCAGGC -3′
	SWS1Rev1	5′- GAGTAGGAGAARATGATGATGG-3′
	SWS1Rev2	5′-GAACTGTTTGTTCATGAAGGCG-3′
SWS2	SWS2Fw1	5′-GYACWATTCAATACAAGAARC-3′
	SWS2Fw3	5′-AGCCTTTGGTCTCTGGCTGTG-3′
	SWS2Rev1	5′-AAAGCARAAGCAGAAGAGGAAC-3′
	SWS2Rev4	5′-CCCGTTGTGTACCAGTCTGG-3′
	SWS2AFw1	5′-GTCCACCCGAGTCATAGAGC-3′
	SWS2ARev2	5′-GCCCACGGTTGTTGACAAC-3′
	SWS2-2Fw2	5′-TCTACACCATGGCTGGATTCAC-3′
	SWS2-2Rev1	5′-GATGGTGGTGAATGGAACAGC-3′
RH2	RH2Fw1	5′- AACTTCTAYATCCCGWTGTCC-3′
	RH2Fw2	5′-TGHTCTTCCTGATCTKCACTGG-3′
	RH2Rev2	5′-GTCTCRTCCTCCACCATGC-3′
	RH2Rev4	5′- TGCGGCATGAGTTCCAGTG-3′
	RH2-2Fw1	5′-CAACAGGACGGGCTGGTGAGG-3′
	RH2-2Rev3	5′-ACCCATTCCAATTGTTGCC-3′
RH1	RH1Fw1	5′-ATGAACGGCACAGAGGGACC-3′
	RH1Fw4	5′-GCAGTGCTCATGCGGAGTC-3′
	RH1Rev2	5′-CCTGTTGCTCCATTTATGCAGG-3′
	RH1Rev4	5′-GCTGGAGGACACAGAAGAGG-3′
LWS	Fw100	5′-GATCCCTTTGAAGGACCAAACT-3′
	Fw1a	5′-TCTTATCAGTCTTCACCAACGG-3′
	Gamma Fw1	5′-TGCTATGCAGCAGATAAATTG-3′
	RevEnd	5′-TTATGCAGGAGCCACAGAGG-3′
	Rev8	5′-GCCCACCTGTCGGTTCATGAAG-3′
	RevEx4	5′- CTTCCACTGAACACATCAGG-3′

Primers were used to amplify sequences from *A. anableps* cDNA and genomic DNA as well as guppy cDNA.

Table 2. Southern blot results.

Probe	Restriction Enzyme	Number of Bands	Band Size (Kb)
LWS	EcoRI	3	4.3, 4.1, 3.8
LWS	BamH1	3	5.0, 4.2, 4.0
SWS1	EcoRI	2	4.0, 2.0
SWS1	BamH1	1	4.1
RH2	EcoRI	1	2.5
RH2	BamH1	0	-
RH1	EcoRI	2	4.5, 4.0
RH1	BamH1	2	4.7, 3.8
SWS2	EcoRI	1	2.0
SWS2	BamH1	1	3.0

Summary of Southern blot analysis results obtained for *A. anableps* opsins probes with genomic DNA hybridized at 41°C. If Southern blot bands outnumbered unique cDNA sequences, we surveyed genomic DNA and sequenced at least five clones. Bands are pictured in Figure S1.

For the LWS opsin subfamily, two rounds of genomic PCR and sequencing were undertaken to supplement the original cDNA screen. The first round amplified the S180α gene that had been retrieved from cDNA and seven novel sequences. However, we suspected several to be mosaics produced during PCR (i.e., template switching) and/or during cloning (e.g., mismatch repair of cloned heteroduplex DNA) [25], [26]. In the second PCR survey of genomic DNA, LWS opsin primers were added at the beginning and then again just before the last PCR cycle in an attempt to eliminate these artefacts [26], [27]. Only genes uncovered in both rounds were considered to be authentic. These genes include LWS S180α, LWS S180β and LWS S180γ and an allele of LWS S180α. Subsequently, LWS S180γ was successfully amplified from cDNA.

Phylogenetic Analyses of A. anableps Opsin Genes

All A. anableps opsin sequences were aligned with representatives of each subfamily from other fish species. Sequences in the alignment were 412 to 819 bp long (Table 3). We used Mega4 [28] to calculate Tamura-Nei genetic distances [29] and to reconstruct a neighbour joining tree (Figure 1). Sequences from each opsin subfamily formed well-supported monophyletic groups, with bootstrap support (500 replicates) ≥97%. Relationships among species within each opsin subfamily were consistent with well-established taxonomy [6], [30]. The root of the tree was positioned along the branch separating the LWS opsins from all others. While no non-opsin out-group sequences were employed in these analyses, the placement of the root between the LWS and all other subfamilies has been well established [31]. The A. anableps sequences occurred in each of the subfamilies confirming that locus-specific primers had amplified the genes they targeted. Phylogenetic analysis revealed an SWS2 gene duplication event that occurred in the common ancestor of bluefin killifish, A. anableps and guppy, although one of the duplicates had not been amplified from guppy and is reported here for the first time.

Variation at Amino Acid Positions Known to Influence Spectral Sensitivity

We hypothesized that A. anableps opsins would contain unique amino acid substitutions to accompany its unusual eye morphology. However, with two exceptions (SWS2A and LWS S180γ), the key-site haplotypes in A. anableps visual opsins also occur in other fish with 'normal' eyes. The residues at all of the 12 key sites in the SWS2A opsin have been seen in other fish, but the entire haplotype found in A. anableps, appears to be unique. This haplotype is unlikely to produce a significant shift in maximal absorption according to mutagenesis analyses [32].

The A. anableps opsin gene, LWS S180γ, also has a unique five key-site haplotype. The fourth key site substitution (T285A) switching SHYTA to SHYAA is predicted to shift the λ_{max} -16 nm [33].

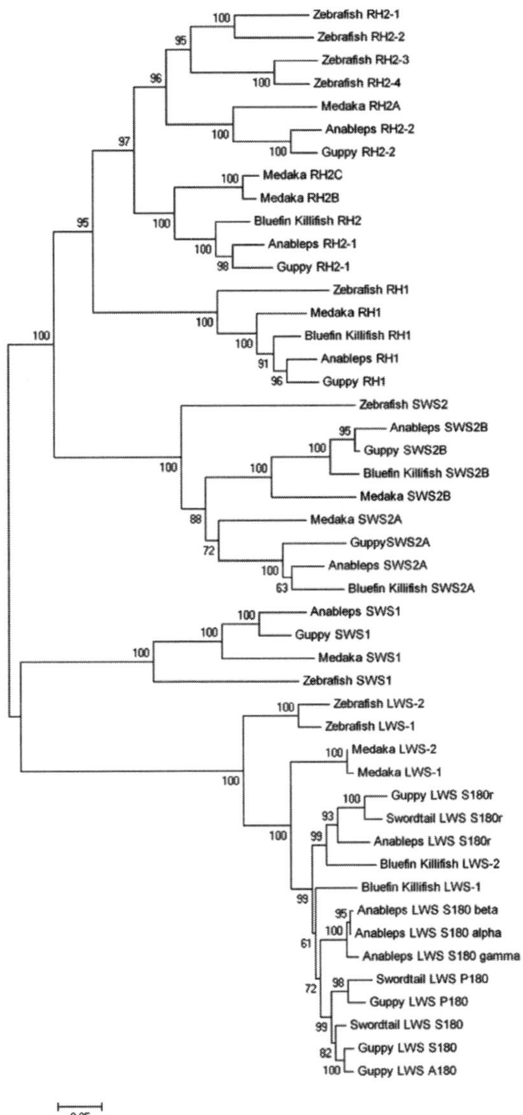

Figure 1. Phylogenetic analysis of A. anableps opsins. A neighbour-joining bootstrap consensus tree of visual opsins from A. anableps and its relatives. The percentage of trees in which the associated taxa clustered together in the bootstrap test (500 replicates) is shown at the nodes. Tamura-Nei algorithm was used and all codon positions were included. Missing nucleotides were treated with pairwise deletion in the analysis. Sequence accession numbers listed in Table 3.

Table 3. Sequences used in phylogenetic analysis.

Common name	Scientific names	Gene name	Accession number
The "Four-eyed" Fish	*Anableps anableps*	LWS S180α	FJ11154
		LWS S180β	FJ11158
		LWS S180γ	FJ11157
		LWS S180r	FJ11155
		SWS1	FJ11153
		SWS2A	FJ11152
		SWS2B	FJ11151
		RH2-1	FJ11149
		RH2-2	FJ11150
		RH1	FJ11156
Guppy	*Poecilia reticulata*	LWS S180	EU329434
		LWS A180	EU329442
		LWS P180	EU329456
		LWS S180r	EU329457
		SWS1	DQ234861
		SWS2A	FJ11159
		SWS2B	DQ234860
		RH2-1	DQ234859
		RH2-2	DQ234858
		RH1	DQ912024
Swordtail	*Xiphophorus pygmaeus*	LWS S180	EU329481
		LWS P180	EU329478
		LWS S180r	EU329479
Bluefin Killifish	*Lucania goodei*	LWS-1	AY296740
		LWS-2	AY296741
		SWS1	AY296735
		SWS2A	AY296737
		SWS2B	AY296736
		RH2-1	AY296739
		RH1	AY296737
Medaka	*Oryzias latipes*	LWS-1	AB223051
		LWS-2	AB223052
		SWS1	AB223058
		SWS2A	AB223056
		SWS2B	AB223057
		RH2a	AB223053
		RH2b	AB223054
		RH2c	AB223055
		RH1	AB180742
Zebrafish	*Danio rerio*	LWS-1	NM131175
		LWS-2	NM001002443
		SWS1	BC060894
		SWS2	NM131192
		RH2-1	NM131253
		RH2-2	NM182891
		RH2-3	NM182892
		RH2-4	NM131254
		RH1	BC05288

Common name, scientific name, gene name and GenBank accession number for all sequences used in phylogenetic analysis.

Discussion

A PCR-based survey of cDNA uncovered eight different opsin genes in Anableps anableps. Southern blotting, which utilized much longer probes than the PCR primers, indicated that additional genes might exist in the RH1, SWS1 and LWS opsin gene subfamilies. A subsequent PCR-based survey of genomic DNA uncovered two additional LWS opsin genes leading us to conclude that A. anableps possess ten visual opsins: one violet-sensitive SWS1 opsin gene, two genes from the blue-sensitive SWS2 subfamily, two genes encoding green-sensitive opsins from the RH2 subfamily, four LWS or red-sensitive opsin genes, and an RH1 gene.

Phylogenetics, Gene Duplication and Key Sites

Phylogenetic analysis showed that the two RH2 opsin genes in A. anableps are orthologs of RH2-1 and RH2-2, duplicates produced in the ancestor of guppy, medaka (Oryzias latipes), pufferfish (Takifugu rubripes), and stickleback (Gasterosteus aculeatus). Although we have no data on opsin gene location in A. anableps, the RH2 gene pair appears to be the product of a tandem duplication event as RH2-1 and RH2-2 are linked in medaka and pufferfish [34], [35]. The single band produced during Southern analysis of RH2 genes might be explained by the lack of a cut site between the tandem duplicates. The A. anableps and guppy SWS2 opsin gene duplicates reported here are orthologs of tandem duplicates found in medaka, called SWS2A and SWS2B [34]. SWS1 appears to be a single-copy gene in nearly all fish, including A. anableps. One exception is ayu (Plecoglossus altivelis), which contains a species-specific SWS1 opsin gene duplication [36]. We sequenced a single RH1 opsin from A. anableps cDNA. Only Conger eel (Conger myriaster) and scabbardfish (Lepidopus fitchi) have RH1 duplicates [37], [38]. The Southern blot analysis showed two bands for RH1 and SWS1. This banding might have been caused by allelic variation in cut site loci or non-specific Southern probe hybridization.

LWS opsin gene duplication events have occurred independently in several fish lineages; the ancestors of zebrafish, medaka, guppy and blind cavefish (Astyanax fasciatus) each experienced independent LWS gene duplication events [11], [34], [39], [40]. In some cases, LWS opsin gene duplication has been followed by amino acid substitutions at sites known to influence spectral sensitivity (human, guppy and zebrafish), whereas in others (medaka, blind cavefish and bluefin killifish) duplicates have the same 5- key-site haplotype [11], [34], [39]–[42]. The single exon LWS opsin gene S180r, which, appears to have been produced by retrotransposition in the ancestor of the livebearers and bluefin killifish [11], has been retained in A. anableps. This gene is one of those that retained the SHYTA five key site haplotype after duplication. LWS P180 (with the key-site haplotype PHFAA) and LWS S180 (with key site haplotype SHYTA) are tandem duplicates that have been uncovered in Poecilia and Xiphophorus [11], two genera in the family Poeciliidae, sister family to Anablepidae [30]. However, the duplication event producing this gene pair appears to pre-date the poeciliid, anablepid divergence. The A. anableps LWS S180α gene is similar to, and has the same key-site haplotype as, guppy LWS S180 and the A. anableps LWS S180γ· gene contains a region at the 3′ end that is remarkably similar to the homologous portion of the poeciliid LWS P180 opsin gene (Figure 2). However, LWS S180γ is not the sister sequence to the poeciliid LWS P180 genes as predicted by the hypothesis that they are products of the same tandem duplication event. We believe this is a consequence of post-duplication gene conversion events within the genus Anableps. Thus it appears that a combination of duplication and conversion have produced a unique opsin (with a SHYAA haplotype) in a fish with unique eye morphology.

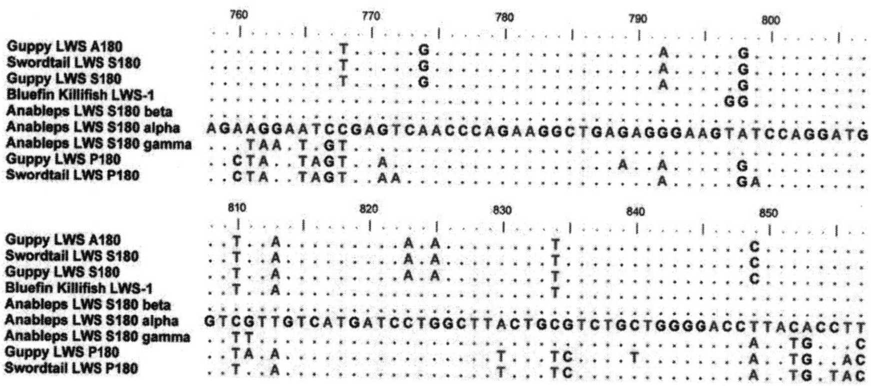

Figure 2. Sequence comparison between LWS genes. An alignment of 100 bp region of interest between between A. anableps LWS S180α and LWS S180γ. Over this area, A. anableps LWS S180γ is more similar to poeciliid P180, than to other A. anableps sequences.

A recent duplicate present in A. anableps and not in close relatives is LWS S180β. It is identical to LWS S180α aside from a short region of sequence variation in the 5′ end. This variation changes the amino acid sequence; however, it does not result in a new key site haplotype as it occurs before amino acid position 180, the first of the five key sites in LWS opsins.

Thus, molecular adaptations for aerial vision, at the primary sequence level, may be confined to one of the most recently created opsin genes, LWS S180γ which has a unique key site haplotype (SHYAA instead of SHYTA), and coincides with the evolution of the unique eye morphology. Alternatively, these observed sequence level changes could be neutral or possibly even mildly deleterious. However, it is likely that spectral tuning through amino acid substitutions is just one part of the adaptations for aerial vision, along with eye morphology, photoreceptor distribution and opsin expression pattern changes [4], [43], [44].

Implications for Expression

Although little is currently known about opsin expression patterns in the A. anableps retina, we predict that the photic contrast between the aerial and aquatic environment will have provided a selective pressure for divergent patterns of expression. The turbid water that A. anableps lives in filters light, allowing the long wavelength light to transmit most readily [45]. It is possible that A. anableps copes with the differences in light composition by using different opsin expression patterns in its two retinal hemispheres. Previous MSP work has attempted to measure pigment differences between retinal hemispheres, but no differences in pigments present were detected [17]. However, as mentioned previously, MSP suggests A. anableps possessed only three visual pigments altogether. Although nine of the ten opsins in this study were recovered from cDNA and thus expressed to some level, there can be extreme variation in opsin expression levels, both between duplicates and during development, therefore it is possible that at any given time only a portion of the repertoire is functionally significant [46], [47]. In future studies, we will use in situ hybridization to examine the mechanism of visual adaptation in A. anableps using probes designed from the opsin repertoire characterized here. By cataloguing its opsin repertoire we have laid the groundwork for much exciting research in not only A. anableps itself, but in the nature of aerial and aquatic vision.

Materials and Methods

RNA Isolation, cDNA Synthesis and DNA Isolation

Live Anableps anableps were obtained from a commercial supplier (The Afishionados, Winnipeg, Manitoba, Canada). One juvenile individual was euthanized in buffered MS222. Total RNA was isolated from one eye using Aurum™ Total RNA Fatty and Fibrous Tissue Pack, immediately after euthanasia and enucleation. RNA was stored at -80°C. cDNA was synthesized using BioRad® iScript Select cDNA Synthesis Kit from total RNA. DNA was isolated from muscle tissue using QIAquick® DNeasy Blood & Tissue Kit.

Primer Design and PCR

PCR primers were developed for eight genes in five visual opsin subfamilies, SWS1, SWS2, RH1, RH2, and LWS (Table 1). These primers were complementary to regions in each opsin gene or subfamily that were conserved in guppy (Poecilia reticulata), and bluefin killifish (Lucania goodei). Two forward and two reverse primers were employed for each gene.

Each primer pair was used to survey cDNA or genomic DNA in PCR reactions using Bio-Rad iProof high-fidelity DNA polymerase in an Eppendorf™ Mastercycler® EP Grad S thermocycler using the following conditions: Initial denaturation at 98°C for 30 seconds, 35 cycles with denaturation at 98°C for 5 seconds, annealing at 50–70°C (in 5°C gradations) for 12 seconds, extension at 72°C for 25 seconds and a final extension at 72°C for 5 minutes. During the second round of genomic DNA screening we added additional primers (1 µl at 10 mM) at the beginning of the last PCR cycle to prevent heteroduplex formation. Guppy cDNA was also surveyed using SWS2A opsin primers designed from conserved regions in A. anableps and bluefin killifish. The guppy PCR templates were obtained from a lab-reared fish descended from samples collected in Cumana, Venezuela (i.e., an Endler's guppy).

Cloning

PCR products were run on 1.5% agarose gel. Amplicons of the predicted size were excised using QIAquick® Gel Extraction Kit. If only one band was observed the portion of the product not run on the gel was purified using QIAquick® PCR Purification Kit. Purified products were A-tailed using Invitrogen™ Taq polymerase and cloned using the Promega® pGEM™ - T Easy Vector System II kit. Clones containing inserts of the correct size were sequenced using labelled M13 forward and reverse primers and a Licor sequencer at the Centre for Biomedical Research at the University of Victoria.

Southern Blotting

A. anableps genomic DNA was extracted from muscle tissue using phenol-chloroform extraction. DNA was digested in two separate reactions for 48 hours at 37°C using restriction enzymes, EcoRI and BamHI. Digestion was followed by overnight ethanol precipitation. Neither EcoRI nor BamHI cut within the region of the opsin genes that were complementary to the probes. 10 to 20 µg of digested DNA was electrophoresed in a 1.5% agarose gel and transferred onto a Bio-Rad® Zeta-Probe nylon membrane using the Bio-Rad® Model 785 Vacuum Blotter. Transferred DNA was immobilized by UV exposure for 5 minutes using a UVP HL-2000 HybriLinker prior to hybridization. DIG-labelled probes complementary to A. anableps opsins were synthesized using a Roche® PCR DIG Probe Synthesis Kit under the following amplification conditions: initial denaturation at 95°C for 2 minutes, 38 cycles with denaturation at 95°C for 30 seconds, annealing at 50–56°C for 30 seconds, extension at 72°C for 40 seconds and a final extension at 72°C for 7 minutes. These LWS, SWS1, SWS2, RH2, RH1 probes were amplified from cloned genomic DNA using the primer sets Fw100/Rev_Ex4, SWS1 Fw1/SWS1 Rev1, SWS2Fw3/SWS2Rev4, RH2 Fw1/RH2 Rev4 and RH1Fw4/RH1Rev4 (Table 1). Southern blot hybridization and detection was conducted according to the protocol provided in the Roche® DIG Application Manual for Filter Hybridization. Overnight hybridization at 41°C was performed in roller bottles using a UVP® HL-2000 HybridLinker. Hybridized membranes were subsequently washed at room temperature for 10 minutes (2×5 minutes) with 2× SSC followed by a 65°C wash for 30 minutes (2×15 minutes) with 0.5× SSC (both solutions contained 0.1% SDS). Roche® sheep Anti-Digoxigenin-AP, Fab fragments conjugated to alkaline phosphatase in conjunction with Roche® CSPD chemiluminescent substrate was used to detect the presence of bound digoxigenin probes. Generated blots were exposed to Roche® Lumi-film Chemiluminescent Detection Film for 3 to 24 hours prior to development.

Phylogenetic Analysis

A phylogenetic tree was reconstructed for the complete set of opsin sequences. It included sequences from guppy, swordtail, bluefin killifish, medaka and zebrafish (Table 3). Phylogenetic trees were constructed using Mega4 utilizing Tamura-Nei algorithm, Neighbour-joining, and support for nodes were estimated using 500 bootstrap reanalyses [28], [29], [48], [49]. Sequences were 412 to 819 bp long.

Acknowledgements

The authors thank Spencer Jack from Afishionados for importing the fish and Roderick Haesevoets at the University of Victoria Centre for Biomedical Research for sequencing assistance.

Author Contributions

Conceived and designed the experiments: GLO JST. Performed the experiments: GLO DJW JM. Analyzed the data: GLO. Wrote the paper: GLO.

References

1. Miller RR (1979) Ecology, Habits and Relationships of the Middle America Cuatro Ojos, Anableps dowi (Pisces: Anablepidae). Copeia 1: 82–91.
2. Zahl PA, McLaughlin JJA, Gomprecht RJ (1977) Visual versatility and feeding of the four-eyed fishes, Anableps. Copeia 4: 791–793.

3. Schwassmann HO, Kruger L (1966) Experimental analysis of the visual system of the four-eyed fish Anableps microlepis. Vision Res 5: 269–281.
4. Sivak J, Howland HC, McGill-Harelstad P (1987) Vision of the Humboldt penguin (Spheniscus humboldti) in air and water. Proc R Soc Lond B Biol Sci 229: 467–472.
5. Leonard DW, Meek KM (1997) Refractive indices of the collagen fibrils and extrafibrillar material of the corneal stroma. Biophys J 72: 1382–1387.
6. Nelson JS (1984) Fishes of the World. New York: John Wiley & Sons.
7. Kawamura S (1995) Phototransduction, excitation and adaptation. Neurobiology and Clinical Aspects of the Outer Retina. Kluwer Academic Publishers.
8. Masland RH (2001) The fundamental plan of the retina. Nat Neurosci 4: 877–886.
9. Bowmaker JK (2008) Evolution of vertebrate visual pigments. Vision Res 48: 2022–2041.
10. Jacobs GH (1993) The distribution and nature of color vision among the mammals. Biol Rev Camb Philos Soc 68: 413–471.
11. Ward MN, Churcher AM, Dick KJ, Laver CR, Owens GL, et al. (2008) The molecular basis of color vision in colorful fish: four long wave-sensitive (LWS) opsins in guppies (Poecilia reticulata) are defined by amino acid substitutions at key functional sites. BMC Evol Biol 8: 210.
12. Yokoyama S, Zhang H, Radlwimmer FB, Blow NS (1999) Adaptive evolution of color vision of the Comoran coelacanth (Latimeria chalumnae). Proc Natl Acad Sci USA 96: 6279–6284.
13. Yokoyama S, Radlwimmer FB (1999) The molecular genetics of red and green color vision in mammals. Genetics 153: 919–932.
14. Bridges CDB (1972) The rhodopsin-porphyropsin visual system. Handbook of sensory physiology 7: 417–480.
15. Meyer-Rochow VB, Coddington PE (2003) Eyes and vision of the New Zealand torrentfish Cheimarrichthys fosteri Von Haast (1874): histology, photochemistry and electrophysiology. Fish Adaptations 337–381.
16. Temple SE, Plate EM, Ramsden S, Haimberger TJ, Roth WM, et al. (2006) Seasonal cycle in vitamin A1/A2-based visual pigment composition during the life history of coho salmon (Oncorhynchus kisutch). J Comp Physiol A Neuroethol Sens Neural Behav Physiol 192: 301–313.
17. Avery JA, Bowmaker JK (1982) Visual pigments in the four-eyed fish, Anableps anableps. Nature 298: 62–63.

18. Applebury ML, Antoch MP, Baxter LC, Chun LL, Falk JD, et al. (2000) The murine cone photoreceptor: a single cone type expresses both S and M opsins with retinal spatial patterning. Neuron 27: 513–523.

19. Hope AJ, Partridge JC, Hayes PK (1998) Switch in rod opsin gene expression in the European eel, Anguilla anguilla (L.). Proc Biol Sci 265: 869–874.

20. Makino CL, Dodd RL (1996) Multiple visual pigments in a photoreceptor of the salamander retina. J Gen Physiol 108: 27–34.

21. Seehausen O, Terai Y, Magalhaes IS, Carleton KL, Mrosso HD, et al. (2008) Speciation through sensory drive in cichlid fish. Nature 455: 620–626.

22. Archer S, Hope A, Partridge JC (1995) The molecular basis for the green-blue sensitivity shift in the rod visual pigments of the European eel. Proc Biol Sci 262: 289–295.

23. Takechi M, Kawamura S (2005) Temporal and spatial changes in the expression pattern of multiple red and green subtype opsin genes during zebrafish development. J Exp Biol 208: 1337–1345.

24. Davies WL, Cowing JA, Carvalho LS, Potter IC, Trezise AE, et al. (2007) Functional characterization, tuning, and regulation of visual pigment gene expression in an anadromous lamprey. FASEB J 21: 2713–2724.

25. Saiki RK, Gelfand DH, Stoffel S, Scharf SJ, Higuchi R, et al. (1988) Primer-directed enzymatic amplification of DNA with a thermostable DNA polymerase. Science 239: 487–491.

26. Odelberg SJ, Weiss RB, Hata A, White R (1995) Template-switching during DNA synthesis by Thermus aquaticus DNA polymerase I. Nucleic Acids Res 23: 2049–2057.

27. Zylstra P, Rothenfluh HS, Weiller GF, Blanden RV, Steele EJ (1998) PCR amplification of murine immunoglobulin germline V genes: strategies for minimization of recombination artefacts. Immunol Cell Biol 76: 395–405.

28. Tamura K, Dudley J, Nei M, Kumar S (2007) MEGA4: Molecular Evolutionary Genetics Analysis (MEGA) software version 4.0. Mol Biol Evol 24: 1596–1599.

29. Tamura K, Nei M (1993) Estimation of the number of nucleotide substitutions in the control region of mitochondrial DNA in humans and chimpanzees. Mol Biol Evol 10: 512–526.

30. Hrbek T, Seckinger J, Meyer A (2007) A phylogenetic and biogeographic perspective on the evolution of poeciliid fishes. Mol Phylogenet Evol 43: 986–998.

31. Okano T, Kojima D, Fukada Y, Shichida Y, Yoshizawa T (1992) Primary structures of chicken cone visual pigments: vertebrate rhodopsins have evolved out of cone visual pigments. Proc Natl Acad Sci USA 89: 5932–5936.

32. Yokoyama S, Takenaka N, Blow N (2007) A novel spectral tuning in the short wavelength-sensitive (SWS1 and SWS2) pigments of bluefin killifish (Lucania goodei). Gene 396: 196–202.

33. Yokoyama S, Yang H, Starmer WT (2008) Molecular basis of spectral tuning in the red- and green-sensitive (M/LWS) pigments in vertebrates. Genetics 179: 2037–2043.

34. Matsumoto Y, Fukamachi S, Mitani H, Kawamura S (2006) Functional characterization of visual opsin repertoire in Medaka (Oryzias latipes). Gene 371: 268–278.

35. Neafsey DE, Hartl DL (2005) Convergent loss of an anciently duplicated, functionally divergent RH2 opsin gene in the fugu and Tetraodon pufferfish lineages. Gene 350: 161–171.

36. Minamoto T, Shimizu I (2005) Molecular cloning of cone opsin genes and their expression in the retina of a smelt, Ayu (Plecoglossus altivelis, Teleostei). Comp Biochem Physiol B Biochem Mol Biol 140: 197–205.

37. Zhang H, Futami K, Horie N, Okamura A, Utoh T, et al. (2000) Molecular cloning of fresh water and deep-sea rod opsin genes from Japanese eel Anguilla japonica and expressional analyses during sexual maturation. FEBS Lett 469: 39–43.

38. Yokoyama S, Tada T, Zhang H, Britt L (2008) Elucidation of phenotypic adaptations: Molecular analyses of dim-light vision proteins in vertebrates. Proc Natl Acad Sci USA 105: 13480–13485.

39. Chinen A, Hamaoka T, Yamada Y, Kawamura S (2003) Gene duplication and spectral diversification of cone visual pigments of zebrafish. Genetics 163: 663–675.

40. Yokoyama R, Yokoyama S (1990) Isolation, DNA sequence and evolution of a color visual pigment gene of the blind cave fish Astyanax fasciatus. Vision Res 30: 807–816.

41. Nathans J, Thomas D, Hogness DS (1986) Molecular genetics of human color vision: the genes encoding blue, green, and red pigments. Science 232: 193–202.

42. Fuller RC, Carleton KL, Fadool JM, Spady TC, Travis J (2004) Population variation in opsin expression in the bluefin killifish, Lucania goodei: a real-time

PCR study. J Comp Physiol A Neuroethol Sens Neural Behav Physiol 190: 147–154.

43. Mass AM, Supin AY (2007) Adaptive features of aquatic mammals' eye. Anat Rec (Hoboken) 290: 701–715.
44. Levenson DH, Ponganis PJ, Crognale MA, Deegan JF, Dizon A, et al. (2006) Visual pigments of marine carnivores: pinnipeds, polar bear, and sea otter. J Comp Physiol A Neuroethol Sens Neural Behav Physiol 192: 833–843.
45. Seehausen O, Alphen JJM, Witte F (1997) Cichlid Fish Diversity Threatened by Eutrophication That Curbs Sexual Selection. Science 277: 1808.
46. Carleton KL, Kocher TD (2001) Cone opsin genes of African cichlid fishes: tuning spectral sensitivity by differential gene expression. Mol Biol Evol 18: 1540–1550.
47. Carleton KL, Spady TC, Streelman JT, Kidd MR, McFarland WN, et al. (2008) Visual sensitivities tuned by heterochronic shifts in opsin gene expression. BMC Biol 6: 22.
48. Saitou N, Nei M (1987) The neighbor-joining method: a new method for reconstructing phylogenetic trees. Mol Biol Evol 4: 406–425.
49. Felsenstein J (1985) Confidence limits on phylogenies: an approach using the bootstrap. Evolution 39: 783–791.

Lateral Transfer of a Lectin-Like Antifreeze Protein Gene in Fishes

Laurie A. Graham, Stephen C. Lougheed, K. Vanya Ewart and Peter L. Davies

ABSTRACT

Fishes living in icy seawater are usually protected from freezing by endogenous antifreeze proteins (AFPs) that bind to ice crystals and stop them from growing. The scattered distribution of five highly diverse AFP types across phylogenetically disparate fish species is puzzling. The appearance of radically different AFPs in closely related species has been attributed to the rapid, independent evolution of these proteins in response to natural selection caused by sea level glaciations within the last 20 million years. In at least one instance the same type of simple repetitive AFP has independently originated in two distant species by convergent evolution. But, the isolated occurrence of three very similar type II AFPs in three distantly related species (herring, smelt and sea raven) cannot be explained by this mechanism. These globular, lectin-like

AFPs have a unique disulfide-bonding pattern, and share up to 85% identity in their amino acid sequences, with regions of even higher identity in their genes. A thorough search of current databases failed to find a homolog in any other species with greater than 40% amino acid sequence identity. Consistent with this result, genomic Southern blots showed the lectin-like AFP gene was absent from all other fish species tested. The remarkable conservation of both intron and exon sequences, the lack of correlation between evolutionary distance and mutation rate, and the pattern of silent vs non-silent codon changes make it unlikely that the gene for this AFP pre-existed but was lost from most branches of the teleost radiation. We propose instead that lateral gene transfer has resulted in the occurrence of the type II AFPs in herring, smelt and sea raven and allowed these species to survive in an otherwise lethal niche.

Introduction

Acquisition of a new gene/trait typically arises from gene duplication and divergence [1]. A classic example of this gradual process is the evolution of a set of pancreatic serine proteases, specifically trypsin, chymotrypsin and elastase, from a common precursor [2], [3]. These paralogs have the same three-dimensional fold and operate by the same enzymatic mechanism, but cleave proteins after different amino acids. The opportunities to short-circuit this process and pass a gene between species by horizontal or lateral gene transfer (LGT) would seem extremely limited, and are largely restricted to prokaryotes. In some bacteria there are established routes (conjugation, transduction and transformation) for the exchange of DNA between strains/species, subject to the strictures of the restriction/modification system in the recipient host. LGT becomes particularly evident where the acquisition of the transferred gene confers a selective advantage on the host, as for example in antibiotic resistance [4], [5]. Here, there is the opportunity to acquire a new gene type within one generation rather than by gradual evolution.

In eukaryotes there is no established mechanism for transferring intact genes between species, although retrovirally processed sequences have been transferred [6]. There is also the added difficulty that genes are packaged within organelles (principally the nucleus) and are therefore less accessible than genes in bacteria. Moreover, in higher eukaryotes there is an additional barrier to transmission in that only LGT to germ-line cells would be passed on. Thus, prior to this report there was no well documented report of a standard eukaryotic gene being passed into or between vertebrate species. Here we provide the first such evidence for LGT in vertebrates. As with antibiotic resistance in bacteria, it has come to light because of the selective advantage the gene confers to a host under intense selective pressure.

The gene in question codes for a type II antifreeze protein (AFP), one of five distinct types that have appeared in fishes. These AFPs stop the growth of seed ice crystals by a surface adsorption-inhibition mechanism and thereby help fish resist freezing in icy seawater [7]. Type II AFPs are homologs (paralogs) of the sugar-binding domain of Ca2+-dependent (C-type) lectins [8]. In C-type lectins [9], [10] such as the rat mannose-binding protein (now named mannan-binding lectin) (Figure 1A), one of the calcium ions is an integral part of the sugar-binding site and makes direct contact with the ligand [11]. Herring and rainbow smelt AFPs require Ca2+ for binding to ice [8], [12]. Again, this metal ion is thought to play a central role in ligand binding because substitution of Ca2+ with other divalent metal ions alters both the antifreeze activity and ice crystal morphology [13]. X-ray crystallography has shown that herring AFP [14] has same fold as rat mannose-binding protein (Figure 1A,B). The more divergent sea raven AFP [15], which is 40% identical to herring and smelt AFPs, is not Ca2+-dependent. Nonetheless, solution structure determination has shown that it too has the same fold as rat mannose-binding protein [16].

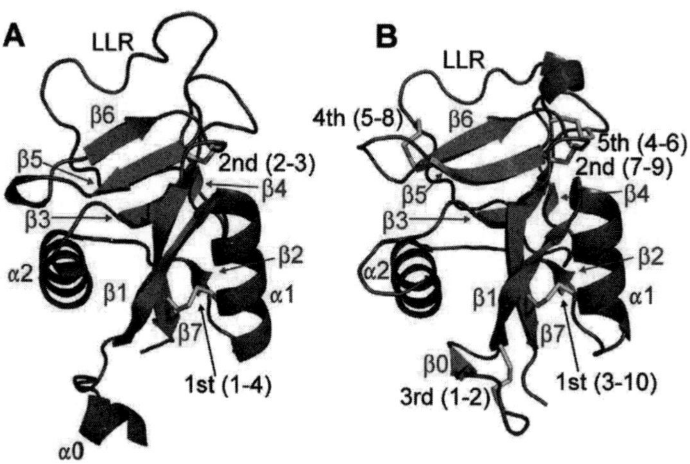

Figure 1. Antifreeze protein - lectin structural comparisons. (A) Rat mannose-binding protein structure (PDB code: 1KWT) and (B) Herring AFP structure (PDB code: 2PY2) showing the location of the 5th disulfide bridge that is peculiar to the type II AFPs. Secondary structure elements (red – helix; green – beta-strand; grey – loop) are numbered from the N terminus. Disulfide bonds are shown in yellow with the linkages of numbered cysteines indicated in brackets. LLR is the long loop region.

The structural feature that the three type II AFPs share, and which distinguishes them from all other C-type lectin domains, is that they have ten cysteines forming five disulfide bridges in identical positions (Figure 2). Most C-type lectins have two or three of the disulfide bridges [9], [10]. One of the two invariant bridges found in all C-type lectins with a long loop region links the first helix

(α1) to the last β-strand (β7) (Figure 1A,B). The other links the start of β5 to the loop between β6 and β7. The 3rd disulfide bridge occurs within the N-terminal extension that is missing from some lectins. A 4th bridge linking the long loop region between β4 and β5 to the start of β6 is comparatively rare but is seen, for example, in some lectins from carp and zebrafish (Figure 2). However, the 5th bridge, linking the loop after β3 to the first Cys of the pair found at the beginning of β3, is peculiar to the type II AFPs.

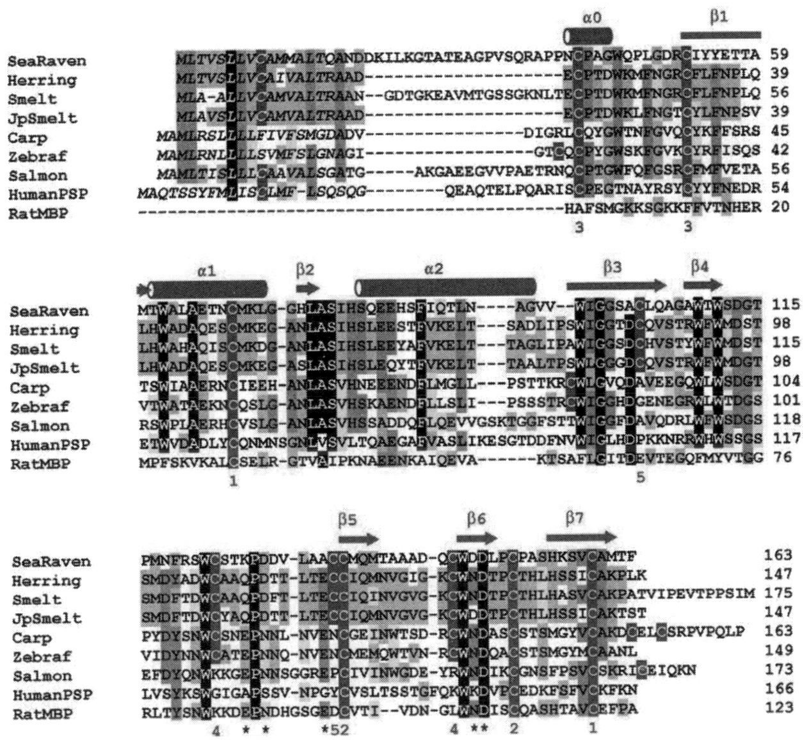

Figure 2. Antifreeze protein - lectin alignments. C-type lectin homolog alignment showing the cysteine pair that makes up the 5th disulfide bond is common to the type II AFPs but is missing in other lectins from the database. Sequences aligned are type II AFPs from the sea raven, Hemitripterus americanus, (SeaRaven, GenBank #AAA49617); the Atlantic herring, Clupea harengus, (Herring, GenBank # AAY60837); the rainbow smelt, Osmerus mordax, (Smelt, GenBank #AAA49442); the Japanese smelt, Hypomesus nipponensis, (JpSmelt (Yamashita et al. 2003)); single lectin domain proteins from the common carp, Cyprinus carpio, (Carp, GenBank #BAA95671); the zebrafish, Danio rerio, (Zebraf, GenBank # XM_001337634); and the Atlantic salmon, Salmo salar, (Salmon, GenBank #AAO43604); as well as the human pancreatic stone protein (PSP, GenBank #NP_002900) and the C-type lectin domain from the rat mannose-binding protein (RatMBP, GenBank #AAA98781). Secondary structure elements from the rat mannose-binding protein (see Figure 1) are displayed above the alignment. Color is used to indicate conservation of the residues between the 9 sequences with matches of 8–9 in black, 6–7 in green, and 3–5 in ochre or yellow. Cysteines are colored red and their pairings are indicated by numbers under the alignment. Sequences are numbered from the start of the signal polypeptide (in italics) except for the rat mannose-binding protein where numbering starts at the beginning of the domain. Residues in the herring and rat mannose-binding protein that are involved in Ca^{2+} coordination are indicated by blue stars below the sequence alignment.

The independent gain of a disulfide bridge in exactly the same place on three separate occasions seems unlikely. In light of this, Liu et al. [14] have proposed that the type II AFPs are derived from a ten-Cys lectin isoform that preexisted in the ancestor to most fishes, but has subsequently been lost from all other branches. We have researched this possibility and find that the evidence, particularly for the herring and smelt AFPs, is overwhelmingly in favour of a different mechanism: LGT. The remarkable conservation of the protein sequences, the unexplained conservation of the intron sequences, the lack of correlation between evolutionary distance and mutation rate, and the pattern of silent vs non-silent codon changes all point to lateral transfer of the gene.

Methods

Unpublished sequences have been deposited in GenBank with the following accession numbers: DQ008165 (Prp8 genomic sequence from rainbow smelt), DQ008166 (Prp8 genomic sequence from Atlantic herring), DQ004949 (AFP genomic sequence from rainbow smelt), DQ003023 (AFP genomic sequence from Atlantic herring).

Isolation of Genomic and Gene-Specific DNAs

Genomic DNAs were isolated from either testes (bowfin Amia calva, rainbow trout Oncorhynchus mykiss, Atlantic cod Gadus morhua, sea raven Hemitripterus americanus, yellow perch Perca flavescens, winter flounder Pseudopleuronectes americanus), liver (Atlantic herring Clupea harengus, rainbow smelt Osmerus mordax), muscle (Pacific herring Clupea pallasi, cisco Coregonus artedi) or whole fish (zebrafish Danio rerio) [17]. All fishes were caught off the Atlantic coast of Canada except the following: rainbow trout (Denmark), zebrafish (local pet store), bowfin, yellow perch and cisco (Lake Ontario), Pacific herring (Pacific coast of Canada).

The primers used to amplify the AFP gene sequences are as follows; rainbow smelt, upstream of the start codon 5′-CAACAGGCTGAAATTGTGCAGACA-3′, ending on the stop codon 5′-TCACATGATTGATGGTGGTGTCAC-3′ and Atlantic herring, upstream of the start codon 5′-CTAAAGGGAAGACAGAGGCAACAG-3′, downstream of the stop codon 5′-TGATTGATGGTGGTGGATGCCTCT-3′. Approximately 200 ng of genomic DNA was amplified using the Expand High-Fidelity PCR system (Roche, Penzberg, Germany). The DNA was denatured for 10 min at 95°C, reagents were added at 80°C, and 30 cycles of PCR were done as follows; 95°C for 1 min, 60°C for 1 min and 72°C for 3 min with 5 sec/cycle added starting at cycle 11, with a final extension of 7 min at 72°C. Products were subcloned into the pCR2.1-TOPO vector (Invitrogen, Carlsbad, CA, U.S.A.) and both strands were sequenced using vector and internal primers.

The primers used to amplify a portion of 16S rDNA from Atlantic herring and rainbow smelt are as follows; 5′-TGAAGACCTGTATGAATGG-3′ and 5′-TTGAACAAACGAACCCTTA-3′. The amplification and subcloning were done as described above but using Taq (Fermentas International, Burlington, ON, Canada) and an annealing temperature of only 50°C.

The primers used to amplify a portion of the Prp8p gene correspond to exonic sequences conserved between zebrafish and pufferfish. They are numbered sequentially from outermost to innermost; #1 sense 5′-CAGCCTGTGAAGGTGCGTGTGTC-3′, #2 sense 5′-TTCCGCTCTTTCAAGGCCACCAA-3′, #3 sense 5′-GGCAT GTACCGCTACAAGTACAA-3′, #1 antisense 5′-CTTCCAAGGAATGTTGGCCTTCCA-3′, #2 antisense 5′-CCGTGTTGGTCCACCAGTCAGCTTT-3′, #3 antisense 5′-GAGATGCT TCAGGTCTTTGCACAT-3′. The rainbow smelt sequence was obtained using primers #1 sense and antisense. The Atlantic herring sequence was obtained in two overlapping segments using #2 sense with #3 antisense and #3 sense with #2 antisense. The amplification, subcloning and sequencing were done as for the AFP genes except that 1 μL of the first PCR reaction was reamplified for 25 cycles and an annealing temperature of 56°C was used.

Phylogenetic Analyses

The evolutionary affinities of the lectin-like AFPs were inferred using both Bayesian and parsimony approaches on the amino acid alignment. Human pancreatic stone protein (PSP) was used as the out-group for both. For Bayesian analysis, the aamodelpr = mixed option was used, where the Markov chain samples each of nine models of evolution according to its probability [18]. Two simultaneous analyses of 1,000,000 generations were run and sampled every 100 generations, until the Potential Scale Reduction Factors [19] for all parameters were very close to one (to the second decimal). Effective sample sizes for all parameters were estimated using TRACER [20] and were all substantially greater than 100, implying effective sampling of the posterior distribution of all parameters. For parsimony analysis, we first performed an exhaustive search in PAUP* [21] with gaps treated as a 21st amino acid, and then evaluated support for the resulting topology using a bootstrap analysis with 1000 pseudoreplicates and ten random additions per replicate.

The phylogenetic relationships between teleosts accepted here are those established by Miya et al. [22] based on complete mitochondrial genome sequences.

However, to orient other intermediate species, particularly AFP-producing ones, to the phylogeny, Bayesian analysis was performed on an alignment of a portion of the 16S rDNA region, corresponding to bases 3081 to 3532 of zebrafish 16S rDNA (AC024175). All sequences were obtained from GenBank, except those obtained above for Atlantic herring and rainbow smelt, and the species not noted elsewhere which include the Japanese pilchard Sardinops melanostictus, common carp Cyprinus carpio, smooth lumpsucker Aptocyclus ventricosus, longhorn sculpin Myoxocephalus octodecemspinosus, Antarctic eelpout Lycodichthys dearborni, Antarctic toothfish Dissostichus mawsoni, dark-banded fusilier Pterocaesio tile, bastard halibut Paralichthys olivaceus and masked triggerfish Sufflamen fraenatus. The.time reversible+I+G model of evolution was selected from among 24 possible models using the Akaike Information Criterion (MrModelTest,) [23]. A Bayesian tree was generated using the program MrBayes [24], with two independent runs each with 1,000,000 generations of MCMC simulations (until the standard deviation of the split frequencies of the two runs was less than 0.01). Trees were sampled every 100 generations beginning at 250,000 generations.

Bioinformatic Analyses

Database searches were done with the complete sequences of rainbow smelt, herring and sea raven AFPs. The protein sequences were used in both protein-protein, position-specific iterated and translated BLAST searches (using default parameters) depending on the database searched (http://www.ncbi.nlm.nih.gov/). The cDNA and genomic sequences were used in BLASTn searches using both default parameters (word size = 11, expect threshold = 10, match = 2, mismatch = –3, gap existence = –5, gap extension = –2, filter and mask on) and with altered parameters (word size = 7, filter and mask off). The following GenBank databases were searched: non redundant, EST, genome survey sequence, high-throughput genomic sequence, patent, whole genome shotgun, sequence tagged site and environmental sequences. Searches of the nr and EST databases included 1) all organisms; 2) just bony fishes (taxid:32443); 3) everything but bony fishes; Species-specific search were performed, as above, on the nr, EST, high-throughput genomic sequence, whole genome shotgun and trace archives for zebrafish (taxid:7955), Takifugu rubripes (taxid:31033) and Tetraodon nigroviridis (taxid:99883). The medaka (Oryzias latipes) BLAST was done at http://dolphin.lab.nig.ac.jp/medaka/ using BLASTn (word size = 7) and tBLASTn (word size = 3, BLOSUM62 scoring matrix) with the filter off and gaps allowed.

The individual intron and exon sequences of herring AFP, as well as intron 2 of smelt AFP, were used for nucleotide-nucleotide searches (BLASTn) against teleost fish sequences (taxid:32443) in the non-redundant database as above.

Analysis of Synonymous and Non-Synonymous Substitutions and Codon Bias

The ratio of non-synonymous substitutions per non-synonymous site to synonymous substitutions per synonymous site (dN/dS) was calculated using the SNAP tool [25]. The portion of AFP sequence compared extends from the first residue of the mature herring AFP (ECP…) to the last residue of the seventh beta strand (…CAK, Figure 2). A section of the sea raven sequence that could not be unambiguously aligned (AGVV in second helix, Figure 2) was excluded in comparisons with this sequence. The portion of Prp8p coding sequence compared corresponds to the overlapping region between the sections of the rainbow smelt and herring sequences cloned in this study.

The codon usage, effective number of codons (ENc), and GC content at the 3rd position of synonymous codons (GC3) was determined for the complete coding sequences of the type II AFPs and the partial coding sequences of the herring and rainbow Prp8p genes using the program codonw [26].

Phylogenetic Analyses Using 16S rRNA

We assembled a 16S dataset for a subset of taxa that included an outgroup (bowfin), four species with the AFP (rainbow and Japanese smelt, Atlantic herring, sea raven), and six other species mentioned above which do not possess the AFP gene according to our Southern blot and on-line searches (zebrafish, rainbow trout, Atlantic cod, winter flounder, Takifugu rubripes and yellow perch). Since a 16S sequence is not yet available for cisco, one of four identical sequences from four species of the same genus (Coregonus peled, DQ399871) was used.

The data were subjected to two analyses to test the admittedly unlikely proposition that smelts, herring and sea raven all possess type II AFPs because they form a monophyletic assemblage. First, a Bayesian analysis was conducted using the GTR+I+G model, as above, selected by MrModeltest [23]. Two independent analyses were run with Metropolis-coupled MCMC using four incrementally heated Markov chains for 1,000,000 generations until the standard deviation of the split frequencies was <0.01. Trees were sampled every 1000 generations, with the first 200 of these discarded as burn-in. A constraint tree was created in MacClade Version 4 [27] with the species possessing the AFP gene constrained to be monophyletic, and then filtered the trees resulting from our Bayesian analysis using PAUP* Version 4.0b10 [21], retaining only those that were consistent with the constraint. Our second approach employed maximum parsimony (gaps treated as missing data). We did two separate exhaustive searches for the most parsimonious tree(s) using PAUP*, one subject to our constraint tree, and the other unconstrained. We then compared the tree lengths for the most parsimonious tree(s) between the two runs.

Southern Blotting

Fish genomic DNAs were digested extensively with PvuII (Fermentas) and 10 μg per lane was resolved on a 0.8% agarose gel. The DNA was transferred to zeta-probe membrane by alkaline capillary blotting as recommended by the manufacturer (Bio-Rad, Richmond, CA, U.S.A.). Probes were labeled using the random primers DNA labeling system (Invitrogen) and consisted of a portion of the rainbow smelt AFP gene (encompassing exons 3 to 6, bases 1023 to 1940 of GenBank #DQ004949), or a portion of a chicken β-tubulin cDNA (from bases 326 to 1423 of GenBank #V00389). Standard blotting techniques were used [28] except that the concentration of Denhardt's was increased to 10×, SDS to 2% and 200 μg/mL sheared and denatured calf thymus DNA was used instead of salmon DNA. All incubations and washes were done at 60°C with a final wash in 0.5% SSC, 1% SDS.

Results

The Ten-Cys Lectin-Like AFPs Have No Close Matches in the Database

Extensive searches of sequences from all organisms, in all relevant GenBank databases, using herring, rainbow smelt and sea raven antifreeze proteins as the queries, revealed no close matches and no ten-Cys lectin other than the type II AFPs. These databases, including the non-redundant and EST databases, contained over 3 million cDNAs from bony fishes. The near completion of the two pufferfish (Fugu rubripes [29] and Tetraodon nigroviridis [30]), medaka (Oryzias latipes) [31]and zebrafish (Danio rerio, http://www.sanger.ac.uk/Projects/D_rerio/) genome sequences provided an opportunity to more thoroughly examine four fish species for a possible progenitor. Again, no close homologs were identified. The highest amino acid sequence identity of mature type II AFP with fish lectin-like proteins is less than 40%. The highest identity with lectins of other vertebrates and invertebrates is 33%. These values are radically different from the 85% identity between the herring and rainbow smelt AFPs.

The High Conservation of the Type II AFP Sequences Belies Their Scattered Distribution in Fish Phylogeny and is Consistent with Lateral Gene Transfer (LGT)

To illustrate the discrepancy between the relatedness of the type II AFP sequences and the relatedness of the fish that produce them, Bayesian and parsimony trees were derived from the protein sequences shown in Figure 2. Since both trees were identical, only the former is presented (Figure 3A). The clustering of the type II AFP producing species (Figure 3A) based solely on the AFP sequences is completely at odds with the phylogenetic tree of teleosts based on ribosomal 16S RNA sequence comparison (Figure 3B). In contrast, the phylogenetic tree in Figure 3B is very similar to those derived from both morphology [32] and complete mitochondrial genome sequences [22]. The high similarity between the herring and rainbow smelt AFPs is amazing given that these fish diverged over 100 million years ago. The Japanese smelt confounds the already remarkable antifreeze sequence similarity, in that its AFP amino acid sequence is about as similar (84%) to that of the Atlantic herring (different superorder) as it is to the rainbow smelt sequence (same family, Figure 3).

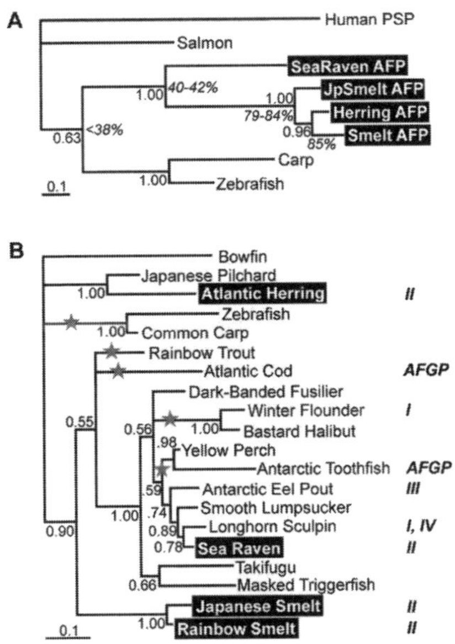

Figure 3. Phylogenetic trees of AFPs and related lectins as well as selected teleost fishes. (A) Bayesian 50% majority rule consensus tree (from 7500 sampled trees) based on the protein sequences aligned in Figure 2. The alignment was truncated to the beginning of the first helix ($\alpha 0$) to the end of the sea raven sequence. Branch lengths represent the expected fraction of changes, and Bayesian posterior probabilities above 50% and selected percent identities (italics) are indicated near the nodes. The highest identity between the AFPs and any other C-type lectin domain is 38%. Herring AFP is 84% and 85% identical to Japanese and rainbow smelt AFPs, respectively. The two smelt sequences share 79% identity. (b) Phylogenetic tree of teleosts based on ribosomal 16S RNA sequence comparison. Species are referred to by their common names. Those that are known to produce AFPs have the type of AFP they produce indicated alongside (AFGP, antifreeze glycoprotein; I, alanine-rich alpha-helical; II, lectin-like; III, beta-clip fold; IV, helix bundle). Those that produce type II AFP are highlighted. The rainbow smelt and herring sequences were double-checked by amplification from the DNAs used in this study and were almost identical to the sequences reported in the database. The non-teleost fish, the bowfin, was used to root the tree and only posterior probabilities >50% are shown. Putative type II gene loses (presuming the presence of the gene in the common ancestor) are indicated by grey stars. A further gene loss on the Takifugu/masked triggerfish lineage is not postulated since these fish group with yellow perch and Antarctic toothfish according to the phylogeny of Miya et al. [22].

The theoretical possibilities that herring and rainbow smelt are much more closely related than previously thought, or that the specimens were misidentified on collection, are negated by our phylogenetic tree using 16S rRNA sequences amplified from the individual Atlantic herring and rainbow smelt used in this study (Figure 3B). The herring clusters with the Japanese pilchard (same subfamily) and the rainbow smelt clusters with the Japanese smelt (same family) as expected. We also incorporated additional AFP-producing species along with their closest relatives from the tree generated by Miya et al. [22] while excluding others, to illustrate the unusual distribution of AFP types in general. Our phylogeny, generated using much less sequence data, is very similar to that of Miya et al. [22]. The two minor exceptions are at trichotomies where trout should be clustered with smelt and winter flounder/halibut should diverge earlier than takifugu/triggerfish.

If, as Liu et al. [14] have postulated, type II AFP existed in the common progenitor to the type II producing species, according to the phylogeny of Miya et al. [22], the gene must have been independently lost on at least five occasions. These theoretical losses are indicated on Figure 3B by grey stars. The absence of the gene in intervening species is supported by the database searches above and Southern blotting below.

Similar Rates of Silent and Missense Mutations in Type II AFP Genes are Inconsistent with Strong Selection for Over 100 Million Years

Another argument against the normal descent/gene loss hypothesis, given the equivalent similarities between the AFPs of the closely related rainbow and Japanese smelts and the distantly related herring, and the greater divergence of the sea raven AFP, is that one would need to postulate starkly contrasting selection pressures on the different fishes at various times. For example, selection must have been much stronger on the herring and smelt sequences than on the sea raven sequence, but only up until the point at which the two smelts diverged. If selection was strong, the estimated ratio of non-synonymous (missense) mutations per non-synonymous site to synonymous (silent) mutation per synonymous site (d_N/d_S) [33] should be much less than one. However, this is not the case, since ratios close to unity are observed in all pairwise AFP comparisons. This is in stark contrast to the values obtained using the highly conserved spliceosomal protein, Prp8p [34] in which d_N/d_S is below 0.02 in all comparisons between herring, rainbow smelt, takifugu and zebrafish sequences. It should be noted a d_N/d_S ratio close to one does not imply a lack of selection for the retention of antifreeze activity in these fishes. Rather, it implies that the majority of the sites within the protein can tolerate substitutions without significantly affecting AFP function. High d_N/d_S ratios (averaging 0.67 and 1.0) have also been observed in the half of the residues (those not involved in ice-binding or structural integrity) of the more structurally-constrained AFP isoforms of two beetle species [35].

Another discrepancy is the differences in the proportion of silent sites that are altered. Fewer than 10% of the synonymous sites differ between the AFP sequences of herring and rainbow smelt. For Prp8p, this value increases to almost 50%, whereas for nonsynonymous sites, the opposite trend (8% for AFP vs 1% for Prp8p) is observed. A low synonymous mutation rate could be the result of selection for particular codons, which has been correlated with both GC content at the 3rd position of synonymous codons (GC3) and expression levels in cyprinid fishes including the common carp [36]. A measure of the variability in codon usage is given by the effective number of codons (ENc), which ranges from 20 for genes which use but a single codon for each amino acid to 61 for genes in which codon usage is random [26]. For type II AFPs, codon usage appears quite random with ENc and GC3 (brackets) values of 60 (38%) for herring, 55 (38%) for rainbow smelt, 59 (39%) for Japanese smelt and 55 (44%) for sea raven. This suggests that codon usage is close to random indicating little or no selection at silent sites. In contrast, the ENc and GC3 values for the Prp8p genes of herring and rainbow smelt are 42 (77%) and 39 (82%) respectively, likely indicative of selection for increased GC content.

Finally, we tested the null hypothesis that the type II AFP gene is the result of normal descent in the absence of gene loss or lateral transfer, by presuming the type II AFP producing species are monophyletic. None of the 800 16S rRNA Bayesian trees was retained after filtering and the five most parsimonious constraint trees were 26 steps longer than the single most parsimonious tree without any constraint (total tree length 524 steps) meaning that monophyly of the type II producing fishes is extremely unlikely, as expected.

Taken together, the discrepancies between the 16S rRNA phylogeny and the conservation pattern of the AFPs, the high ratio (0.9) of the rate of missense to silent mutations, which suggests that the amino acid sequences of herring and rainbow smelt AFPs are not under strong selection pressure, and the low rate of silent substitution in the absence of an appreciable codon bias, are totally inconsistent with normal descent of the type II AFP gene from a common ancestor over 100 million years ago. This contrasts with the Prp8p gene, which shows a much lower rate of missense to silent mutations along with a five-fold higher rate of silent substitution with selection. An alternate and more plausible explanation for these data is that the type II AFP gene was laterally transferred into or between the herring and smelt lineages not long before the divergence of the two smelt species.

LGT probably occurred on at least two occasions: in an earlier event to the ancestor of the sea raven, and more recently, to or between the herring and smelts.

The Conservation of Non-Coding Sequences Also Supports LGT

To further test the LGT hypothesis, we cloned and sequenced the introns and exons of both Atlantic herring and rainbow smelt AFPs and aligned these with the previously known sea raven sequence. All three genes have five introns in identical positions (Figure 4A,B). The second intron in the rainbow smelt AFP gene is interrupted by a mini-exon that codes for an N-terminal extension to the mature protein. But, this exon might be of very recent origin because its sequence is not present in the closely related Japanese smelt (Hypomesus nipponensis) [37]. BLAST searches, using both isolated exons and complete cDNA sequences, detected only two matches (55/65 and 50/59) with an expect value less than 10^{-3}. Both correspond to sequences encoding low-complexity signal peptides, so their significance is doubtful. This paucity of sequences related to the AFP gene suggests close homologues or recognizable pseudogenes are absent from all fish and non-fish genomes sequenced to date.

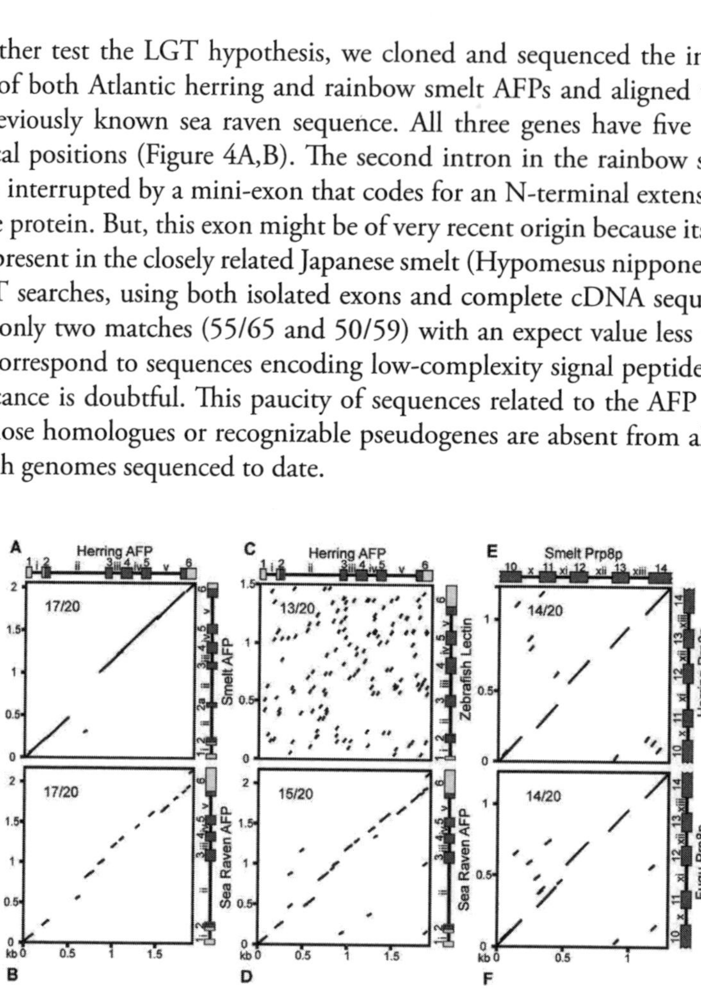

Figure 4. Dot matrix comparisons of lectin-like AFP and control genes. (A,B) Atlantic herring type II AFP gene (Herring, GenBank #DQ003023) compared to (A), the rainbow smelt type II AFP gene (Smelt, GenBank #DQ004949), and (B), the sea raven type II AFP gene. The genes are draw to scale with exons (boxes) numbered 1–6 and color-coded, where yellow represents untranslated regions, red the signal polypeptide, green other coding regions, and blue the extra exon in the rainbow smelt AFP gene. Introns are designated i to v. Data points represent a 17/20 nucleotide match. (C,D) Atlantic herring type II AFP gene compared to (C), zebrafish lectin gene at a 13/20 nucleotide match, and to (D), the sea raven type II AFP gene at a 15/20 nucleotide match. (E,F) Dot matrix comparison of a segment (exons 10–14) of the rainbow smelt Prp8p gene compared to (E), the Atlantic herring Prp8p gene at a 14/20 nucleotide match and to (F), the pufferfish (Fugu) Prp8p gene at a 14/20 nucleotide match.

Consistent with this LGT hypothesis, the AFP gene introns reveal a remarkable degree of identity of up to 97% between rainbow smelt and herring (Figure 4A; Table 1). In the dot matrix analysis, where 17 out of 20 (17/20) bases were matched for each data point, the only significant break in the alignment occurs in intron 2. Elsewhere, intron and exon sequences are equally well conserved. Although conservation of branching points and regulatory elements could account for some limited conservation between introns, this degree of intron sequence identity in fishes belonging to different superorders is unusual. Nucleotide and translated BLAST searches, using the entire gene sequence and each individual intron, only detected one additional match with an expect value less than 10–3. This match, of 60 out of 78 positions with two gaps, is between an uncharacterized zebrafish genomic sequence and intron 2 from herring. We do not consider this significant because it only covers 12% of the intron, there are no other matches within this contig, and this portion of the intron is not conserved between herring and rainbow smelt. As well, the only exons predicted using the gene prediction program GENSCAN [38] corresponded to the AFPs. Taken together, this suggests that these introns are unlikely to contain functional or regulatory domains unless they are specific to the AFP genes themselves.

Table 1. Percent identities between each intron and exon in the herring (H), rainbow smelt (S) and sea raven (SR) AFP gene sequences.

Identities – Pairwise Exclusion of Gaps – 100% = identical

Region	SR/H % Identity		SR/H (bp)	SR/S % Identity		SR/S (bp)	H/S % Identity		H/S (bp)
	No Gaps	Gaps		No Gaps	Gaps		No Gaps	Gaps	
Exon 1	73.3	64.7	60/68	78.8	71.9	52/57	90.7	86.0	54/57
Intron 1	71.3	52.8	94/127	71.4	47.2	84/127	97.1	88.7	105/115
Exon 2	78.0	67.6	91/105	75.1	62.9	88/105	87.5	84.6	88/91
Exon 3	58.8	34.3	80/137	61.3	35.8	80/137	98.8	98.8	80/80
Intron 3	74.5	67.0	98/111	78.6	69.4	98/111	96.1	94.2	102/104
Exon 4	64.1	59.5	117/126	65.8	61.1	117/126	88.1	88.1	126/126
Intron 4	72.8	64.4	92/104	75.0	66.3	92/104	97.1	97.1	104/104
Exon 5	60.6	60.6	109/109	62.4	62.4	109/109	89.9	89.9	109/109
Intron 5	84.8	56.5	230/345	83.0	58.4	230/327	96.5	90.8	317/337
Exon 6	78.2	36.4	165/354	73.0	30.7	148/352	90.8	83.1	162/177

The columns labeled with % are the identities. The first (no gaps) is calculated by pair wise exclusion of gaps and the second (gaps) is calculated over the length of the alignment, gaps included. The column labeled with bp shows the total length of each alignment (first no gaps/second with gaps). The second intron is not included as large portions of the sequence are probably not homologous.

The herring and sea raven AFP genes also share similarity throughout their length, and the dot matrix analyses with at least 15 (or even 17) matches in a 20 base window, show again that there is conservation of both intron and exon sequences (Figure 4B,D) ranging from 34–68% identity (Table 1). In contrast, the next best match to a fish lectin sequence (zebrafish) shows no pattern of alignment for a dot matrix plot even when based on a 13/20 base match (Figure 4C). As a control, the single-copy gene sequences for a well-conserved spliceosomal protein

(Prp8p) showed continuous 14/20 base matches within the exon sequences in comparisons between the rainbow smelt, herring and pufferfish (Fugu) genes, but no matches within the introns (Figure 4E,F). This lack of intron sequence identity in a gene from distant species is normal, even in one coding for a highly conserved protein. It helps make the point that the remarkable intron sequence similarity between the herring and rainbow smelt AFP genes is consistent with LGT and would be hard to explain by another mechanism.

Genomic Southern Blots Confirm the Absence of Type II AFP Gene Homologs in Other Fishes

To experimentally illustrate the sequence conservation of the type II AFP genes, and at the same time to confirm the absence of homologs in more closely related fishes, we have probed a genomic Southern blot of Pvu II-digested DNA from 11 species arrayed in order of their taxonomic relationships (Figure 5A). When the blot was probed with the 3′-half of the rainbow smelt AFP gene, encompassing both exons and introns, there was strong hybridization to a 7.8 Kb band of rainbow smelt DNA, and to multiple bands in the Atlantic and Pacific herring DNAs ranging from 3.5 to >10 Kb. There was also hybridization to the sea raven DNA at ~2.5 Kb. However, there was no sign of hybridization to any of the other DNAs, despite the ease of detection of highly diluted control DNA at a concentration equivalent to a single gene copy. This confirms the results of the database search and illustrates that failure to find a close homolog is not due to a defect in the search strategy or a gap in the coverage of DNA sequences. When the same blot was stripped and reprobed with beta-tubulin cDNA there were signals from multiple genes in all species, illustrating that there was hybridizable DNA in each lane (Figure 5B). When the blot was reprobed with the Prp8p gene (single-copy), there were one or two hybridizing bands in each DNA-containing lane, again showing that single copy sequences can readily be detected on the blot (not shown).

Discussion

The isolated occurrence of type II AFP in three distant branches of the teleost radiation is extraordinary. These lectin homologs are the only ones to have a fifth disulfide bridge in a specific location, and they are far more similar to each other than to any other lectin homolog. Their resemblance extends to the DNA sequence level, where even the introns are up to 97% identical. This sequence similarity is independently demonstrated by genomic Southern blotting, which also confirms the absence of type II AFP homologs in other fishes, some of which are quite closely related to the type II AFP-producing species. The most likely explanation for up to 85% amino acid sequence identity, low silent mutation rate, and extreme conservation of number, position and sequence of introns is that the type II AFP gene has been laterally transferred. Nevertheless, we have considered other possible explanations.

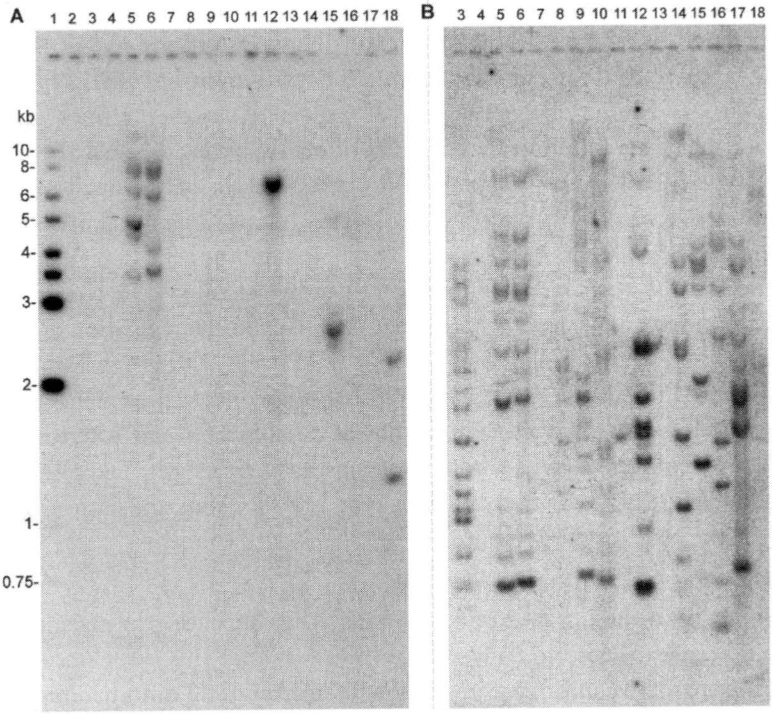

Figure 5. Southern blot of fish genomic DNAs. (A) Blot probed with rainbow smelt AFP genomic DNA from exon 3 to 6. (B) same blot probed with chicken beta-tubulin cDNA. Lane 1 contains markers. Lanes 2, 4, 7, 11 and 13 are blank. Other lanes contain Pvu II-digested genomic DNAs (10 μg) from the following species: 3, bowfin; 5, Atlantic herring; 6, Pacific herring; 8, zebrafish; 9, rainbow trout; 10, cisco; 12, rainbow smelt; 14, Atlantic cod; 15, sea raven; 16, yellow perch; 17, winter flounder; and 18, zebrafish. Lane 18 also contains plasmid DNAs for the herring and rainbow smelt AFP genes at a single gene copy loading.

In the first scenario, that of gene loss, the ten-Cys type II AFP lectin homolog would have been present in the common ancestor to herring, smelt and sea raven. Since herring and smelt belong to different superorders (and in some phylogenetic schemes to different infradivisions), this ancestor would be the progenitor of nearly all teleosts. For the gene to have disappeared from those other species surveyed in the data bases and on the genomic Southern blot would require at least five gene deletion events. Taken alone, this might not be totally unexpected as it appears that notothenoid fishes that do not live in the icy Antarctic seas have lost many or all of their antifreeze glycoprotein genes [39]. But what can account for the conservation of coding sequences in the absence of strong selection as indicated by the near equivalent rates of missense and silent mutations and the low rate of silent mutations in the absense of codon selection? And how can introns that are up to 97% identical between herring and smelt after >100 million year of separation be explained?

Highly conserved sequence segments, termed ultraconserved elements, have been revealed by genome-wide comparisons between various species, including humans and fish, and some of these elements lie within introns [40]. Type II AFPs are unlikely to belong to this category of sequence, however, as most ultraconserved elements are found in the genomes of many species, whereas the distribution of type II AFP genes is extremely limited. As well, only a small proportion of introns have been shown to contain conserved noncoding elements. These can be up to several hundred base pairs in length and are thought to regulate expression of either the genes in which they lie or nearby genes [41]. Moreover, they are mainly found in and around genes that are involved in the regulation of development, which is not the function of the type II AFP gene. Although a small subset of genes may contain more than one type of ultraconserved element, these elements tend to be interspersed with regions of variable sequence, whereas the type II AFPs are highly conserved throughout most of their length. Taken together, it seems unlikely that the conservation of type II AFP exons and introns over the length of the gene can be attributed to ultraconserved elements.

Another scenario is convergent evolution of type II AFPs from lectin homologs. At least one instance of very similar AFPs appearing in divergent fishes [42] has been attributed to convergent evolution [43]. This is the occurrence of the highly repetitive antifreeze glycoproteins in Antarctic nototheniids and the unrelated Arctic cods. The former appear to have arisen de novo from expansion of a tripeptide sequence within the trypsinogen gene [44], [45]. A different example of convergent AFP evolution rests with the insect AFPs, where moth and beetle AFPs [46], [47] have ended up with nearly identical ice-binding sites consisting of two parallel ranks of equally spaced threonines despite being derived from very different beta-helical folds, one left-handed and the other right-handed [48]. Although one could imagine the 5th disulfide bridge having been independently evolved on three separate occasions, especially if it had some functional role in ice binding, there is no way that convergent evolution could account for the overall amino acid sequence similarity and the similarity in both the third codon position and intron sequences.

Although many suggested cases of LGT, particularly between bacteria and higher eukaryotes, have been discounted [49] there is more robust evidence for LGT of mitochondrial DNA in plants [50]. Certainly, LGT between bacteria is well established and occurs frequently when there is selective pressure, as for example in the acquisition of antibiotic resistance [51]. Kurland et al. [52] have emphasized two criteria that bear on the success of LGT. One is that the alien sequences should not spoil the efficiency of an integrated system that has co-evolved to be optimal in that organism. The other is that for alien sequences to be perpetuated in the genome they must be adaptive. Both of these criteria are met here because 1) the antifreeze protein is presumably a single gene trait that is additional to, and largely independent of, existing systems, and 2) it is clearly of adaptive value. We suggest that the considerable selective pressure for survival in icy seawater in the face of past climate change [53], [54] has revealed the lateral transfer between fish species of a nuclear gene for freeze resistance. Indeed, the massive gene amplification that has accompanied the acquisition of AFP genes [17] is indicative of the intense selective pressure to produce adequate amounts of AFP to survive in icy seawater resulting from the Cenozoic glaciations of the last 10–20 million years [55]. Species acquiring antifreeze genes would not only have had resistance to freezing during glacial episodes but they would have faced less competition with, and predation from, non-resistant species.

There are a number of possible mechanisms that could explain LGT between species of fish, such as transfer by shared parasites, viruses or transposable elements. However, a much simpler scenario is possible. Sperm-mediated LGT is based on the ability of sperm to absorb foreign DNA from solution, and partial uptake of DNA by the sperm nucleus has been observed for many species, including zebrafish [56]. Transgenic offspring have been generated in this manner for a variety of species ranging from bees and sea urchins to fish, birds, mammals and other vertebrates (reviewed in Smith and Spadafora [57]). The exogenous DNA usually persists extrachromosomally for some time, but chromosomal integration has been observed in certain cases, such as with the fish, Labeo rohita [58].

Naturally-occurring sperm-mediated LGT has not yet been documented but is much more feasible for vertebrates with external fertilization, such as fish, for several reasons. During active spawning, particularly in the case of herring, the water over a huge area is often visibly discolored due to the massive release of sperm [59]. Lysis of sperm is observed in seawater [60], releasing large amounts of DNA into the water column. Although DNAses are abundant in seawater, extracellular DNA still has a half-life of several hours [61]. Also, fish eggs have a hole in their chorion (micropyle) through which the sperm and any attached DNA can enter the egg. Therefore, it is feasible that foreign DNA could be taken up by fish eggs naturally, but in most cases, it would not be retained due to failure to meet the criteria of Kurland et al. [52] mentioned above. However, because an AFPs gene has the potential to independently confer a strong selective advantage, it could become established in the population.

Acknowledgements

We thank Alana Nguyen and Robert L. Campbell for assistance, and Andrew Roger, David Irwin, Gary Scott and Mike Reith for their constructive criticisms. P.L.D. holds a Canada Research Chair in Protein Engineering. This is NRC publication number 2005-42487.

Author Contributions

Conceived and designed the experiments: PD LG KE. Performed the experiments: LG. Analyzed the data: PD LG SL KE. Wrote the paper: PD LG SL KE.

References

1. Ohno S (1970) Evolution by gene duplication. Heidelberg: Springer-Verlag.
2. Hartley BS, Brown JR, Kauffman DL, Smillie LB (1965) Evolutionary similarities between pancreatic proteolytic enzymes. Nature 207: 1157–1159.
3. Neurath H (1984) Evolution of proteolytic enzymes. Science 224: 350–357.
4. Levy SB, Marshall B (2004) Antibacterial resistance worldwide: causes, challenges and responses. Nat Med 10: S122–129.
5. Tenover FC (2006) Mechanisms of antimicrobial resistance in bacteria. Am J Infect Control 34: S3–10. discussion S64–73.
6. Kordis D, Gubensek F (1998) Unusual horizontal transfer of a long interspersed nuclear element between distant vertebrate classes. Proc Natl Acad Sci USA 95: 10704–10709.
7. Raymond JA, DeVries AL (1977) Adsorption inhibition as a mechanism of freezing resistance in polar fishes. Proc Natl Acad Sci USA 74: 2589–2593.
8. Ewart KV, Rubinsky B, Fletcher GL (1992) Structural and functional similarity between fish antifreeze proteins and calcium-dependent lectins. Biochem Biophys Res Commun 185: 335–340.
9. Drickamer K (1999) C-type lectin-like domains. Curr Opin Struct Biol 9: 585–590.
10. Zelensky AN, Gready JE (2003) Comparative analysis of structural properties of the C-type-lectin-like domain (CTLD). Proteins 52: 466–477.
11. Weis WI, Drickamer K, Hendrickson WA (1992) Structure of a C-type mannose-binding protein complexed with an oligosaccharide. Nature 360: 127–134.

12. Ewart KV, Fletcher GL (1993) Herring antifreeze protein: primary structure and evidence for a C-type lectin evolutionary origin. Mol Mar Biol Biotechnol 2: 20–27.

13. Ewart KV, Yang DS, Ananthanarayanan VS, Fletcher GL, Hew CL (1996) Ca2+-dependent antifreeze proteins. Modulation of conformation and activity by divalent metal ions. J Biol Chem 271: 16627–16632.

14. Liu Y, Li Z, Lin Q, Kosinski J, Seetharaman J, et al. (2007) Structure and evolutionary origin of Ca2+-dependent herring type II antifreeze protein. PLoS ONE 2: e548.

15. Ng NF, Hew CL (1992) Structure of an antifreeze polypeptide from the sea raven. Disulfide bonds and similarity to lectin-binding proteins. J Biol Chem 267: 16069–75.

16. Gronwald W, Loewen MC, Lix B, Daugulis AJ, Sonnichsen FD, et al. (1998) The solution structure of type II antifreeze protein reveals a new member of the lectin family. Biochemistry 37: 4712–4721.

17. Scott GK, Hew CL, Davies PL (1985) Antifreeze protein genes are tandemly linked and clustered in the genome of the winter flounder. Proc Natl Acad Sci USA 82: 2613–2617.

18. Huelsenbeck JP, Ronquist F (2005) Bayesian analysis of molecular evolution using MrBayes. In: Nielsen R, editor. Statistical Methods in Molecular Evolution. New York: Springer. pp. 183–232.

19. Gelman A, Rubin DB (1992) Inference from iterative simulation using multiple sequences. Stat Sci 7: 457–472.

20. Rambaut A, Drummond AJ (2003) Tracer, version 1.2.1. Program distributed by the author. Institute of Evolutionary Biology, Ashworth Laboratories, Kings Buildings, West Mains Road. Edinburgh, Scotland. Available: http://tree.bio.ed.ac.uk/software/tracer/ .

21. Swofford DL (2002) PAUP*. Phylogenetic analysis using parsimony (* and other methods). Version 4. Sunderland, Massachusetts: Sinauer Associates.

22. Miya M, Takeshima H, Endo H, Ishiguro NB, Inoue JG, et al. (2003) Major patterns of higher teleostean phylogenies: a new perspective based on 100 complete mitochondrial DNA sequences. Mol Phylogenet Evol 26: 121–138.

23. Nylander JAA (2006) MrModeltest v2.2. Program distributed by the author. Sweden: Department of Systematic Zoology, Evolutionary Biology Centre, Uppsala University. Available: http://www.abc.se/nylander/.

24. Ronquist F, Huelsenbeck J (2003) Mrbayes 3: bayesian phylogenetic inference under mixed models. Bioinformatics 19: 1572–1574.

25. Korber B (2000) HIV sequence signatures and similarities. In: Rodrigo AG, Learn GH, editors. Computational and evolutionary analysis of HIV molecular sequences. Dordrecht, Netherlands: Kluwer Academic Publishers. pp. 55–72. Available: http://www.hiv.lanl.gov/content/hiv-db/SNAP/WEBSNAP/SNAP.

26. Peden JF (1999) Analysis of codon usage. PhD Thesis, Department of Genetics, University of Nottingham, UK.

27. Maddison DR, Maddison WP (2000) MacClade, Version 4.0. Sunderland, Massachusetts: Sinauer Associates.

28. Sambrook J, Fritsch EF, Maniatis T (1989) Molecular Cloning: A Laboratory Manual. Cold Spring Harbour: Cold Spring Harbour Laboratory Press.

29. Aparicio S, Chapman J, Stupka E, Putnam N, Chia JM, et al. (2002) Whole-genome shotgun assembly and analysis of the genome of Fugu rubripes. Science 297: 1301–1310.

30. Jaillon O, Aury JM, Brunet F, Petit JL, Stange-Thomann N, et al. (2004) Genome duplication in the teleost fish Tetraodon nigroviridis reveals the early vertebrate proto-karyotype. Nature 431: 946–957.

31. Kasahara M, Naruse K, Sasaki S, Nakatani Y, Qu W, et al. (2007) The medaka draft genome and insights into vertebrate genome evolution. Nature 447: 714–719.

32. Nelson JS (1984) Fishes of the World. New York: John Wiley and Sons.

33. Nei M, Gojobori T (1986) Simple methods for estimating the numbers of synonymous and nonsynonymous nucleotide substitutions. Mol Biol Evol 3: 418–426.

34. Luo HR, Moreau GA, Levin N, Moore MJ (1999) The human Prp8 protein is a component of both U2- and U12-dependent spliceosomes. RNA 5: 893–908.

35. Graham LA, Qin W, Lougheed SC, Davies PL, Walker VK (2007) Evolution of hyperactive, repetitive antifreeze proteins in beetles. J. Mol. Evol. 64: 387–398.

36. Romero H, Zavala A, Musto H, Bernardi G (2003) The influence of translational selection on codon usage in fishes from the family Cyprinidae. Gene 317: 141–147.

37. Yamashita Y, Miura R, Takemoto Y, Tsuda S, Kawahara H, et al. (2003) Type II antifreeze protein from a mid-latitude freshwater fish, Japanese smelt (Hypomesus nipponensis). Biosci Biotechnol Biochem 67: 461–466.

38. Burge C, Karlin S (1997) Prediction of complete gene structures in human genomic DNA. J Mol Biol 268: 78–94.

39. Cheng CH, Detrich HW (2007) Molecular ecophysiology of Antarctic notothenioid fishes. Philos Trans R Soc B 362: 2215–2232.

40. Bejerano G, Pheasant M, Makunin I, Stephen S, Kent KW, et al. (2004) Ultraconserved elements in the human genome. Science 304: 1321–1325.

41. McEwen GK, Woolfe A, Goode D, Vavouri T, Callaway H, et al. (2006) Ancient duplicated conserved noncoding elements in vertebrates: a genomic and functional analysis. Genome Res. 16: 451–465.

42. Davies PL, Ewart KV, Fletcher GL (1993) The diversity and distribution of fish antifreeze proteins: new insights into their origins. In: Hochachka PW, editor. Fish biochemistry and molecular biology. pp. 279–291.

43. Chen L, DeVries AL, Cheng CH (1997) Convergent evolution of antifreeze glycoproteins in Antarctic notothenioid fish and Arctic cod. Proc Natl Acad Sci USA 94: 3817–3822.

44. Cheng CH, Chen L (1999) Evolution of an antifreeze glycoprotein. Nature 401: 443–444.

45. Chen L, DeVries AL, Cheng CH (1997) Evolution of antifreeze glycoprotein gene from a trypsinogen gene in Antarctic notothenioid fish. Proc Natl Acad Sci USA 94: 3811–3816.

46. Graether SP, Kuiper MJ, Gagne SM, Walker VK, Jia Z, et al. (2000) Beta-helix structure and ice-binding properties of a hyperactive antifreeze protein from an insect. Nature 406: 325–328.

47. Liou YC, Tocilj A, Davies PL, Jia Z (2000) Mimicry of ice structure by surface hydroxyls and water of a beta-helix antifreeze protein. Nature 406: 322–324.

48. Davies PL, Baardsnes J, Kuiper MJ, Walker VK (2002) Structure and function of antifreeze proteins. Philos. Trans. R. Soc. Lond. B Biol. Sci. 357: 927–935.

49. Andersson JO (2005) Lateral gene transfer in eukaryotes. Cell Mol Life Sci 62: 1–16.

50. Mower JP, Stefanovic S, Young GJ, Palmer JD (2004) Plant genetics: gene transfer from parasitic to host plants. Nature 432: 165–166.

51. Ochman H, Lawrence JG, Groisman EA (2000) Lateral gene transfer and the nature of bacterial innovation. Nature 405: 299–304.

52. Kurland CG, Canback B, Berg OG (2003) Horizontal gene transfer: a critical view. Proc Natl Acad Sci USA 100: 9658–9662.

53. Scott GK, Fletcher GL, Davies PL (1986) Fish antifreeze proteins:recent gene evolution. Can J Fish Aquat Sci 43: 1028–1034.

54. Cheng CH (1998) Evolution of the diverse antifreeze proteins. Curr Opin Genet Dev 8: 715–720.

55. Moran K, Backman J, Brinkhuis H, Clemens SC, Cronin T, et al. (2006) The Cenozoic palaeoenvironment of the Arctic Ocean. Nature 441: 601–605.

56. Patil JG, Khoo HW (1996) Nuclear internalization of foreign DNA by zebrafish spermatozoa and its enhancement by electroporation. J Exp Zool 274: 121–129.

57. Smith K, Spadafora C (2005) Sperm-mediated gene transfer: applications and implications. Bioessays 27: 551–562.

58. Venugopal T, Anathy V, Kirankumar S, Pandian TJ (2004) Growth enhancement and food conversion efficiency of transgenic fish Labeo rohita. J Exp Zoolog A Comp Exp Biol 301: 477–490.

59. Hourston AS, Rosenthal H (1976) Sperm density during active spawning of pacific herring Clupea harengus pallasi. J Fish Res Board Can 33: 1788–1790.

60. Dundas IED (1985) Fate and possible effects of excessive sperm released during spawning. Mar Ecol Prog Ser 30: 287–290.

61. Lorenz MG, Wackernagel W (1994) Bacterial gene transfer by natural genetic transformation in the environment. Microbiol Rev 58: 563–602.

Use of Number by Fish

Christian Agrillo, Marco Dadda, Giovanna Serena and Angelo Bisazza

ABSTRACT

Background

Research on human infants, mammals, birds and fish has demonstrated that rudimentary numerical abilities pre-date the evolution of human language. Yet there is controversy as to whether animals represent numbers mentally or rather base their judgments on non-numerical perceptual variables that co-vary with numerosity. To date, mental representation of number has been convincingly documented only for a few mammals.

Methodology/Principal Findings

Here we used a training procedure to investigate whether mosquitofish could learn to discriminate between two and three objects even when denied access to non-numerical information. In the first experiment, fish were trained to discriminate between two sets of geometric figures. These varied in shape, size, brightness and distance, but no control for non-numerical variables was made. Subjects were then re-tested while controlling for one non-numerical variable at a time. Total luminance of the stimuli and the sum of perimeter of figures appeared irrelevant, but performance dropped to chance level when stimuli were matched for the cumulative surface area or for the overall space occupied by the arrays, indicating that these latter cues had been spontaneously used by the fish during the learning process. In a second experiment, where the task consisted of discriminating 2 vs 3 elements with all non-numerical variables simultaneously controlled for, all subjects proved able to learn the discrimination, and interestingly they did not make more errors than the fish in Experiment 1 that could access non-numerical information in order to accomplish the task.

Conclusions/Significance

Mosquitofish can learn to discriminate small quantities, even when non-numerical indicators of quantity are unavailable, hence providing the first evidence that fish, like primates, can use numbers. As in humans and non-human primates, genuine counting appears to be a 'last resort' strategy in fish, when no other perceptual mechanism may suggest the quantity of the elements. However, our data suggest that, at least in fish, the priority of perceptual over numerical information is not related to a greater cognitive load imposed by direct numerical computation.

Introduction

Abilities such as recording the number of events, enumerating items in a set, or comparing two different sets of objects, can be adaptive in a number of ecological contexts. Lyon [1], for instance, reported a spontaneous use of numerical information (egg recognition and counting) in a natural context as a strategy to reduce the costs of conspecific brood parasitism in American coots.

McComb and co-workers [2] using playback experiments found that wild lions based the decision whether or not to attack a group of intruders on a comparison of the number of roaring intruders they had heared and the number and composition of their own group. Over the last two decades or so, extensive laboratory research carried out on monkeys and apes [3]–[6] has revealed the existence of non-verbal systems of numerical representation that non-human primates apparently share both with human infants and with human adults tested in comparable conditions [7], [8]. In recent years, rudimentary numerical abilities have been reported in several other mammalian and avian species, among others, elephants, dolphins, dogs, cats, robins and chicks [9]–[14].

Recently we found [15] that fish, seeking safety from predators, display a rudimentary numerical ability in selecting the largest shoal. Interestingly, the limits shown by fish in this task closely resemble those that have been reported for primates. These data in particular suggest the existence, as in primates, of two independent pre-verbal systems: one for counting a small quantity (≤4) precisely, and the other for estimating large quantities (>4) approximately. These findings suggest the possibility that all extant vertebrates share similar quantificational mechanisms, which may have an ancient phylogenetic origin, at least predating the divergence of the tetrapod lineage.

Nonetheless, before concluding that the same systems are involved, it is necessary to understand whether similar limits really reflect identical underlying mechanisms. In particular, it has been extensively demonstrated that both humans and nonhuman animals can discriminate between two quantities without necessarily counting the number of objects. Numerosity normally co-varies with several other physical attributes, and organisms can use the relative magnitude of continuous variables such as the total area of the stimuli or the sum of their contour, to estimate which group is larger/smaller [16]–[18]. Discriminations based on number or on continuous extent often yield comparable results and therefore carefully controlled experiments are necessary to show that an animal is really using numerical information. Experiments of this type demonstrate that, when selecting the larger shoal, mosquitofish (Gambusia holbrooki) spontaneously use non-numerical cues, namely the sum of areas of the shoals and the overall quantity of movements of the individuals within the shoal [15]. This does not necessarily imply that fish are unable to discriminate two groups on the basis of the numerosity alone. Overall perceptual cues may simply be the easiest indicators of numerosity in this task. Indeed, there is persuasive evidence that even humans and non-human primates, which have the capacity to represent number, in many circumstances base their quantity judgment primarily on proxy measures such as area, contour or density of elements and that they use number as a last resort, when there are no other available cues [19], [20].

In the present work we investigated whether fish can discriminate between two quantities when access to non-numerical cues was prevented. The procedure used in our previous studies with fish did not easily permit a fine-grained manipulation of stimuli and an efficient control of continuous perceptual variables that correlate with number. Therefore we adopted a procedure modelled on carefully-controlled experiments conducted on non-human primates, which consisted of training the subject to discriminate between sets containing different numbers of geometric figures while controlling for the perceptual non-numerical variables [3], [21].

The first experiment aimed to determine which cues mosquitofish used spontaneously when both numerical information and continuous physical attributes are available. Subjects learned a discrimination between 2 and 3 objects in the absence of any manipulation of the stimuli; after animals had achieved learning criterion they were tested without reward while controlling for one perceptual non-numerical variable at a time. In the second experiment we trained fish to discriminate between 2 and 3 objects while we simultaneously controlled for non-numerical variables, in order to determine whether fish could discriminate quantities by using only numerical information as shown for mammals.

Results

Experiment 1.a. Cues Spontaneously Used by Fish to Discriminate Between Quantities

Ten female mosquitofish were placed in an unfamiliar tank and trained to discriminate between two doors in order to re-join their social group (Fig. 1). Doors were associated with a pair of stimuli consisting of two or three small figures (Fig. 2). These figures were randomly selected with replacement from a pool of approximately 100, and no control for non-numerical variables was operated in the learning phase. Subjects were given six trials per day for a maximum of ten days. Once a subject had reached the learning criterion, it was admitted to the test phase and was examined in the same apparatus without reward (no possibity to re-join the conspecifics) while controlling for one perceptual non-numerical variable at time. We controlled those variables that were shown to be relevant in previous studies with vertebrates, namely the total luminance of the two stimuli, the sum of perimeter of the figures, the cumulative surface area, and the overall space occupied by the arrays. Since operant conditioning is normally a stressful procedure for fish, we adopted a pre-training procedure that consisted of exposing the subjects, in the seven days preceding the training, to the choice of similar pairs of stimuli in order to move from one compartment to the other of their home-tank.

Figure 1. Apparatus used to train fish. Subjects were singly placed in the middle of a test chamber provided with two doors (one associated to three and the other associated to two elements) placed at two opposite corners. Subject could pass through the reinforced door to rejoin shoal mates in the outer tank (not shown).

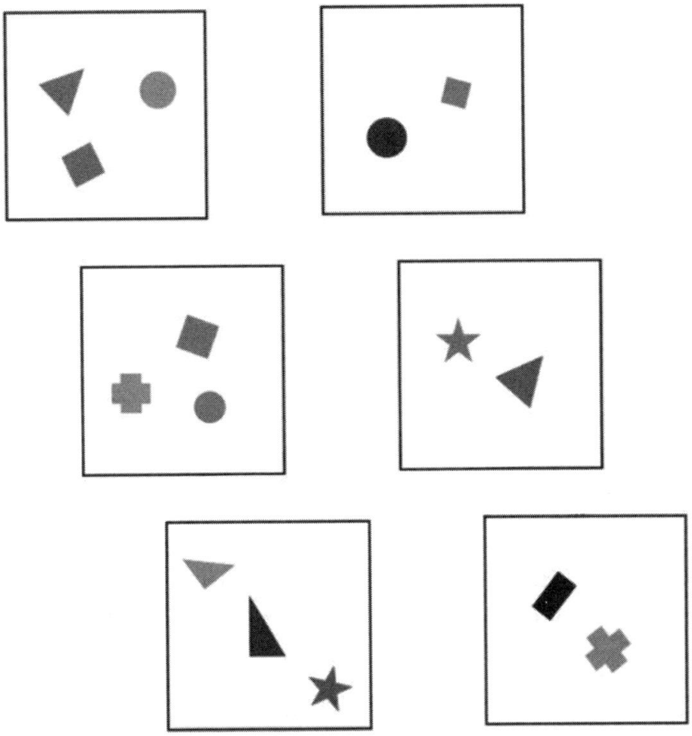

Figure 2. Schematic representation of the stimuli used in experiment 1. Each pair was composed of one set of two and one of three elements. Elements varied in shape, size, brightness and position, and were randomly selected from a large pool.

All ten subjects reached the learning criterion in the training phase, but one was excluded from the subsequent test phase due to poor health, and hence nine started the test phase. We reported no difference in the proportion of correct choices between fish trained with three (mean±std. dev.: 0.753±0.065) and those trained with two figures (0.678±0.028; $t(7) = 2.337$, $p = 0.052$). In the test phase a significant discrimination was observed when no perceptual cue was controlled for ($t(8) = 2.449$, $p = 0.020$) and when the total luminance was controlled for ($t(8) = 2.310$, $p = 0.025$); no significant choice toward the trained quantity was found when the sum of perimeter of the figures ($t(8) = 1.316$, $p = 0.225$), the cumulative surface area ($t(8) = -1.512$, $p = 0.169$), and the overall space occupied by the arrays ($t(8) = -0.373$, $p = 0.719$) were controlled for (Fig. 3a).

However, since area and perimeter of the figures are strictly related to each other, in this experiment by controlling one variable we inevitably affected the other, so that it was not possible to conclude whether one or both variables were important in the discrimination.

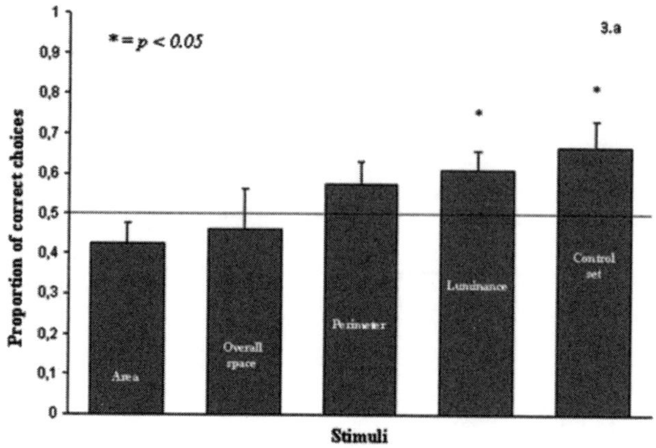

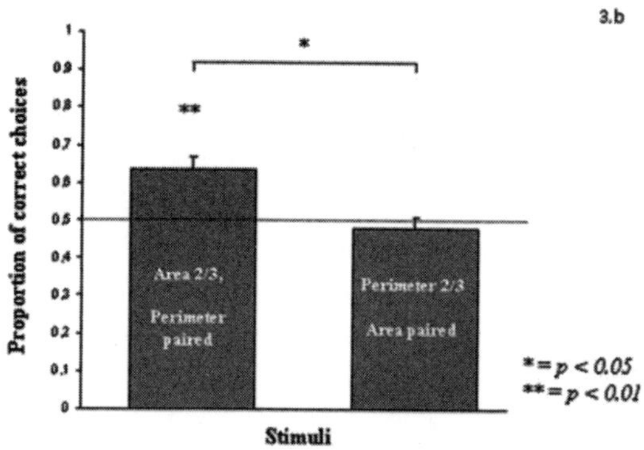

Figure 3. Results of Experiment 1. (a) Proportion of correct choices when area, space, perimeter and luminance were singly controlled for (exp 1 a). (b) Proportion of correct choices when area and perimeter were controlled for (exp 1b).

Experiment 1.b. Area vs. Perimeter

We accordingly set up another experiment with ten subjects, using the same procedure as before in pre-training and training phases, whereas in the test phase fish were presented with only two different sets of stimuli: one set in which the cumulative surface area was paired whereas the sum of perimeter was not (i.e. the perimeter could suggest the exact ratio between the quantities), and one set in which the sum of the perimeter was paired whereas the cumulative surface area was not (i.e. the area could suggest the exact ratio between the quantities).

We reported no difference in the proportion of correct choices between fish trained with three (mean±std. dev.: 0.788±0.066) and those trained with two figures (0.753±0.065; $t(8) = 0.832$, $p = 0.429$). When the relative ratio of the areas (but not the perimeter) was predictive of the numerical ratio, we observed a significant choice toward the trained quantity ($t(9) = 3.786$, $p = 0.004$) whereas no significant choice was observed in the condition in which the perimeter, but not the area, could be used to distinguish between two quantities ($t(9) = -0.653$, $p = 0.530$). The difference between the two conditions was significant (paired t-test, $t(9) = 2.865$, $p = 0.019$, Fig. 3b).

On the whole, results of Experiment 1 showed that fish were found to base their discrimination on the cumulative surface area occupied by figures and on the overall space occupied by the arrays, while they did not use the sum of perimeter, the total luminance of the stimuli or the number of items. Interestingly, in Experiment 1.a we observed a negative correlation between the proportion of correct choices when cumulative surface area was paired and when the overall space of the arrays was paired (Spearman test, $\rho s = -0.734$, $p = 0.024$) indicating that there was an individual variability in the cues used with some subjects relying on the cumulative surface area for discriminating and not being affected by the overall space, while others used the overall space but not area to solve the task.

Experiment 2. Discrimination of Small Quantities using only Numerical Information

In this experiment we trained fourteen fish to discriminate between 2 and 3 objects while we simultaneously controlled stimuli for their non-numerical variables in both the pre-training and the training phase, with the aim of determining whether fish could learn the discrimination using only numerical information. Using the same geometric figures as the previous experiment, we designed pairs of stimuli in which the total luminance, the cumulative surface area, and the overall space occupied by the arrays were paired between the groups with two and three elements. We found no difference in the proportion of correct choices between fish trained with three (mean±std. dev.: 0.690±0.037) and those trained with two figures (0.651±0.070; $t(12) = 1.328$, $p = 0.209$). All 14 fish reached the criterion (chi square test, $p<0.05$), proving thus able to select the trained numerosity. Overall the choice for the trained stimuli is highly significant ($t(13) = 11.103$, $p<0.001$).

As a by-product of controlling for three perceptual variables, stimuli differed for two other non-numerical variables that the fish could have used instead of number to learn the discrimination. The by-product of pairing the cumulative surface area between sets with two and three elements was that in the latter sets smaller-than-average in area figures were more frequent. The by-product of pairing the overall space occupied by configuration was that figures were more spaced out in the sets containing two figures. After reaching criterion, fish were thereby subjected to a test phase without reinforcement using pairs of stimuli composed of figures of identical size and similarly spaced.

Results showed that fish still significantly selected the trained numerosity, even when all the elements were equal to each other and the density of the elements was controlled for ($t(13) = 4.397$, $p = 0.001$).

When we compared the number of trials necessary to reach criterion in Experiment 1 (when all numerical and non-numerical cues were available) and Experiment 2 (where only numerical cues were available), we found no difference between experiments (trials in Experiment 1: 25.2±11.7; trials in Experiment 2: 29.14±9.7; $F(1,33) = 1.064$, $p = 0.170$; power = 0.170).

Discussion

Our experiments show that the ability of mosquitofish to discriminate among sets containing a different number of elements is not limited to the socio-sexual context [15], [22], [23], but also applies to sets of abstract elements. They also indicate that mosquitofish can accomplish this task when all non-numerical perceptual variables are matched between the stimuli, thus strongly suggesting that teleosts, like mammals, possess true counting abilities, at least in the domain of small numbers.

Cues Spontaneously used by Fish to Discriminate between Quantities

The first experiment showed that during the extinction phase a good performance was maintained when no control of perceptual variables was operated or after the total luminance of the stimuli and the sum of perimeter of figures were matched. Conversely mosquitofish were unable to select the learned numerosity when stimuli were matched for the cumulative surface area or for the overall space occupied by the arrays, thus suggesting that these two cues had been used during the learning process. Previous works have already demonstrated that such variables play an important role as proxies of numerosity in humans and other mammals [12], [18], [19]. It is interesting to notice that the finding of Experiment 1.a, that fish are influenced by two different non-numerical cues, is somehow a paradox. In fact, because this experiment only controlled for one variable at a time, the fish should have succeeded in all conditions. That is, when the cumulative surface area was controlled for, the overall space of the arrays was not, and vice versa. One explanation for this apparent conflict is that fish combined two different non-numerical cues to learn the numerical discrimination, and that, consequently, the absence of one of the two cues was sufficient to worsen their performance. A second possibility is that fish used only one non-numerical cue, but that there was individual variation in the cue they adopted to learn the discrimination. Our data are more in accordance with the latter hypothesis. In Experiment 1.a the performance of subjects in trials when the cumulative surface area was controlled for correlated negatively with the performance in trials when the overall space occupied by the arrays was controlled for, suggesting that fish that were more influenced by the cumulative surface area were unaffected by manipulation of the overall space of the arrays, whereas subjects that relied on this latter cue did not use the cumulative surface area during the learning process. However, this evidence is based on the examination of a small number of subjects, and caution should be exercised before drawing firm conclusions on this question.

Discrimination of Small Quantities using only Numerical Information

Results of the second experiment showed that fish can use numbers when perceptual variables that correlate with numerosity were excluded. To date, this is the first evidence that a lower vertebrate can really represent and compare numbers. The capacity to discriminate among sets containing different numbers of objects by using numerical information only, previously reported for six month old infants [24], primates, dolphins and dogs [4], [10], [11], [25], is here extended to include a species, the Eastern mosquitofish, which is phylogenetically very distant from mammals and has a much smaller brain size compared with the former species.

Observations made in this study were limited to a single quantity discrimination, 2 vs 3. This was shown to be the upper limit in the capacity of discrimination of six-month old infants [18]. In experiments with continuous variables controlled for, non-human primates successfully discriminate between 3 and 4 objects [4], and non-verbal counting abilities of human adults can be even better [26]. Mosquitofish have been shown to discriminate a shoal of three fish from one of four in two different contexts [15], [22], but in these experiments access to continuous extent of the stimuli could not be prevented. In one of these studies [22], shoals were in two distinct compartments of the apparatus so that they could not be seen and compared simultaneously. This implies that whatever information, numerical or continuous, the mosquitofish encoded, they were able to maintain it temporarily in working memory. Further experiments are necessary to determine whether the upper limit of discrimination of fish also matches that of mammals when access to continuous extent of the stimuli is prevented.

Many authors now agree that there are two distinct non-verbal systems for representing numerosity in animals, adults and human infants [7], [17]. The first mechanism proposed is the one most likely investigated by us in this study. It is an object-tracking system that operates on a small number of items by keeping track of individual objects [27], [28]. It is precise but, due to the limited number of available indexes, it is supposed to allow for the parallel representation of up to 3–4 elements only [29]. The second is an analog magnitude system of numerical representations that allows approximate discrimination of large quantities. It obeys Weber's Law, which holds that as numerical magnitude increases, a larger disparity is needed to obtain the same level of discrimination [8], [30]. Fish have shown to rival primates in their ability to discriminate large quantities approximately [15], [22], [31], [32]. However, while controlled experiments have shown that six month babies and non-human primates can perform large number discrimination using only numerical information, no such evidence exists for fish. Future research should assess if the analog magnitude system can operate in fish when access to continuous extent of the stimuli is prevented.

Number as a Last Resort?

Comparison of the two experiments suggests that although mosquitofish are capable of using both number and continuous extent, they spontaneously use the latter to estimate quantities. Similar results have been reported to occur in dolphins, macaques, six month old infants and human adults [10], [16], [18], [19], [27]. For instance, a bottlenose dolphin trained to distinguish between two quantities spontaneously used overall surface area of the elements or brightness for performing the discrimination [10]. However, controlling for non-numerical cues, these authors demonstrated that the dolphin could discriminate the stimuli solely on the basis of the numerosity feature and that eventually it was able to successfully transfer the discrimination to novel numerosities outside the former range.

Traditionally, the explanation for these results is that number requires more effortful processing compared with continuous extent, and therefore counting represents a 'last resort' strategy, when no other perceptual mechanism may suggest the quantity of the elements [10], [18], [33]–[35]. However, recent studies have questioned this assumption, showing that adult humans, pre-verbal children, chimpanzees and macaques spontaneously and automatically encode information about continuous extent and numerosity simultaneously, and that the relative salience of these two dimensions depends on factors such as type of task, numerosity ratio and previous experience [17], [36]–[39].

Recently, Burr and Ross [40] have provided evidence for a putative physiological mechanism underlying this capacity. After being exposed for 30 sec to a large number of spots in one portion of their visual field, the subjects of this experiment tended to underestimate by three times the number of spots being subsequently presented in the same region of retina. The presence of a retinotopic adaptation clearly indicates that the visual system is able to extract, at an early stage, the numerical information from a visual scene, just as it extracts other 'primary visual properties' such as color, size, orientation and spatial frequency.

Our study was not designed to specifically investigate this issue. However we found that learning was equally effective in the first experiment when subjects could use all physical properties of the stimuli and in the second, when they had access only to the numerical cues, suggesting that the precedence of perceptual cues is not determined, at least in fish, by a greater cognitive effort when numerical computation is involved.

Why do mosquitofish preferentially use continuous extent over numerical information given that the two alternatives are similar in cognitive demand? One possibility is that quantity information is ecologically more relevant for this species. For example, in foraging contexts animals often tend to maximise the amount of resources acquired with a minimum of energy expenditure [41], [42]. Even though number of items and total amount of resource gained frequently correlate, sometimes this does not occur, for example when there is a large variation in the size of food items. Selection for optimising food intake could have favoured mechanisms based on continuous extent, such as area, as they are more reliable indicators of the resource potentially gained [43], [44]. Alternatively, perceptual cues of the stimuli may simply be the quickest indicator of the numerosity, for example because they involve earlier stages in neural visual or auditory processing. Mosquitofish use quantity discrimination in fitness related contexts, such as choosing the safer social group or the larger number of potential mates [15], [23], in which speed of decision is often crucial. Mechanisms based on continuous extent may have been favoured in this species since they allow choosing the best option in the fastest way. One recent study with adult humans [38] has provided evidence that the extraction of a representation of continuous extent, such as the area of stimuli, in most cases proceeds more rapidly than the extraction of a representation of discrete quantity. There is some evidence that this may be the case for rats and pigeons too [34], [45], suggesting that it may represent a common property of vertebrate visual system.

In summary, this study provides a new insight into the evolution of cognitive abilities of vertebrates. Many authors have proposed the existence of shared mechanisms for non-verbal numerical discrimination in humans, non-human mammals, and birds [4], [14], [27], [46]. The present results provide further evidence that is coherent with previous works [15], [22], [31], [47], indicating a fundamental similarity of mechanisms underlying non-verbal numerical abilities in distantly related vertebrates and reinforcing the idea that numerical systems may be more ancient than we had previously assumed.

Materials and Methods

Experiment 1.a. Cues Spontaneously Used by Fish to Discriminate Between Quantities

Subjects

Ten female Eastern mosquitofish (Gambusia holbrooki) were used as subjects of this experiment. Fish were collected from Valle Averto, a system of brackish water ponds and ditches in the Venetian lagoon basin (northern Italy), returned to the laboratory and initially maintained in small mixed-sex groups (12–15 fish, approx. 1:1 sex ratio) kept in 70-l glass aquaria with abundant vegetation (Vesicularia dubyana and Ceratophyllum demersum), lit by a 15 W fluorescent lamp (16L:8D) and with a water temperature that was maintained at 25±2°C. Subjects were used once; companion females, on the other hand, were used more than once.

Apparatus and Stimuli

Pre-training phase. One week before the training, fish were placed in a 68×68×38 cm tank, divided into four equal sectors by white plastic partitions (Fig. 4). The tank was lit by four fluorescent lamps positioned around the borders, and water was maintained at a temperature of 25°±2°C. The bottom was covered with natural gravel, and vegetation (Vesicularia dubyana) was provided as well as aquarium filters.

Figure 4. Pre-training apparatus. Aerial (a), and lateral view (b). Eight pairs of equal doors allowed fish to move between the four compartments. Stimuli (3 figures or 2) were placed above each door and only the door below the reinforced quantity permitted the passage.

To move between sectors, each partition was provided with two doors of equal size (2.5×3.5×1 cm) closed by a flexible plastic material and located 12.5 cm from the floor of the tank, with a distance of 8 cm between them. Above each of the two doors we placed two identical stimuli, each occupying a 3×3 cm area. Each stimulus set contained one exemplar with two elements and one with three. Elements were geometric figures differing in shape, size and luminance, randomly chosen from a set of approximately 100 elements and positioned on a white background. The average distance among elements in stimuli containing two or three elements was the same (see examples in Fig. 2).

Only the door below the reinforced quantity permitted them to pass from one sector to the other. This was achieved by gluing the transparent material on the top of the door, so that fish could easily bend it and pass through the door. On the other door the transparent material was glued also at the bottom, so that fish could not pass through. An openable door could be traversed in both directions, and pairs of stimuli were placed on both sides of the partition so that a total of 8 different pairs were presented inside the tank at the same time. These stimuli were changed daily, and a total of 56 different pairs of stimuli were used during the pre-training phase.

The experimental apparatus (Fig. 1) was used in the training phase and in the test phase. It consisted of a small white test chamber (16×16×16 cm) inserted in a larger tank (60×26×36 cm) to provide a comfortable area with vegetation and food where the test fish was placed together with other three companion females, 10 minutes before starting the training session. The tank was inserted in a dark room and covered with a one-way screen to eliminate extra-tank cues. Female mosquitofish are highly social and spontaneously tend to join the other females when placed in an uncomfortable environment [15]. Previous work has shown that this procedure provides motivation for social reinstatement in fish [48].

At two corners of the chamber, two small tunnels (3×4×2.5 cm, located 2 cm from the floor of the tank) made from white plastic material were inserted, allowing the fish to pass through it to rejoin conspecifics in the outer tank. At the end of each tunnel there was a door similar to that used in the pre-training tank. As previously, one door was blocked, while the other could be opened by bending the flexible plastic material.

Sixty new pairs of stimuli were used, with the same characteristics of those used in the pre-training phase. As the elements of the stimuli were randomly selected, during pre-training and the training phase fish could learn to distinguish between two quantities by using both number and non-numerical information that correlated with number, such as the cumulative surface area or the overall space occupied by the arrays. Conversely, in the control test, five different sets of stimuli were presented. In four, we controlled for one continuous variable at time, namely the cumulative surface area of the elements, the total luminance of the stimuli, the sum of perimeter, and the overall space occupied by the arrays. The fifth was a control set of stimuli, in which no control for non-numerical variables was performed. All stimuli were created by using Microsoft Office 2003 and the area, perimeter and luminance was controlled using TpsDig software.

Procedure

Three different steps were planned: the pre-training, the following training phase and the test phase. Half of the subjects were trained toward the larger quantity (three), whereas the second half were trained toward the smaller one (two). In the first step, two subjects were kept for 7 days inside the pre-training tank. All the couples of stimuli were changed daily and fish were left free to swim inside the four sectors without any interference from the experimenter for the whole period.

At the beginning of day 8, all fish commenced the training phase in the experimental apparatus: fish were singly tested each day (6 trials per day) from a minimum of three to a maximum of ten days. During the trials, fish were brought to the test tank by inserting them into a transparent plastic cylinder (4.5 cm in diameter) and placing it in the centre of the test chamber. After 30 seconds, the cylinder was removed, leaving the fish in the middle of the test chamber. The first door they initially reached was recorded until the fish was able to exit and rejoin conspecifics (the maximum time allowed to exit was 20 minutes). Inter-trial intervals lasted 5 minutes, during which the fish was allowed to shoal with the conspecifics; in the meantime the experimenter changed the pair of stimuli. The location of the trained quantity was exchanged at any successive trial. Furthermore, since the subject was disoriented between successive trials and no external cue was available, the two corners were equivalent from the point of view of the fish, reducing any possibility that fish may have preferentially chosen one door on the basis of the geometrical information of the environment [48].

The learning criterion was a statistically significant frequency of correct choice estimated with chi square test. Starting from day 3, we statistically analysed the daily performance of the subject, and once discrimination reached significance it was admitted the next day to the following test phase. Procedure for this phase was similar to that used during the training phase, with the exception that we adopted an extinction procedure by keeping both doors blocked. The first choice was recorded until a maximum period of 2 minutes. After this period, fish were released outside the test tank and could join their conspecifics; 5 minutes later, the subject was re-inserted into the test chamber in the presence of a new pair of stimuli. This phase lasted 5 days, with 6 trials per day, for a total of 30 overall trials, 6 for each set of stimuli. The five sets were randomly intermingled during each daily session. Statistical tests were conducted using SPSS 15.0.

Experiment 1.b. Area vs. Perimeter

Ten female mosquitofish were used as subjects. Experimental apparatus and procedure were the same described in Experiment 1.a. The same stimuli described in Experiment 1.a were used in pre-training and training for this experiment. For the test phase, fish were presented with only two sets of stimuli. In one, the cumulative surface area of the stimuli was exactly paired, whereas the relative ratio of the perimeter between the groups - 3 and 2 elements - was equal to 3/2 (the perimeter could then suggest the exact ratio between the quantity whereas the area could not); in the other set we used an opposite pattern, controlling for the perimeter but having the area that could suggest the exact ratio between the quantities (3/2). In both cases we paired stimuli for the overall space occupied by the arrays. During each daily session, half of the trials presented the former set whereas the remaining presented the latter set. The two sets were randomly intermingled within each session.

Experiment 2. Discrimination of Small Quantities using only Numerical Information

Subjects and Apparatus

A total of 14 female mosquitofish were used as subjects. Apparatus was the same as for the previous experiment.

Stimuli and Procedure

The procedure for this experiment was similar to the previous one, with the exception that during pre-training and training phases we used pairs of stimuli in which the cumulative surface area, the total luminance and the overall space occupied by the arrays were simultaneously controlled for. The key phase for this experiment was the training phase, since we aimed at determining whether fish could learn the discrimination in the absence of non-numerical cues. During the training phases of this experiment all subjects received the same number of trials, 36, comprising 6 trials per day for a total of 6 days. As before, the criterion for discrimination was a statistically significant frequency of correct choices during the training phase.

By pairing the cumulative surface area and the overall space occupied by the arrays we could have provided subjects with two additional non-numerical cues.

In each pair, the stimulus with the larger number of elements (three) tended, inevitably, to contain small elements more often than the corresponding stimulus with two elements. By occupying the same overall space, the stimulus with three elements also tended to have a shorter distance between the elements. Both cues could, in principle, be used by fish to learn the discrimination. We therefore added a test phase, in which we presented with an extinction procedure pairs of stimuli in which all elements were identical in size and shape (all circles, all stars, etc.) and were similarly spaced. Fish received a total of 24 trials (6 trials per day, for 4 days).

Acknowledgements

The authors would like to thank Jonathan Daisley and Peter Kramer for their comments and Barbara Bonaldo for her help conducting the experiments. The reported experiments comply with all laws of the country (Italy) in which they were performed.

Author Contributions

Conceived and designed the experiments: CA AB. Performed the experiments: CA MD GS. Analyzed the data: GS AB. Contributed reagents/materials/analysis tools: MD. Wrote the paper: CA AB.

References

1. Lyon BE (2003) Egg recognition and counting reduce costs of avian conspecific brood parasitism. Nature 422: 495–499.
2. McComb K, Packer C, Pusey A (1994) Roaring and numerical assessment in contests between groups of female lions, Panthera leo. Animal Behavior 47: 379–387.
3. Brannon EM, Terrace HS (1998) Ordering of the numerosities 1 to 9 by monkeys. Science 282: 746–749.
4. Hauser MD, Carey S, Hauser LB (2000) Spontaneous number representation in semi-free-ranging rhesus monkeys. Proceedings of the Royal Society B: Biological Sciences 267: 829–833.
5. Matsuzawa T (1985) Use of numbers by a chimpanzee. Nature 315: 57–59.

6. Rumbaugh DM, Savage-Rumbaugh S, Hegel MT (1987) Summation in the chimpanzee (Pan troglodytes). Journal of Experimental Psychology: Animal Behavior Processes 13: 107–115.

7. Feigenson L, Dehaene S, Spelke E (2004) Core systems of number. Trends in Cognitive Sciences 8: 307–314.

8. Xu F (2003) Numerosity discrimination in infants: Evidence for two systems of representations. Cognition 89: B15–B25.

9. Irie-Sugimoto N, Kobayashi T, Sato T, Hasegawa T (2009) Relative quantity judgment by Asian elephants (Elephas maximus). Animal Cognition 21: 193–199.

10. Kilian A, Yaman S, Von Fersen L, Güntürkün O (2003) A bottlenose dolphin discriminates visual stimuli differing in numerosity. Learning and Behavior 31: 133–142.

11. West R, Young R (2002) Do domestic dogs show any evidence of being able to count? Animal Cognition 5: 183–186.

12. Pisa PE, Agrillo C (2009) Quantity discrimination in felines: a preliminary investigation of the domestic cat (Felis silvestris catus). Journal of Ethology 27: 289–293.

13. Hunt S, Low J, Burns KC (2008) Adaptive numerical competency in a food-hoarding songbird. Proceedings of the Royal Society B: Biological Sciences 275: 2373–2379.

14. Rugani R, Regolin L, Vallortigara G (2007) Rudimental numerical competence in 5-day-old domestic chicks (Gallus gallus): identification of ordinal position. Journal of Experimental Psychology: Animal Behavior Processes 33: 21–31.

15. Agrillo C, Dadda M, Serena G, Bisazza A (2008) Do fish count? Spontaneous discrimination of quantity in female mosquitofish. Animal Cognition 11: 495–503.

16. Clearfield MW, Mix KS (1999) Number versus contour length in infants' discrimination of small visual sets. Psychological Science 10: 408–411.

17. Cordes S, Brannon EM (2008) Quantitative competencies in infancy. Developmental Science 11: 803–808.

18. Feigenson L, Carey S, Spelke E (2002) Infants' discrimination of number vs. continuous extent. Cognitive Psychology 44: 33–66.

19. Durgin FH (1995) Texture density adaptation and the perceived numerosity and distribution of texture. Journal of Experimental Psychology: Human Perception and Performance 21: 149–169.

20. Vos PG, van Oeffelen MP, Tibosch HJ, Allik J (1988) Interactions between area and numerosity. Psychological Research 50: 148–154.
21. Judge PG, Evans TA, Vyas DK (2005) Ordinal representation of numeric quantities by brown capuchin monkeys (Cebus apella). Journal of Experimental Psychology: Animal Behavior Processes 31: 79–94.
22. Agrillo C, Dadda M, Bisazza A (2007) Quantity discrimination in female mosquitofish. Animal Cognition 10: 63–70.
23. Agrillo C, Dadda M, Serena G (2008) Choice of female groups by male mosquitofish (Gambusia holbrooki). Ethology 114: 479–488.
24. Xu F, Spelke ES (2000) Large number discrimination in 6-month-old infants. Cognition 74: B1–B11.
25. Beran MJ (2008) Monkeys (Macaca mulatta and Cebus apella) track, enumerate, and compare multiple sets of moving items. Journal of Experimental Psychology: Animal Behavior Processes 34: 63–74.
26. Barth H, Kanwisher N, Spelke E (2003) The construction of large number representations in adults. Cognition 86: 201–221.
27. Feigenson L, Carey S, Hauser M (2002) The representations underlying infants' choice of more: object files versus analog magnitudes. Psychological Science 13: 150–156.
28. Trick LM, Pylyshyn ZW (1994) Why are small and large numbers enumerated differently? A limited-capacity preattentive stage in vision. Psychological Review 101: 80–102.
29. Pylyshyn ZW, Storm RW (1988) Tracking multiple independent targets: evidence for a parallel tracking mechanism. Spatial Vision 3: 179–197.
30. Jordan KE, Brannon EM (2006) Weber's Law influences numerical representations in rhesus macaques (Macaca mulatta). Animal Cognition 9: 159–172.
31. Buckingham JN, Wong BBM, Rosenthal GG (2007) Shoaling decisions in female swordtails: How do fish gauge group size? Behavior 144: 1333–1346.
32. Binoy VV, Thomas KJ (2004) The climbing perch (Anabas testudineus Bloch), a freshwater fish, prefers larger unfamiliar shoals to smaller familiar shoals. Current Science 86: 207–211.
33. Davis H, Memmott J (1982) Counting behavior in animal: a critical evalutation. Psychological Bullettin 92: 547–571.
34. Roberts WA, Coughlin R, Roberts S (2000) Pigeons flexibly time or count on cue. Psychological Science 11: 218–222.

35. Seron X, Pesenti M (2001) The number sense theory needs more empirical evidence. Mind and Language 16: 76–88.

36. Cantlon JF, Brannon EM (2007) How much does number matter to a monkey (Macaca mulatta)? Journal of Experimental Psychology: Animal Behavior Processes 33: 32–41.

37. Cordes S, Brannon EM (2008) The difficulties of representing continuous extent in infancy: Using number is just easier. Child Development 79: 476–489.

38. Hurewitz F, Gelman R, Schnitzer B (2006) Sometimes area counts more than number. Proceedings of the National Academy of Sciences of the United States of America 103: 19599–19604.

39. Tomonaga M (2008) Relative numerosity discrimination by chimpanzees (Pan troglodytes): evidence for approximate numerical representations. Animal Cognition 11: 43–57.

40. Burr D, Ross J (2008) A Visual Sense of Number. Current Biology 18: 425–428.

41. Kamil AC, Krebs JR, Pulliam HR (1987) Foraging Behavior. New York and London: Plenum Press.

42. McArthur RH, Pianka ER (1966) On the optimal use of a patchy environment. American Naturalist 100: 603–609.

43. Beran MJ, Evans TA, Harris EH (2008) Perception of food amounts by chimpanzees based on the number, size, contour length and visibility of items. Animal Behavior 75: 1793–1802.

44. Stevens JR, Wood JN, Hauser MD (2007) When quantity trumps number: discrimination experiments in cotton-top tamarins (Saguinus oedipus) and common marmosets (Callithrix jacchus). Animal Cognition 10: 429–437.

45. Breukelaar JWC, Dalrymple-Alford JC (1998) Timing ability and numerical competence in rats. Journal of Experimental Psychology: Animal Behavior Processes 24: 84–97.

46. Brannon EM (2005) What animals know about numbers. In: Campbell JID, editor. Handbook of Mathematical Cognition. New York: Psychology Press.

47. Agrillo C, Dadda M (2007) Discrimination of the larger shoal in the poeciliid fish Girardinus falcatus. Ethology Ecology and Evolution 19: 145–157.

48. Sovrano VA, Bisazza A, Vallortigara G (2002) Modularity and spatial reorientation in a simple mind: encoding of geometric and nongeometric properties of a spatial environment by fish. Cognition 85: B51–B59.

Copyrights

Index

A

D

E

F

N

R

S

T

U

V

W

X

Y

Z